# Discrete Structures

## an introduction to mathematics for computer science

Fletcher R. Norris
University of North Carolina at Wilmington

Prentice-Hall, Inc., Englewood Cliffs, New Jersey 07632

*Library of Congress Cataloging in Publication Data*

NORRIS, FLETCHER R. (date)
Discrete structures.

Bibliography: p.
Includes index.
1. Electronic data processing—Mathematics.
I. Title.
QA76.9.M35N67 1985 519.4 84-24837
ISBN 0-13-215260-6

Editorial/production supervision: Kathryn Gollin Marshak
Interior design: Maureen Eide and Lee Cohen
Cover design: Maureen Eide
Manufacturing buyer: Gordon Osbourne



Printed in the United States of America

10 9 8 7 6 5 4 3 2 1

ISBN 0-13-215260-6

Prentice-Hall International, Inc., *London*
Prentice-Hall of Australia Pty. Limited, *Sydney*
Editora Prentice-Hall do Brasil, Ltda., *Rio de Janeiro*
Prentice-Hall Canada Inc., *Toronto*
Prentice-Hall of India Private Limited, *New Delhi*
Prentice-Hall of Japan, Inc., *Tokyo*
Prentice-Hall of Southeast Asia Pte. Ltd., *Singapore*
Whitehall Books Limited, *Wellington, New Zealand*

*To Janis*

Contents

## 2 Sets 39



## 3 Boolean Algebra 83



## 4 The Algebra of Switching Circuits 105

## 5 *Functions, Recursion, and Induction* 136

## *Relations and Their Graphs* 167



## 7 *Applications of Graph Theory* 202

# 8 Discrete Counting: An Introduction to Combinatorics 242



# 9 Posets and Lattices 252

# The Binary Number System 264



# B Matrices: An Introduction 279

# Preface

This book presents a first course in discrete mathematical structures with an emphasis on applications to computer science. The general topic of discrete structures has existed as a somewhat unified body of mathematical content roughly since the publication of the ACM Curriculum '68. This report, and other more recent ones, identifies a body of mathematical knowledge that is deemed essential for the mastery of some of the higher-level computer science courses, especially those relating to computer organization, logic design and switching circuits, data structures, and those with a theoretical orientation.

Subsequent to the Curriculum '68 report, several good textbooks appeared on the market as a direct result. Inevitably, many of these books assumed a level unattainable by many entering computer science students. This was unfortunate, since the material is needed at an early stage in college by computer science majors. The understanding of discrete mathematical topics is fundamental for academic work in computer science and should be done early. Most of the needed materials can be made accessible to students with only a strong background from college algebra and possibly one introductory programming course. This book is intended for just such an audience and is suitable for the freshman-sophomore level. It should be viewed as a first college-level mathematics course of substance for computer science majors.

A second objective I have addressed is an attempt to unify the seemingly disparate topics into a single treatment. The approach is that of an introductory survey of discrete mathematics, much as calculus textbooks are unified introductory treatments of continuous mathematics. The notion of the discreteness and finiteness of computer calculations and algorithms and the underlying mathematical models are stressed throughout.

The book was written with the student in mind. The topics chosen have been shown to be relevant to a computer science curriculum, and no attempt has been made to write a sophisticated book that is unreadable for the average beginning college student.

Even though this textbook is intended to be the first college-level mathematics course for computer science majors, it can be used for other purposes as well. The text provides excellent collateral materials for mathematics majors who wish to broaden their knowledge. The background of such students is richer, so more of the material can be covered and at greater depth. Other mathematics courses, such as the calculus, will inevitably be more meaningful with the maturity gained from this first course in discrete mathematics, even though there is no direct connection intended between the two.

## To the Instructor

Presumably, this course will be taught by both mathematicians and computer scientists. The approach I have taken is to be mathematically precise but not rigorous. I have attempted to give the students the flavor of discrete topics without burdening them with a highly theoretical orientation. The students want to know how the mathematics involved will help them in computer science. A mathematical approach to computer science has been proved to be sound, but at a beginning level it is of primary importance that the topics be relevant and meaningful.

Chapter 1 presents the fundamental model of discrete systems, the mathematics of logical propositions. In Chapter 2 the algebra of sets is presented as another representation of the model of Chapter 1. This development, as well as that of binary relations, is derived exclusively from the results of Chapter 1. Much of these materials should be covered rapidly, as they may be review for many students. Experience has shown, however, that this may not, in fact, be the case.

Chapter 3 presents the most important algebraic structure for the computer scientist: Boolean algebra. The presentation is that this is simply another model of the one covered in the first two chapters. In Chapter 4 the all-important application of Boolean algebra, that of switching circuits, is presented.

In Chapter 5 the modern terminology of functions is introduced together with the development of Polish notation as a function of two variables. Recursive functions and recursion in general are carefully motivated. This treatment is followed by its close parallel, mathematical induction, the basic proof method of discrete systems.

Chapter 6 covers the elements of graph theory both undirected and directed as pictorial representation of relations. Section 6.5 is devoted entirely to trees since this is such an important graph-theoretic topic for computer scientists. Most of the applications of graphs and trees are collected together in Chapter 7. These were put in one place to give the instructor the opportunity to choose the ones most relevant to his or her own curriculum. Strategies on various coverages are given in the Instructor's Manual.

Chapter 8 is an introductory treatment of combinatorics and discrete counting methods. This is a follow-up to earlier treatments of the topics which are at appropriate places throughout the beginning chapters. The chapter is not intended to be extensive since presumably an entire course in the curriculum will be devoted to this important topic. Chapter 9 is a somewhat theoretical treatment of partial orders and lattices as algebraic systems of considerable immediate importance to computer science majors.

The appendices on the binary number system and matrices are intended for supplementary reading for those students unfamiliar with the topics. Parts of Appendix A should be discussed before the treatment of switching circuits in Chapter 4. Some students will be familiar with the binary number system, but for those who are not, a relatively extensive treatment of binary arithmetic is given. The coverage of matrices in Appendix B is needed before they are used in Chapter 6 on graph theory.

You will probably find that there is more material than can be covered in one semester. This obviously gives you the freedom to stress what is important in your own curriculum.

An Instructor's Manual accompanies the textbook and will be available for those teaching the course. In it are given the goals and purpose of the book; a general philosophy of the importance of an early coverage of discrete mathematics by computer science students; and a chapter-by-chapter commentary together with solutions to selected exercises that are not solved in the book itself.

## To the Student

Many students will find they have familiarity with some of the topics in this book: for instance, truth tables of logical propositions, elements of set theory, and basic notions of functions and mathematical induction. The approach taken is that the mathematics of logical propositions is the underlying model of discrete systems and that each subsequent topic relies heavily on this beginning. Ample motivation is given to assure that the entire range of topics is meaningful and relevant for further study in computer science.

## Acknowledgments

I would like to express my gratitude to all those who have helped to make this enjoyable undertaking come to fruition. To the hundreds of students who studied and learned from the several initial drafts and who pointed out the early inconsistencies go my heartfelt thanks. Special thanks go to Sandy Plummer,

Mike Brown, and Kathy Kellog, who as student assistants made my life considerably easier by suggesting ideas, working exercises, and checking solutions in the book, although any errors that remain are, of course, mine and mine alone.

The assistance given me by my colleagues, Drs. Barbara Ann Greim, Sandra McLaurin, and Bill Etheridge, who taught from earlier versions of the book, has been especially valuable. Each is aware of the special needs of beginning as well as advanced students and their insights have been taken to heart by me in this book.

Dr. Barbara Ann Greim deserves special credit also for her thorough and meticulous editing of the final version. There are far fewer errors of mine remaining in the book due to her careful and alert reading. Many thanks to her!

I want to thank Pat Joseph and Judy McKee, who at times thought the typing would never end as the manuscript grew, for their unflagging competence and humor.

To the editors at Prentice-Hall, Jim Fegen and Kathy Marshak, who expressed confidence, provided encouragement, and who politely pushed, go my gratitude.

I would like to acknowledge the encouragement given to me in the early stages of the project by my late department chairman and friend, Dr. Fred Toney. However, the same encouragement and enthusiastic support has been given by Dr. Douglas Smith, the current chairman of the department of Mathematical Sciences. I take this opportunity to express my sincere appreciation to him for his understanding of the nature and scope of the project.

How is it possible to show appreciation to a wife and family who allowed this book to happen? Only someone who has undertaken such a project as this in his "spare time" can understand. I dedicate this book to Janis.

FLETCHER R. NORRIS

# Propositions and Logic

# 1

## 1.1 Introduction

The simplest of all discrete systems are the ones that give rise to simple yes/no responses. These are evidenced in a variety of settings. A light bulb can be *on* or *off*. The dimness or brightness is not under consideration. There is no in between; either the light is on or it is off. Another example is a test question that is to be answered as true or false. There is *no* "maybe"; the statement is either true or it is false and you are graded accordingly. Still another example is the IF-THEN decision alternative in programming. Conditional statements can be answered as yes or no and the appropriate branch is taken.

In this chapter we examine the nature of this two-part discrete system. The structure of this system and its associated laws and rules are fundamental to all discrete systems and thus form the basis for our study.

## 1.2 Propositions and Statements

The fundamental element, the building block of our study, will be the *proposition*. A proposition is easily defined as: a simple declarative sentence that can be classified (by everyone) as being either true or false, but not both. That is, it has a *truth value* (true or false) which is easily determined and is not ambiguous in any sense.

Certain statements that *are* propositions are:

George Washington was the first president of the United States.
Andrew Jackson is president of the United States now.
Minnesota is in Alaska.
The number for this course is 273.
Today is the 23rd of August.
Some doctors are rich.
7 is an odd number.
7 is an even number.

Some that are *not* propositions are:

How old are you?
Shut the door!
This is a great book.
This is a large book.
A pink thought jumped in a raisin.
$x + y = z$. (What are the values of $x$, $y$, and $z$?)
Colorless green ideas sleep furiously.
Millard Fillmore is president. (President of what?)

In addition to those propositions that have a known truth value, there are others that have truth values that are not easily verified, or even values that are impossible to determine. Examples of such are:

Snow fell on Iceland in the year 1000 B.C.
There were 14,000 BASIC programs run yesterday.
Abraham Lincoln slept for 7 hours and 25 minutes on the night of August 15, 1849.
There are 100,000,000,000 grains of sand at Wrightsville Beach, N.C.

It would be very hard indeed to verify the truth or falsity of the statements above. We should not let this distract us from declaring them to be propositions, since we would certainly agree that each is either true or false.

Just as arithmetic deals with numbers, the study of logic deals with propositions. As we undertake this study, our immediate task will be to discover what operations can be performed on propositions. We also look at how these propositions can be manipulated and combined, that is, what can be *calculated* using propositions, and how to perform these calculations. In short, we will look at a calculus of propositions. Thus the name of our study is, quite appropriately, *propositional calculus*.

When you learned arithmetic you examined the real numbers and discovered that they could be manipulated and combined through the operations of addition, multiplication, subtraction, and division. Other numbers were obtained by means of these operations. Quite obviously, these are not the appropriate operations for propositions. (What does it mean to say "'7 is odd' + 'Today is Monday'"?)

As noted earlier, a proposition is a simple declarative statement with a truth value. As we combine propositions with operations, we get statements perhaps not so simple but having truth values. To clarify matters, let us designate *variable propositions* by the letters $p$, $q$, and $r$, as was done in algebra when we used $x$, $y$, and $z$ as variables to represent numbers.

Thus, if $p$ is the proposition "$2 + 2 = 5$," then $p$ is *false;* and if $q$ is the proposition "Washington, D.C., is the capital of the United States," then $q$ is *true.* A more complicated statement would be (through combining $p$ and $q$)

"Washington, D.C., is the capital of the United States and $2 + 2 \neq 5$."

What is the truth value of this new statement? Notice that we can write the statement as "$q$ and not $p$."

When propositions are thus combined to form more complex statements we have what is known as a *statement form* or simply a *statement.* We define a statement form as (1) a proposition or (2) any combination of statement forms with the appropriate operation symbols, which we will look at in the next sections.

## EXERCISES

State which of the following are propositions and which are not.

1. $x = 2 + 7$.
2. She is a beautiful person.
3. Ban the bomb.
4. Time flies.
5. This is a silly sentence.
6. This sentence contains no u's.
7. BASIC is an easy programming language to learn.
8. Aren't we through yet?
9. All primes are odd numbers.
10. Some odd numbers are prime.
11. Harry Truman was a good president.
12. I like cold coffee.
13. Peaches are blue.
14. You can see forever.

15. $10 + 10 = 100$. (What base?)

16. $x + y = 5$.

17. There exist real numbers $x$ and $y$ such that $x + y = 5$.

18. We are in the midst of a recession.

19. They are failing students.

20. That's incredible.

## 1.3 Operations on Propositions and Statements

The operations we will define on propositions are commonly known as *connectives* since they will connect propositions and statements to produce new statements.

Negation ($\neg$) The simplest operation that can be applied to statements is that of *negation.* We observed this earlier when we considered $p$ as the proposition: $2 + 2 = 5$, and then used it later as $2 + 2 \neq 5$. The latter form is "not $p$." The symbol we will use for negation will be $\neg$. If proposition $p$ is true, then $\neg p$ will be false; and if $p$ is false, then $\neg p$ will be true. Thus $2 + 2 \neq 5$ can be written $\neg(2 + 2 = 5)$.

The relation between the truth values of $p$ and $\neg p$ can best be illustrated by a diagram called a *truth table*, as shown in Table 1.1. It will be noted that the symbol $\neg$ operates on a single proposition or statement. Thus it is called a *unary* operator. This is similar to the unary minus (or negation) of numbers, which converts a number to the negative of itself.

**TABLE 1.1** Truth Table for Negation

| $p$ | $\neg p$ |
|---|---|
| T | F |
| F | T |

**EXAMPLE 1** (a) Let $p$ be the proposition: It is raining.
Then $\neg p$ is: It is not raining.

(b) Let $q$ be the proposition: $2 + 3 < 5$.
Then $\neg q$ is: $2 + 3 \not< 5$ or $(2 + 3 \geq 5)$ or $\neg(2 + 3 < 5)$.

Conjunction ($\wedge$) Perhaps the next-simplest connector is one that operates on *two* propositions. If we are given two propositions, $p$ and $q$, this connector combines them into the single statement "*p and q*." This operator is known as the *conjunction* operator. The truth value for this new statement will have the mean-

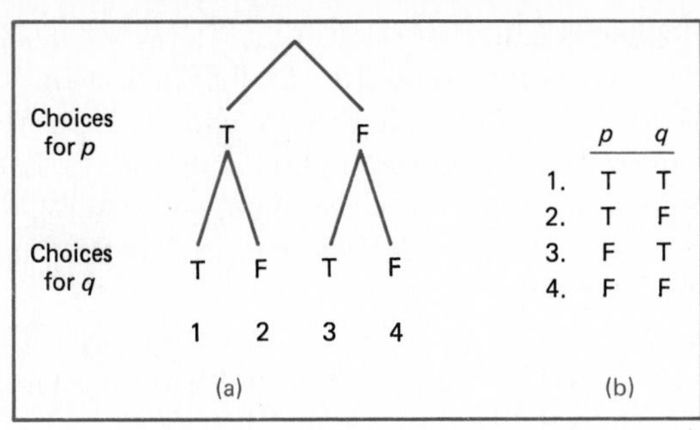


**FIGURE 1.1** Possible combinations of truth value for two propositions.

ing "*p and q*" and will be true when *both p* is true *and q* is true. In all other cases "*p and q*" will be false. Thus it is false when $p$ is false or when $q$ is false, or when both are false.

We will use the symbol $\wedge$ for conjunction. The statement "*p and q*" will thus be symbolized as $p \wedge q$. The operator $\wedge$ is *binary*. That is, it operates on two propositions or statements. Thus it is analogous to addition and multiplication of two real numbers.

To illustrate the truth table for $p \wedge q$, we have to determine how many rows the truth table will have. Now the first variable, $p$, can assume each of the values true and false; and for *each* of these choices, the variable $q$ can also be true and false. This means that there will be four possible combinations of the truth values, as shown in Figure 1.1(a). These values yield the four rows as shown in Figure 1.1(b).

Figure 1.1(a) is commonly known as a "tree diagram," with its "branches" showing the possible combinations of values. This technique of "counting" the number of distinct units of a discrete structure will be used frequently in this book. Trees themselves are an important discrete structure and will be examined rather thoroughly in Chapters 6 and 7.

As stated above, the only row in which $p \wedge q$ has the truth value true (T) is the first row, that is, when both $p$ and $q$ have truth values T. The other rows will be false (F). This is shown in Table 1.2.

**TABLE 1.2** Truth Table for Conjunction

| $p$ | $q$ | $p \wedge q$ |
|---|---|---|
| T | T | T |
| T | F | F |
| F | T | F |
| F | F | F |

**EXAMPLE 2** Construct the truth table for the statement $p \wedge \neg q$. (The unary negation takes precedence over the binary conjunction. Thus this statement is the same as $p \wedge (\neg q)$).

To illustrate the formation of the truth table of such a statement, first, we merely copy the original truth values for $p$ and $q$. Table 1.3(a) shows this phase where the recopied columns are numbered ① and ②. Next we form the negation of $q(\neg q)$. This is shown as column ③ in Table 1.3(b). Last [Table 1.3(c)] we form the conjunction of $p$ and $\neg q$, obtaining column (4), which is the final column computed. We will adopt the convention that the final column of any truth table will be underlined twice.

We see that the only time $p \wedge \neg q$ has the truth value, T, is in the second row of the truth table, that is, when $p$ is true and when $q$ is false (or $\neg q$ is true).

**TABLE 1.3** Forming the Truth Table of $p \wedge (\neg q)$

(a)

| $p$ | $q$ | $p \wedge \neg q$ |
|---|---|---|
| T | T | T   T |
| T | F | T   F |
| F | T | F   T |
| F | F | F   F |
| | | ①   ② |

(b)

| $p$ | $q$ | $p \wedge \neg q$ |
|---|---|---|
| T | T | T   F T |
| T | F | T   T F |
| F | T | F   F T |
| F | F | F   T F |
| | | ①   ③② |

(c)

| $p$ | $q$ | $p \wedge \neg q$ |
|---|---|---|
| T | T | T F F T |
| T | F | T T T F |
| F | T | F F F T |
| F | F | F F T F |
| | | ①④③② |

**EXAMPLE 3** Construct the truth table for $\neg(\neg p \wedge \neg q)$.

First we copy the original truth values for $p$ and $q$, obtaining columns ① and ② in Table 1.4(a). Next, we form the negation of these two columns, getting columns ③ and ④ of Table 1.4(b). The third step will be to form the

**TABLE 1.4** Formation of Truth Table for $\neg(\neg p \wedge \neg q)$

(a)

| $p$ | $q$ | $\neg(\neg p \wedge \neg q)$ |
|---|---|---|
| T | T | T   T |
| T | F | T   F |
| F | T | F   T |
| F | F | F   F |
| | | ①   ② |

(b)

| $p$ | $q$ | $\neg(\neg p \wedge \neg q)$ |
|---|---|---|
| T | T | F T   F T |
| T | F | F T   T F |
| F | T | T F   F T |
| F | F | T F   T F |
| | | ③①   ④② |

(c)

| $p$ | $q$ | $\neg(\neg p \wedge \neg q)$ |
|---|---|---|
| T | T | F T F F T |
| T | F | F T F T F |
| F | T | T F F F T |
| F | F | T F T T F |
| | | ③①⑤④② |

combine ③ and ④

(d)

| $p$ | $q$ | $\neg(\neg p \wedge \neg q)$ |
|---|---|---|
| T | T | T F T F F T |
| T | F | T F T F T F |
| F | T | T T F F F T |
| F | F | F T F T T F |
| | | ⑥ ③①⑤④② |

negation of column ⑤

conjunction of these two negations: column ⑤ of Table 1.4(c). Finally, we perform the negation that is outside the parentheses. This is the negation of the conjunction at column ⑤. The final result, underlined twice, appears in column ⑥ of Table 1.4(d).

In both of these examples, columns ① and ② are merely copies of the original truth values for $p$ and $q$. They are recopied here for convenience to show that they are needed for further operations. They need not be recopied, since you have them for reference at the left.

## EXERCISES

1. Construct truth tables for each of the following.
   (a) $p \wedge (q \wedge p)$ (b) $p \wedge \neg(p \wedge q)$
   (c) $(\neg p \wedge q) \wedge \neg p$ (d) $p \wedge (q \wedge \neg(p \wedge \neg q))$
   (e) $(\neg p \wedge \neg q) \wedge p$ (f) $\neg(\neg p \wedge \neg(\neg q))$

**Disjunction ($\vee$) and X-OR ($\oplus$)** The next statement we will consider is "*p or q*." The word "or" in English leads to a bit of ambiguity. The sense of "This object is a book *or* it is a chair" is that it cannot be both: one *or* the other, but not both. In this sense, we use "or" as an *exclusive* connector, which we will call X-OR (exclusive-or).

On the other hand, the statement "Jan is smart *or* she studies hard" should not exclude the possibility that Jan can, say, be smart enough to study hard. The sense of *or* in this context is *inclusive*. In modern mathematics and for applications in computer science, this inclusive relation is the one that is used most frequently.

Mathematically, there should be no fuzziness or ambiguity about the meaning of a word or concept, so we adopt two symbols for the different or's. The *inclusive or* will be symbolized as $\vee$. When we state "$p$ or $q$" in the inclusive sense, we write $p \vee q$. For the *exclusive or* we will use $p \oplus q$. The inclusive or ($p \vee q$) is the one most commonly used and is given the name *disjunction*.

The truth tables for both disjunction and X-OR are given in Table 1.5. Note that both are true when either one of $p$ or $q$ is true (lines 2 and 3). The

**TABLE 1.5** Truth Tables for (a) Disjunction ($\vee$) and (b) X-OR ($\oplus$)

| $p$ | $q$ | $p \vee q$ | $p \oplus q$ |
|---|---|---|---|
| T | T | T | F |
| T | F | T | T |
| F | T | T | T |
| F | F | F | F |

(a) (b)

disjunction ($p \vee q$) is also true when both $p$ and $q$ are true, whereas the X-OR ($p \oplus q$) is not.

**EXAMPLE 4** Construct the truth table for $(p \wedge q) \wedge (\neg p \vee q)$.

This truth table is constructed as shown in Table 1.6.

**TABLE 1.6**

| $p$ | $q$ | $(p \wedge q)$ | $\vee$ | $(\neg p$ | $\vee$ | $q)$ |
|---|---|---|---|---|---|---|
| T | T | T | T | F | T | T |
| T | F | F | F | F | F | F |
| F | T | F | T | T | T | T |
| F | F | F | T | T | T | F |
| | | ① | ⑤ | ③ | ④ | ② |

⑤ ← disjunction of ① ∨ ④

**EXAMPLE 5** Construct the truth table for $\neg(p \wedge \neg(q \vee r))$.

Note that in this example we have three variable propositions. How many rows will be needed for the truth table? Referring to Figure 1.1(a), we saw that we needed four rows for two propositions. For each of these rows, we need one for *each* of the truth values of the third proposition [see Figure 1.2(a)]. This means that there are $2^3 = 8$ possible combinations of truth values. Consequently, we need eight rows for the truth table.

Observe in Figure 1.2(b) the systematic way in which the truth values for $p$, $q$, and $r$ are organized. The final column computed is ⑦.

**FIGURE 1.2**

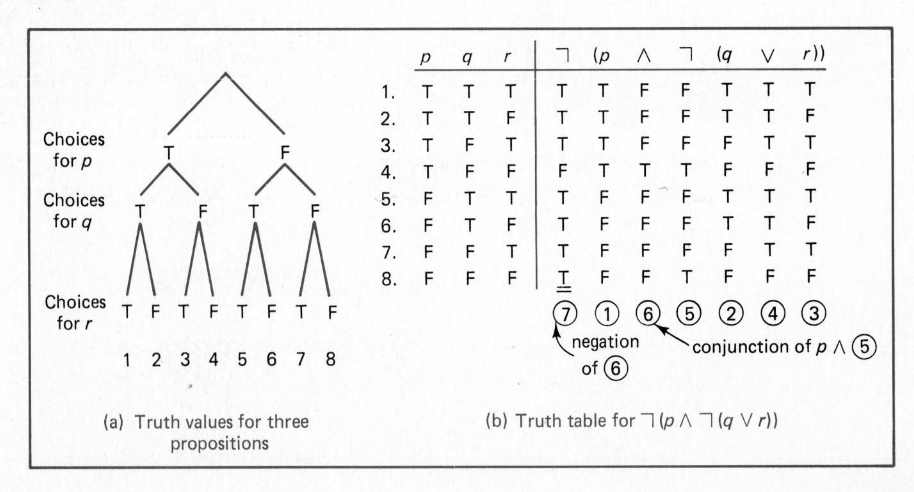


| | $p$ | $q$ | $r$ | $\neg$ | $(p$ | $\wedge$ | $\neg$ | $(q$ | $\vee$ | $r))$ |
|---|---|---|---|---|---|---|---|---|---|---|
| 1. | T | T | T | T | T | F | F | T | T | T |
| 2. | T | T | F | T | T | F | F | T | T | F |
| 3. | T | F | T | T | T | F | F | F | T | T |
| 4. | T | F | F | F | T | T | T | F | F | F |
| 5. | F | T | T | T | F | F | F | T | T | T |
| 6. | F | T | F | T | F | F | F | T | T | F |
| 7. | F | F | T | T | F | F | F | F | T | T |
| 8. | F | F | F | T | F | F | T | F | F | F |
| | | | | ⑦ | ① | ⑥ | ⑤ | ② | ④ | ③ |

(a) Truth values for three propositions

(b) Truth table for $\neg(p \wedge \neg(q \vee r))$

2. Construct truth tables for each of the following statements.

(a) $p \wedge \neg(q \vee \neg p)$

(b) $\neg p \vee (q \wedge p)$

(c) $p \wedge (p \vee q)$

(d) $p \vee (p \wedge q)$ [check your result with statement (c)]

(e) $\neg(\neg p \vee \neg q)$ (check your result with $p \wedge q$)

(f) $\neg(\neg p \wedge \neg q)$ (check your result with $p \vee q$)

(g) $p \wedge (q \wedge r)$

(h) $(p \wedge q) \wedge r$ [check your result with statement (g)]

(i) $p \vee (p \oplus q)$

(j) $\neg q \oplus (p \wedge q)$

3. How many rows will be needed for a truth table on *four* variable propositions? *Five* variable propositions? *n* variable propositions?

4. For the propositions

$p$: Snow fell in Anchorage.

$q$: Abraham Lincoln is alive.

$r$: The Braves won the pennant.

(a) Write the symbolic form for each of the following statements.
   (i) The Braves won the pennant or Abraham Lincoln is not alive.
   (ii) Either Abraham Lincoln is not alive or snow fell in Anchorage and the Braves won the pennant.
   (iii) Neither did it snow in Anchorage nor is Abraham Lincoln alive or the Braves won the pennant.

(b) Interpret in English each of the following symbolic expressions.
   (i) $p \wedge \neg(q \vee \neg r)$
   (ii) $\neg(r \wedge q) \vee (p \wedge r)$
   (iii) $(p \vee \neg r) \wedge \neg q$

Application to Programming Many of you have had some experience with programming languages. One of the important applications of propositions and their connectors is in programming. In most languages, the IF statement in a program represents a conditional branch to another part of the program.

For example, the proposition (statement) "X is less than 100" (X < 100) will be either true or false at any given time in any program in which X is defined. If the programmer wishes to make an alternative decision when X < 100 is true, the code would be similar to

IF(X < 100)THEN

Sometimes the branching can be done based on the truth values of more than one statement. If the programmer wants to perform some segment when X < 100 *and* at the same time a variable Y is >12, the code could be

IF(X < 100 *AND* Y > 12)

The *AND* in this statement is a *conjunction* of the two statements X < 100 and Y > 12. The segment will be performed when *both* are true, namely when the conjunction is true.

If, on the other hand, the branch were to be made when either one or the other or both were true, the disjunction *OR* would be used:

IF(X < 100 *OR* Y > 12)

The negation of a statement is usually represented by *NOT*. The X-OR connective is not conveniently represented and a combination of *AND*'s and *OR*'s is needed. For example, if one wants to branch when X < 100 or when Y > 12 but not when both are true, the code would have to mimic the X-OR truth table. That is, one would explicitly have to code lines 2 and 3 of Table 1.5(b). Line 2 ($p \wedge \neg q$) would be represented by

X < 100 *AND NOT* Y > 12

Line 3 ($\neg p \wedge q$) would be represented by

*NOT* X < 100 *AND* Y > 12

These two statements would then have to be connected with the *OR* operator, yielding

(X < 100 *AND NOT* Y > 12) *OR* (*NOT* X < 100 *AND* Y > 12)

Of course, a simpler representation can be written:

(X < 100 *AND* Y ≤ 12) *OR* (X ≥ 100 *AND* Y > 12)

Conditional (→) The reference made above to the IF-THEN programming statement leads us to another statement that can be formed from two propositions. It is the one that customarily takes the form: *if p, then q*. This logical connector is called the *conditional*, and is symbolized by $p \rightarrow q$. The statement $p$ is called the *premise, hypothesis*, or *antecedent;* and $q$ is referred to as the *conclusion* or *consequence*. Some authors use the term *implication* for this connector; we will use this for a later concept concerning logical arguments.

The conditional, like conjunction and disjunction, is a binary operation. But unlike them, some sort of causal relationship is usually expected between $p$ and $q$, such as:

*If* it is 11:00 P.M., *then* it is dark outside.
*If* it is April 15, *then* taxes are due.
*If* you study hard, *then* you will make an A.
As the song goes, *if* the day is clear, *then* you can see forever.

whereas such statements as

*If* taxes are due, *then* you study hard.
*If* $3 + 4 = 9$, *then* the radio is on.
*If* it is 11:00 P.M., *then* peaches are blue.

are deemed to be nonsensical. In each case the premise has no apparent relationship with or influence over the conclusion.

The conditional is an especially important statement in mathematics. Most mathematical theorems are of this form:

*If* a number $x$ is prime, *then* $x$ is odd or $x = 2$.
*If* a figure is a rectangle, *then* it is a parallelogram.
*If* $xy = 0$, *then* $x = 0$ or $y = 0$ (in the inclusive sense).

Also, as discussed earlier, most computer programs with any degree of difficulty utilize the *if-then* statement for purposes of making alternative decisions or for branching to various segments of the program. Thus it behooves us to formulate a method for assigning truth values to this important statement. We must bear in mind that as with conjunction and disjunction, we are merely combining two statements to make a third. According to our conventions, this third statement must have truth values assigned, even though the statements used may bear no apparent relationship. For all possible combinations of T-F values of the hypothesis and conclusion we need to arrive at a meaning, or truth value, for the conditional.

To motivate matters, let us consider a conditional that "has a sensical meaning." To this end let

$p$ be the proposition: The time is midnight.

and

$q$ be the proposition: It is dark outside.

and the conditional

$p \rightarrow q$: *If* it is midnight, *then* it is dark outside.

Now we have to consider the four possible combinations of truth values for $p$ and $q$; these are shown in Table 1.7. The conditional at line 1 must be true (T), for it is true that if it is midnight, then it is dark outside; and line 2 must be false (F), for there is no way for it to be midnight and *not* be dark outside at the same time (exclusive, of course, of artificial lighting or the aurora

**TABLE 1.7**

| $p$ | $q$ | $p \to q$ |
|---|---|---|
| 1. The time is midnight (T) | It is dark outside (T) | ? |
| 2. The time is midnight (T) | It is *not* dark outside (F) | ? |
| 3. It is *not* midnight (F) | It is dark outside (T) | ? |
| 4. It is *not* midnight (F) | It is *not* dark outside (F) | ? |

borealis). Thus we can fill in lines 1 and 2 in the partial truth table shown in Table 1.8.

**TABLE 1.8** Partial Truth Table for Conditional

| $p$ | $q$ | $p \to q$ |
|---|---|---|
| 1. T | T | T |
| 2. T | F | F |
| 3. F | T | ? |
| 4. F | F | ? |

Lines 3 and 4 present us with more difficulty in interpretation. In both cases "It is *not* midnight." Line 3 then states that "It is dark outside." This is entirely possible, so we have to say that this case does not have to be false. The case at line 4: "If it is *not* midnight, *then* it is *not* dark outside" can in no way be claimed to be false either.

Logically, and mathematically, in order to have assurance of no ambiguity, if it is *possible* for the conditional to be true, we assert that the statement is true. This leads to a plausible set of truth values and as we shall see later, conforms to established theories of inference.

Thus we arrive at the completed truth table for the conditional (Table 1.9).

**TABLE 1.9** Truth Table for Conditional

| $p$ | $q$ | $p \to q$ |
|---|---|---|
| 1. T | T | T |
| 2. T | F | F |
| 3. F | T | T |
| 4. F | F | T |

**EXAMPLE 6** Another way to interpret the conditional is to consider it as a *promise*, and ask: When can we say that the promise is broken? This will be equivalent to saying that the statement is false.

To illustrate:

Let $p$ be: You do your homework.
Let $q$ be: You will make an A in the course.

and consider a teacher's "promise":

$p \to q$: *If* you do your homework, *then* you will make an A.

When is the promise broken? Surely it is broken when at the same time we have $p$: You do your homework and $\neg q$: You are not given an A. This is the second line of the truth table (see Table 1.9):

| | $p$ | $q$ | $p \to q$ |
|---|---|---|---|
| 1. | T | T | |
| 2. | T | F | F |
| 3. | F | T | |
| 4. | F | F | |

Now look at the other lines.

*Line 1:* You do your work; you receive an A. Surely the promise is carried out. Thus line 1 is *true.*

*Lines 3 and 4:* You do *not* do your work. *Nothing* has been promised in the case where you do not do the work, so in *no way* can these lines be considered to be false, whether you get an A or not. Consequently, they are *true.*

Thus we have lines 1, 3, and 4 being *true,* corresponding to the conditional truth table of Table 1.9.

These two examples are meant to motivate only. They are statements with some attached "meaning." This, of course, does not have to be the case. For example, "*If* $2 + 2 = 5$, *then* the earth is square" is a conditional statement. Our reasoning is that *regardless* of the meaning of or relationship between the statements joined, the truth values of the conditional remain the same as with the examples given; the only time that it is false is when the *premise is true* and the *conclusion is false.* Thus, although it may be disconcerting, the statement "If $2 + 2 = 5$, *then* the earth is square" is true, and "*If* $3 + 2 = 5$, *then* the earth is square" is false. These represent lines 4 and 2, respectively, in the truth table of the conditional (Table 1.9).

There are many synonyms for the *if-then* statement in English and a number of these are used equivalently in mathematical reasoning. $p \to q$ can be expressed as:

1. *If* $p$, *then* $q$.
2. $q$ *if* $p$.

3. *p only if q.*
4. *q* follows from *p*.
5. *q* provided *p*.
6. *q* whenever *p*.
7. *q* is a necessary condition for *p* (*q* follows necessarily from *p*).
8. *p* is a sufficient condition for *q*.

Two points to make note of regarding the conditional are:

1. The conditional is *true* whenever the hypothesis is false ($\neg p$) [see Table 1.10(a)].
2. The conditional is *true* whenever the conclusion, *q*, is true [see Table 1.10(b)].

**TABLE 1.10**

(a) Conditional is true when *p* is false

| $p$ | $q$ | $p \to q$ | |
|---|---|---|---|
| T | T | T | |
| T | F | F | |
| F | T | T | *p* is false |
| F | F | T | *p* is false |

(b) Conditional is true when *q* is true

| $p$ | $q$ | $p \to q$ | |
|---|---|---|---|
| T | T | T | *q* is true |
| T | F | F | |
| F | T | T | *q* is true |
| F | F | T | |

What we can say in this regard, then, is that the truth values of the conditional will be the same as when $\neg p$ is true [Table 1.10(a)], *or* when *q* is true [Table 1.10(b)]. This combination is the same as the statement $\neg p \vee q$. Table 1.11 shows that $\neg p \vee q$ has the same truth table as that for $p \to q$.

**TABLE 1.11** $\neg p \vee q$ Has Same Truth Table as That for $p \to q$

| $p$ | $q$ | $\neg p \vee q$ | $p \to q$ |
|---|---|---|---|
| T | T | F T T | T |
| T | F | F F F | F |
| F | T | T T T | T |
| F | F | T T F | T |
| | | ① ③ ② | |

Determining Statements from Truth Tables The foregoing analysis with the conditional allows us a means of determining a statement for any given truth table. Suppose, for example, that we are given Table 1.12 and wish to determine a statement with this particular truth function. We note from line 3 that "not $p$ and $q$," $\neg p \wedge q$, produces a value of true (T). Line 4, $\neg p \wedge \neg q$, also produces a true value. Thus we get a value of T either from line 3 or from line 4 (or both). Thus the required statement is formed from the *disjunction* of the two rows:

$$(\neg p \wedge q) \vee (\neg p \wedge \neg q)$$

**TABLE 1.12**

| | $p$ | $q$ | ? |
|---|---|---|---|
| 1. | T | T | F |
| 2. | T | F | F |
| 3. | F | T | T |
| 4. | F | F | T |

**EXAMPLE 7** Find a statement having the truth values as indicated by Table 1.13.

**TABLE 1.13**

| | $p$ | $q$ | ? |
|---|---|---|---|
| 1. | T | T | F |
| 2. | T | F | T |
| 3. | F | T | T |
| 4. | F | F | F |

We see that the statement is true when line 2 is true $(p \wedge \neg q)$ or when line 3 is true $(\neg p \wedge q)$. Thus the required statement is

$$(p \wedge \neg q) \vee (\neg p \wedge q)$$

Compare this truth table with the one for X-OR in Table 1.5(b).

**EXAMPLE 8** Find the statement having the truth value as indicated in Table 1.14.

**TABLE 1.14**

| | $p$ | $q$ | $r$ | ? |
|---|---|---|---|---|
| 1. | T | T | T | T |
| 2. | T | T | F | F |
| 3. | T | F | T | F |
| 4. | T | F | F | T |
| 5. | F | T | T | F |
| 6. | F | T | F | F |
| 7. | F | F | T | T |
| 8. | F | F | F | T |

The required statement is made from the disjunction of the four rows with value (T), namely rows 1, 4, 7, and 8. We get

$$(p \wedge q \wedge r) \vee (p \wedge \neg q \wedge \neg r) \vee (\neg p \wedge \neg q \wedge r) \vee (\neg p \wedge \neg q \wedge \neg r)$$

## EXERCISES

5. Construct truth tables for each of the following.
   (a) $\neg p \rightarrow (p \vee q)$ (b) $q \rightarrow \neg(p \wedge q)$ (c) $(p \wedge q) \rightarrow (p \vee \neg q)$
   (d) $p \rightarrow (q \oplus \neg r)$ (e) $(p \rightarrow q) \wedge \neg(q \rightarrow p)$ (f) $p \rightarrow (\neg q \rightarrow \neg(p \wedge r))$
   (g) $(r \vee q) \oplus \neg(q \vee \neg r)$ (h) $p \rightarrow (q \rightarrow r)$ (i) $(p \rightarrow q) \rightarrow r$

6. Reduce the following sentences to statement forms containing $\wedge$, $\vee$, $\neg$, or $\rightarrow$.
   (a) If today is Tuesday, then this must be Belgium.
   (b) Either John is lying or Fred is telling the truth or Mary is happy.
   (c) Grass will grow only if there is sunshine.
   (d) If no aid is present, then the people will starve.
   (e) College professors are either happy or foolish.
   (f) The sun will stop shining whenever 2 + 3 = 5.
   (g) A necessary condition for the mayor to be reelected is that he carry District 11.
   (h) A sufficient condition for the drought to end is that it rain.

7. Given the propositions

   $p$: Doctors are rich.

   $q$: Teachers are foolish.

   $r$: Lawyers are smart.

   (a) Write each sentence in symbolic form.
      (i) Either doctors are rich or lawyers are smart when teachers are not foolish.
      (ii) It is not true that if lawyers are smart, then teachers are foolish and doctors are not rich.
   (b) Write each symbolic expression as a sentence in English.
      (i) $(p \vee r) \rightarrow q$
      (ii) $\neg(p \wedge q) \rightarrow \neg r$

8. Find statements corresponding to each truth table.

(a)

| $p$ | $q$ | ? |
|---|---|---|
| T | T | F |
| T | F | F |
| F | T | F |
| F | F | T |

(b)

| $p$ | $q$ | $r$ | ? |
|---|---|---|---|
| T | T | T | F |
| T | T | F | F |
| T | F | T | T |
| T | F | F | T |
| F | T | T | F |
| F | T | F | T |
| F | F | T | F |
| F | F | F | F |

(c)

| $p$ | $q$ | $r$ | ? |
|---|---|---|---|
| T | T | T | T |
| T | T | F | T |
| T | F | T | T |
| T | F | F | F |
| F | T | T | T |
| F | T | F | T |
| F | F | T | T |
| F | F | F | F |

**Biconditional ($\leftrightarrow$)** It should also be noted that $p \rightarrow q$ does not have the same meaning as $q \rightarrow p$. The latter is said to be the converse of the conditional, $p \rightarrow q$, and more will be said about it in Section 1.7.

If, however, it is known that both $p \rightarrow q$ and $q \rightarrow p$ are true, we know that $q$ follows from $p$ *and* $p$ follows from $q$. This situation gives rise to our next operation.

Let's examine the case where we have both $p \rightarrow q$ and $q \rightarrow p$. Here we will look at the truth table for the statement $(p \rightarrow q) \wedge (q \rightarrow p)$, shown in Table 1.15. We see that our statement is true when both $p$ and $q$ are true $(p \wedge q)$; also, it is true when they are both false $(\neg p \wedge \neg q)$. Another way to state this is that the statement is true when $p$ and $q$ have the *same* truth value; and false otherwise.

When this is the case we use the symbol $p \leftrightarrow q$. This logical operator is known as the *biconditional*. Mathematical synonyms are: (1) $p$ if and only if $q$ (this is sometimes shortened to $p$ *iff* $q$) and (2) $p$ is a necessary and sufficient condition for $q$. (Some authors use the term *equivalence* for this connector.)

**TABLE 1.15** Truth Table for $[(p \rightarrow q) \wedge (q \rightarrow p)]$ or $p \leftrightarrow q$

| $p$ | $q$ | $(p \rightarrow q)$ | $\wedge$ | $(q \rightarrow p)$ |
|---|---|---|---|---|
| T | T | T | T | T |
| T | F | F | F | T |
| F | T | T | F | F |
| F | F | T | T | T |
| | | ① | ② | ② |

Many theorems and definitions in mathematics take this form.

1. A number $>1$ is prime *iff* the number has no divisors other than itself and 1.
2. An algebraic equation $ax^2 + bx + c = 0$ has real roots *iff* the discriminant, $b^2 - 4ac$, is greater than zero.
3. In Euclidean geometry two lines $l$ and $m$ are parallel *iff* they lie in the same plane and have no points in common.

**EXAMPLE 9** Construct the truth table for $(\neg p \vee q) \leftrightarrow (p \wedge q)$.

This table is constructed as shown in Table 1.16.

**TABLE 1.16**

| $p$ | $q$ | $(\neg p$ | $\vee$ | $q)$ | $\leftrightarrow$ | $(p \wedge q)$ |
|---|---|---|---|---|---|---|
| T | T | F | T | T | T | T |
| T | F | F | F | F | T | F |
| F | T | T | T | T | F | F |
| F | F | T | T | F | F | F |
| | | ① | ③ | ② | ⑤ | ④ |

## EXERCISES

9. Construct the truth tables for each statement.
   (a) $(p \rightarrow q) \leftrightarrow (q \rightarrow \neg p)$
   (b) $\neg p \leftrightarrow (p \rightarrow (p \rightarrow q))$
   (c) $(p \vee r) \wedge (\neg q \leftrightarrow p)$
   (d) $\neg(p \leftrightarrow q)$ (compare with $p \oplus q$ in Table 1.5)
10. Show that $p \leftrightarrow q$ has the same truth table as that for $(p \wedge q) \vee (\neg p \wedge \neg q)$.

## 1.4 Tautologies and Contradictions

Consider the truth table for $(p \rightarrow q) \leftrightarrow (\neg p \vee q)$. As we see from Table 1.17, the final column computed is all T's. A statement in which this is the case—that is, it has the final value, "true" for all possible true/false combinations for the variables—is known as a *tautology*. Three other examples of tautologies are shown in Table 1.18.

**TABLE 1.17**

| $p$ | $q$ | $(p \rightarrow q)$ | $\leftrightarrow$ | $(\neg p \vee q)$ |
|---|---|---|---|---|
| T | T | T | T | T |
| T | F | F | T | F |
| F | T | T | T | T |
| F | F | T | T | T |
| | | ① | ③ | ② see Table 1.11 |

**TABLE 1.18** Examples of Tautologies

| $p$ | $q$ | (1) $p \vee \neg p$ | (2) $p \rightarrow (p \vee q)$ | (3) $(p \wedge q) \rightarrow q$ |
|---|---|---|---|---|
| T | T | TT FT | T T T | T T T |
| T | F | TT FT | T T T | F T F |
| F | T | FT TF | F T T | F T T |
| F | F | FT TF | F T F | F T F |
| | | ①③②① | ① ③ ② | ① ③② |

A *contradiction* is a statement that always takes on the value "false." Hence a statement $P$ is a contradiction if and only if (iff) $\neg P$ is a tautology, and $P$ is a tautology iff $\neg P$ is a contradiction. Three examples of contradictions are shown in Table 1.19.

**TABLE 1.19** Examples of Contradictions

| $p$ | $q$ | (1) $q \wedge \neg q$ | (2) $((p \vee q) \wedge \neg p) \wedge (\neg q)$ | (3) $p \wedge (\neg p)$ |
|---|---|---|---|---|
| T | T | TF FT | T FF F F | TF F |
| T | F | FF TF | T FF F T | TF F |
| F | T | TF FT | T TT F F | FF T |
| F | F | FF TF | F FT F T | FF T |
| | | ①③②① | ① ③② ⑤④ | ①③ ② |

## EXERCISES

1. Show that $(p \wedge (p \vee q)) \leftrightarrow p$ is a tautology.
2. Show that $((p \rightarrow q) \rightarrow p) \rightarrow p$ is a tautology.
3. Show that $((p \wedge q) \vee (\neg p \wedge q)) \vee ((p \wedge \neg q) \vee (\neg p \wedge \neg q))$ is a tautology.
4. Show that $(p \wedge (p \rightarrow q)) \wedge (p \rightarrow \neg q)$ is a contradiction.
5. Show that $\neg[(p \wedge q) \vee (\neg p \vee \neg q)]$ is a contradiction.
6. Determine whether each statement is a tautology or a contradiction or neither.
   (a) $\neg(p \rightarrow (p \vee q))$ (b) $(p \wedge \neg q) \rightarrow (p \vee q)$
   (c) $p \rightarrow (p \rightarrow \neg q)$ (d) $\neg(p \vee q) \vee (q \rightarrow p)$
   (e) $(p \wedge (p \rightarrow q)) \rightarrow q$

## 1.5 Logical Equivalence

If two statements $P$ and $Q$ have the same truth table, they are said to be *logically equivalent*. When this is the case, we use the symbol $P \Leftrightarrow Q$.

A pair of equivalent statements was examined in Table 1.11 in connection with the conditional operator. There, we noted that $p \rightarrow q$ has the same truth table as $\neg p \vee q$ (see Table 1.20). Thus we state that $(p \rightarrow q) \Leftrightarrow (\neg p \vee q)$.

**TABLE 1.20**

| $p \rightarrow q$ | $\neg p \vee q$ |
|---|---|
| T | T |
| F | F |
| T | T |
| T | T |

same truth tables

It is evident that these two truth tables have the *same* truth value in every instance. But this is precisely how we defined the biconditional ($\leftrightarrow$): The biconditional between two statements is *true* when and only when the truth values of the two are the *same*. In our example they are the same everywhere; therefore, the biconditional will be *all T's*: a tautology. This leads to the following definition:

**Definition** Two statements $P$ and $Q$ are *logically equivalent* if and only if the biconditional $P \leftrightarrow Q$ is a *tautology*.

**EXAMPLE 1** $(p \wedge q) \to r$ is logically equivalent to $p \to (q \to r)$. This can be seen from Table 1.21, where $[(p \wedge q) \to r] \leftrightarrow [p \to (q \to r)]$ is seen to be a tautology.

**TABLE 1.21**

| $p$ | $q$ | $r$ | $[(p \wedge q)$ | $\to r]$ | $\leftrightarrow$ | $[p$ | $\to$ | $(q \to r)]$ |
|---|---|---|---|---|---|---|---|---|
| T | T | T | T | T | T | T | T | T |
| T | T | F | T | F | T | T | F | F |
| T | F | T | F | T | T | T | T | T |
| T | F | F | F | T | T | T | T | T |
| F | T | T | F | T | T | F | T | T |
| F | T | F | F | T | T | F | T | F |
| F | F | T | F | T | T | F | T | T |
| F | F | F | F | T | T | F | T | T |
| | | | ① | ② | ⑥ | ③ | ⑤ | ④ |

There are many examples of logical equivalence in the study of propositional calculus that are of considerable importance. We will examine a few in detail.

To motivate this concept of equivalence, consider the propositions

$p$: John is smart.
$q$: John is lucky.

and the statement

$p \vee q$: John is smart or he is lucky.

It should make equal sense to state that

John is lucky or John is smart (i.e., $q \vee p$).

What this intuitively implies is that $p \vee q$ should be logically equivalent to $q \vee p$. The fact that, indeed, $(p \vee q) \Leftrightarrow (q \vee p)$ is evident from Table 1.22, where it is seen that the biconditional is a tautology. If the order in which the variables of a binary operation are combined does not matter, the operation is said to be *commutative*. Thus we see that disjunction is commutative.

**TABLE 1.22**

| $p$ | $q$ | $(p \vee q)$ | $\leftrightarrow$ | $(q \vee p)$ |
|---|---|---|---|---|
| T | T | T | T | T |
| T | F | T | T | T |
| F | T | T | T | T |
| F | F | F | T | F |
| | | ① | ③ | ② |

In like fashion, equivalence between the statements $(p \wedge q)$ and $(q \wedge p)$ can be demonstrated (see Exercise 1) and we establish that conjunction is commutative also. Thus we have established the following two equivalencies or identities:

1. $(p \wedge q) \Leftrightarrow (q \wedge p)$ conjunction is commutative
2. $(p \vee q) \Leftrightarrow (q \vee p)$ disjunction is commutative

These laws are, of course, analogous to the commutative laws of addition and multiplication of real numbers: $x + y = y + x$ and $xy = yx$.

In a similar fashion (see Exercise 1), the *associative* laws of conjunction and disjunction can be demonstrated:

3. $p \wedge (q \wedge r) \Leftrightarrow (p \wedge q) \wedge r$ conjunction is associative
4. $p \vee (q \vee r) \Leftrightarrow (p \vee q) \vee r$ disjunction is associative

Another important logical equivalence is $p \Leftrightarrow (p \wedge p)$ and its accompanying statement $p \Leftrightarrow (p \vee p)$, as seen from Table 1.23, in which we see that each column has the same truth values.

**TABLE 1.23**

| $p$ | $p \wedge p$ | $p \vee p$ |
|---|---|---|
| T | T | T |
| F | F | F |

Here we see a departure from our association with addition and multiplication of real numbers in that the only number $x$ for which $x = x + x$ and $x = x \cdot x$ is $x = 0$. It should not be too surprising to see some departure from the usual laws of arithmetic, since we are dealing with propositions and statements, not numbers. This rule thus established is known as *idempotence.*

5. $p \Leftrightarrow (p \wedge p)$ idempotence of conjunction
6. $p \Leftrightarrow (p \vee p)$ idempotence of disjunction

Still another equivalence that does not follow the patterns set forth by arithmetic is the *absorption* law. This is a law that combines conjunction and disjunction. As do the other laws discussed above, this one takes two forms also. The absorption laws are

7. $p \Leftrightarrow [p \wedge (p \vee q)]$ absorption
8. $p \Leftrightarrow [p \vee (p \wedge q)]$ absorption

We will demonstrate one of them here with Table 1.24. You will be asked to show the other absorption law in Exercise 1.

**TABLE 1.24**

| $p$ | $q$ | $p \wedge (p \vee q)$ |
|---|---|---|
| T | T | T T T |
| T | F | T T T |
| F | T | F F T |
| F | F | F F F |

① ③ ②

*same table as that for proposition p*

Another important law, this one also combining conjunction and disjunction, is the *distributive* law. You know this one as the "factoring" rule of algebra: $x(y + z) = xy + xz$; that is, "multiplication distributes over addition." You also know that addition does *not* distribute over multiplication $(x + (y \cdot z)) \neq (x + y)(x + z)$. However, in the algebra of logical propositions, *both* distributive laws are true:

9. $p \wedge (q \vee r) \Leftrightarrow (p \wedge q) \vee (p \wedge r)$ distributive law
10. $p \vee (q \wedge r) \Leftrightarrow (p \vee q) \wedge (p \vee r)$ distributive law

Still another equivalence that demands special attention is *De Morgan's law.* This law combines conjunction and disjunction along with negation. It, like the others, has two forms:

11. $\neg(p \wedge q) \Leftrightarrow (\neg p \vee \neg q)$ De Morgan's law
12. $\neg(p \vee q) \Leftrightarrow (\neg p \wedge \neg q)$ De Morgan's law

More informally, these two De Morgan's laws state that

**11.** The negation of the conjunction of two statements is logically equivalent to the *disjunction* of the negations of the statements.

**12.** The negation of the disjunction of two statements is logically equivalent to the *conjunction* of the negations of the statements.

Another law to take note of is the law of *double negation*, or *involution*. This states simply that

**13.** $p \Leftrightarrow \neg\neg p \Leftrightarrow \neg(\neg p)$ law of double negation (involution)

Other laws and their accompanying names are shown below in (14). These are of importance equal to the previous laws and the reader should examine the truth tables displayed to prove these equivalencies.

**14.** Given that a statement $T$ is a tautology and statement $F$ is a contradiction, then for any statement $p$, we have the following:

identity laws:

(a) $p \wedge T \Leftrightarrow p$

| $p$ | $p \wedge T$ |
|---|---|
| T | TTT |
| F | FFT |

(b) $p \vee F \Leftrightarrow p$

| $p$ | $p \vee F$ |
|---|---|
| T | TTF |
| F | FFF |

laws of boundness (or dominance):

(c) $p \wedge F \Leftrightarrow F$

| $p$ | $p \wedge F$ |
|---|---|
| T | TFF |
| F | FFF |

(d) $p \vee T \Leftrightarrow T$

| $p$ | $p \vee T$ |
|---|---|
| T | TTT |
| F | FTT |

laws of complement:

(e) $p \wedge \neg p \Leftrightarrow F$

| $p$ | $p \wedge \neg p$ |
|---|---|
| T | TF F |
| F | FF T |

(f) $p \vee \neg p \Leftrightarrow T$

| $p$ | $p \vee \neg p$ |
|---|---|
| T | TT F |
| F | FT T |

The list of logical equivalencies that we have established is shown in Table 1.25. The student would be well advised to learn this table of equivalencies soon, as it will be cited often in the next three chapters.

**TABLE 1.25** Logical Equivalencies

| | *Rule* | *Conjunction* $\wedge$ | *Disjunction* $\vee$ |
|---|---|---|---|
| (1)–(2) | Commutative | $p \wedge q \Leftrightarrow q \wedge p$ | $p \vee q \Leftrightarrow q \vee p$ |
| (3)–(4) | Associative | $p \wedge (q \wedge r) \Leftrightarrow (p \wedge q) \wedge r$ | $p \vee (q \vee r) \Leftrightarrow (p \vee q) \vee r$ |
| (5)–(6) | Idempotence | $p \Leftrightarrow p \wedge p$ | $p \Leftrightarrow p \vee p$ |
| (7)–(8) | Absorption | $p \Leftrightarrow p \wedge (p \vee q)$ | $p \Leftrightarrow p \vee (p \wedge q)$ |
| (9)–(10) | Distribution | $p \wedge (q \vee r) \Leftrightarrow (p \wedge q) \vee (p \wedge r)$ | $p \vee (q \wedge r) \Leftrightarrow (p \vee q) \wedge (p \vee r)$ |
| (11)–(12) | De Morgan | $\neg(p \wedge q) \Leftrightarrow (\neg p \vee \neg q)$ | $\neg(p \vee q) \Leftrightarrow (\neg p \wedge \neg q)$ |
| (13) | Double negation | $p \Leftrightarrow \neg\neg p \Leftrightarrow \neg(\neg p)$ | |
| (14) (a)–(b) | Identity | $p \wedge T \Leftrightarrow p$ | $p \vee F \Leftrightarrow p$ |
| (c)–(d) | Boundness | $p \wedge F \Leftrightarrow F$ | $p \vee T \Leftrightarrow T$ |
| (e)–(f) | Complement | $p \wedge \neg p \Leftrightarrow F$ | $p \vee \neg p \Leftrightarrow T$ |

## EXERCISES

1. Use truth tables to establish the following equivalencies.
   (a) Commutativity of $\wedge$ [rule (1)]
   (b) Associativity of $\wedge$ and $\vee$ [rules (3) and (4)]
   (c) $p \Leftrightarrow [p \vee (p \wedge q)]$ absorption [rule (8)]
   (d) The distributive laws [rules (9) and (10)]
   (e) De Morgan's Laws [rules (11) and (12)]
   (f) Involution [rule (13)]
   (g) $(p \wedge q) \Leftrightarrow \neg(\neg p \vee \neg q)$
   (h) $(p \vee q) \Leftrightarrow \neg(\neg p \wedge \neg q)$
   (i) $\neg p \wedge (p \vee q) \Leftrightarrow \neg p \wedge q$
   (j) $p \vee (\neg p \wedge q) \Leftrightarrow p \vee q$

Using the Laws of Logical Equivalence The list of equivalencies in Table 1.25 can be used to establish other equivalencies. The *transitive* rule of logical equivalence states that: Given three statements $p$, $q$, and $r$, if $p \Leftrightarrow q$ and $q \Leftrightarrow r$, then $p \Leftrightarrow r$. This rule is stated informally at this time, but will be studied later in greater depth.

**EXAMPLE 2** Let us show that $(p \wedge q) \Leftrightarrow [(p \vee \neg q) \wedge q]$ without using truth tables.

It should be noted that it is best to start with the more complicated statement and attempt to reduce it to the simpler by a series of equivalencies. The resulting statements will be equivalent to the first one because of the transitive rule.

| | |
|---|---|
| $(p \vee \neg q) \wedge q \Leftrightarrow q \wedge (p \vee \neg q)$ | by commutativity, rule (1) |
| $\Leftrightarrow (q \wedge p) \vee (q \wedge \neg q)$ | by distribution, rule (9) |
| $\Leftrightarrow (q \wedge p) \vee F$ | by complement, rule (14e) |
| $\Leftrightarrow q \wedge p$ | by identity, rule (14b) |
| $\Leftrightarrow p \wedge q$ | by commutativity, rule (1) |

Thus the two statements are equivalent.

**EXAMPLE 3** Show $((p \wedge q) \vee \neg q) \Leftrightarrow (p \vee \neg q)$ without using truth tables.

| | |
|---|---|
| $(p \wedge q) \vee \neg q \Leftrightarrow \neg q \vee (p \wedge q)$ | by commutativity, rule (2) |
| $\Leftrightarrow (\neg q \vee p) \wedge (\neg q \vee q)$ | by distribution, rule (10) |
| $\Leftrightarrow (\neg q \vee p) \wedge T$ | by complement, rule (14f) |
| $\Leftrightarrow \neg q \vee p$ | by identity, rule (14a) |
| $\Leftrightarrow p \vee \neg q$ | by commutativity, rule (1) |

**EXAMPLE 4** Show $\neg(p \vee r) \vee (p \rightarrow q) \Leftrightarrow (\neg p \vee q)$ without using truth tables.

| | |
|---|---|
| $\neg(p \vee r) \vee (p \rightarrow q) \Leftrightarrow \neg(p \vee r) \vee (\neg p \vee q)$ | see equivalence at Table 1.11 |
| $\Leftrightarrow (\neg p \wedge \neg r) \vee (\neg p \vee q)$ | by De Morgan's law, rule (12) |
| $\Leftrightarrow ((\neg p \wedge \neg r) \vee \neg p) \vee q$ | by associativity, rule (4) |
| $\Leftrightarrow (\neg p \vee (\neg p \wedge \neg r)) \vee q$ | by commutativity, rule (2) |
| $\Leftrightarrow \neg p \vee q$ | by absorption, rule (8) |

**EXAMPLE 5** Use the other rules in Table 1.25 to establish the absorption law (7): $p \wedge (p \vee q) \Leftrightarrow p$. Do not use truth tables.

A starting point is

| | |
|---|---|
| $p \wedge (p \vee q) \Leftrightarrow (p \wedge p) \vee (p \wedge q)$ | by distribution, rule (9) |
| $\Leftrightarrow p \vee (p \wedge q)$ | by idempotence, rule (5) |

which turns out to be absorption law (8). Obviously, reapplying the distributive law becomes cyclic, and thus we get nowhere. This is one of those cases where a "trick" is needed. Such a trick is found by using the identity law (14b). So

| | |
|---|---|
| $p \wedge (p \vee q) \Leftrightarrow (p \vee F) \wedge (p \vee q)$ | by identity, rule (14b) |
| $\Leftrightarrow p \vee (F \wedge q)$ | by distribution, rule (10) |
| $\Leftrightarrow p \vee F$ | by boundness, rule (14c) |
| $\Leftrightarrow p$ | by identity, rule (14b) |

One may also use Table 1.25 to show that one statement may be expressed in another form and we are not sure of the exact nature of this other form. As an example, we can demonstrate that the conjunction $(p \wedge q)$ can be expressed directly in terms negation $(\neg)$ and disjunction $(\vee)$. In establishing such a conjecture, there is, in a sense, no goal: $(p \wedge q) \Leftrightarrow ?$. So we cannot use truth tables. We use the following sequence of equivalencies:

$$\begin{aligned} p \wedge q &\Leftrightarrow \neg\neg(p \wedge q) && \text{by double negation, rule (13)} \\ &\Leftrightarrow \neg(\neg(p \wedge q)) && \text{by double negation, rule (13)} \\ &\Leftrightarrow \neg(\neg p \vee \neg q) && \text{by De Morgan's law, (11)} \end{aligned}$$

Thus we have the desired result. $\wedge$ is expressed directly in terms of $\neg$ and $\vee$.

## EXERCISES

2. Use the laws of Table 1.25 to establish the equivalencies. State the rules used.
   (a) $(p \vee q) \Leftrightarrow (p \wedge \neg q) \vee q$
   (b) $(p \wedge \neg q) \Leftrightarrow \neg q \wedge (p \vee q)$
   (c) $p \vee (p \wedge q) \Leftrightarrow p$ [do not use rule (8)]
   (d) $\neg(p \wedge \neg q) \Leftrightarrow p \rightarrow q$ (see also Table 1.11)
   (e) $\neg p \vee (q \rightarrow \neg r) \Leftrightarrow \neg p \vee \neg q \vee \neg r$ (see also Table 1.11)
   (f) $p \wedge (\neg p \vee q) \Leftrightarrow p \wedge q$
3. Use truth tables or Table 1.25 to determine whether the following pairs of statements are equivalent.
   (a) $\neg p \wedge q, \neg(p \wedge q)$
   (b) $p \rightarrow (q \wedge r), (p \rightarrow q) \wedge (p \rightarrow r)$
   (c) $p \vee (q \rightarrow r), (p \vee q) \rightarrow (p \rightarrow r)$
   (d) $(p \wedge \neg q) \vee (\neg p \wedge q) \vee (p \wedge q); p \vee q$
4. Show that $p \vee q$ can be expressed in terms of $\neg$ and $\wedge$.
5. Show that $p \rightarrow q$ can be expressed in terms of $\neg$ and $\vee$.
6. Show that $p \rightarrow q$ can be expressed in terms of $\neg$ and $\wedge$.
7. Show that $p \leftrightarrow q$ can be expressed in terms of $\neg$ and $\vee$.
8. Show that $p \leftrightarrow q$ can be expressed in terms of $\neg$ and $\wedge$.
9. Show that $p \oplus q$ can be expressed in terms of
   (a) $\neg$ and $\wedge$ (b) $\neg$ and $\vee$

## 1.6 Duality

It should be observed that for the most part, the rules of equivalence in Table 1.25 occur in pairs. For each rule of conjunction there is a corresponding rule of disjunction, and vice versa. In each pair from rules (1)–(12) the corresponding rule can be obtained merely by interchanging every instance of $\vee$ with $\wedge$ and every $\wedge$ with $\vee$. For example, both distribution laws hold:

$$p \wedge (q \vee r) \Leftrightarrow (p \wedge q) \vee (p \wedge r)$$

$$p \vee (q \wedge r) \Leftrightarrow (p \vee q) \wedge (p \vee r)$$

Similarly, observe that the pairs in rule (14) can be obtained the same way, except that additionally in these rules, every instance of $T$ is replaced by $F$ and every $F$ is replaced by $T$ in addition to interchanging the $\wedge$'s and $\vee$'s.

For example, note the identity rules (14a) and (14b):

$$p \wedge T \Leftrightarrow p$$

$$p \vee F \Leftrightarrow p$$

Examine the table carefully and verify that this is the case for every pair.

When such interchanges as above are made in a logical statement we say that we have obtained the *dual* of the statement. Thus the dual of the statement $\neg(p \wedge q) \vee (\neg r \vee T)$ is $\neg(p \vee q) \wedge (\neg r \wedge F)$.

We say that rule (13) (involution) is self-dual.

Suppose that from our laws and rules we are able to derive a certain equivalence. Then the dual of the equivalence can be derived in the same manner as the original equivalence by using dual statements and dual laws at each step. This means that we may conclude the truth of the dual equivalence without having to write out any additional proof.

This concept leads to what is known as the *principle of duality.*

□ Principle of Duality  If two statements $P$ and $Q$ can be shown to be equivalent, then another pair of equivalent statements can be obtained by replacing every instance of $\wedge$ with $\vee$, every $\vee$ with $\wedge$, every $T$ with $F$ and every $F$ with $T$; that is, by exchanging $\wedge$ and $\vee$, and exchanging $T$ and $F$.

In Example 3 of Section 1.5 we showed that $((p \wedge q) \vee \neg q) \Leftrightarrow (p \vee \neg q)$. By using the principle of duality, we immediately have the equivalence

$$((p \vee q) \wedge \neg q) \Leftrightarrow (p \wedge \neg q)$$

[See Exercise 2(b) of Section 1.5.] This principle is very important in that once we have derived a statement of equivalence, we have, in effect, derived another one, the dual of the first.

## EXERCISES

**1.** Find the dual statement of each of the following.

(a) $(p \vee q) \vee \neg(T \wedge q)$

(b) $\neg(p \wedge q \wedge r) \vee \neg(T \vee p)$

(c) $(p \wedge q) \vee (\neg p \wedge q) \vee (p \wedge \neg q) \vee (\neg p \wedge \neg q)$

(d) $(p \wedge q \wedge F) \vee (\neg p \wedge \neg q) \wedge (F \vee p)$

2. Dualize the proof at Example 4 of Section 1.5 to establish

$$\neg(p \wedge r) \wedge (\neg p \wedge q) \Leftrightarrow (\neg p \wedge q)$$

3. Which of the following pairs of statements are duals?

(a) $p \wedge (T \vee \neg q)$; $p \vee (F \wedge q)$

(b) $(p \wedge q) \vee (q \wedge \neg p)$; $(q \wedge p) \vee (p \wedge \neg q)$

(c) $(p \wedge r) \vee \neg(q \wedge r)$; $(p \vee r) \wedge \neg(q \vee r)$

(d) $(p \vee q) \rightarrow (\neg p \wedge T)$; $(p \wedge q) \rightarrow (\neg p \vee F)$

## 1.7 Converse, Inverse, and Contrapositive

We noted earlier (Section 1.3) that even though the commutative law holds for conjunction and disjunction, it does not hold for the conditional. For, observe in Table 1.26 that $p \rightarrow q$ and $q \rightarrow p$ do not have the same truth table.

**TABLE 1.26**

| $p$ | $q$ | $p \rightarrow q$ | $q \rightarrow p$ |
|---|---|---|---|
| T | T | T | T |
| T | F | F | T |
| F | T | T | F |
| F | F | T | T |

$$p \rightarrow q \not\Leftrightarrow q \rightarrow p$$

For a given conditional statement $p \rightarrow q$, the statement $q \rightarrow p$ is said to be its *converse*.

**EXAMPLE 1**

| | *Conditional* | *Converse* |
|---|---|---|
| (a) | *If* it is cold, *then* I will wear my coat. | *If* I wear my coat, *then* it is cold. |
| (b) | *If* you fall in the pond, *then* you will get wet. | *If* you are wet, *then* you fell in the pond. |
| (c) | I'll kiss you *if* the moon is made of green cheese. | *If* I kiss you, *then* the moon is made of green cheese. |

It should be obvious in the examples above that the converse does not have the same meaning as the conditional.

The *inverse* of the conditional $p \to q$ is the statement $\neg p \to \neg q$. Like the converse, the inverse is not equivalent to the conditional, but it *is* equivalent to the converse, as shown in Table 1.27.

The *contrapositive* of the conditional $p \to q$ is the statement $\neg q \to \neg p$. As shown in Table 1.27, the contrapositive *is* equivalent to the given conditional.

**TABLE 1.27**

| $p$ | $q$ | *Conditional* $p \to q$ | *Contrapositive* $\neg q \to \neg p$ | *Converse* $q \to p$ | *Inverse* $\neg p \to \neg q$ |
|---|---|---|---|---|---|
| T | T | T | T | T | T |
| T | F | F | F | T | T |
| F | T | T | T | F | F |
| F | F | T | T | T | T |

↑ *same truth table* ↑ (Conditional, Contrapositive) ↑ *same truth table* ↑ (Converse, Inverse)

conditional ⇔ contrapositive

converse ⇔ inverse

**EXAMPLE 2** The converse of $(p \vee \neg q) \to \neg(r \wedge p)$ is $\neg(r \wedge p) \to (p \vee \neg q)$. The inverse is $\neg(p \vee \neg q) \to (r \wedge p)$, which is equivalent to $(\neg p \wedge q) \to (r \wedge p)$ by using De Morgan's law. The contrapositive is $(r \wedge p) \to (\neg p \wedge q)$.

**EXAMPLE 3** Consider the following conditional statement:

"I will use BASIC *if* the programming assignment is easy."

If we let $b$ be the proposition: "I will use BASIC" and $a$ be the proposition: "The programming assignment is easy," then our statement is of the form $a \to b$. The converse, $b \to a$, is

"If I use BASIC, then the programming assignment is easy."

The inverse, $\neg a \to \neg b$, is

"If the programming assignment is difficult (not easy), then I will not use BASIC."

The contrapositive, $\neg b \to \neg a$, is

"If I do not use BASIC, then the programming assignment is not easy."

**Proof by Contradiction** As noted earlier, mathematical theorems are often stated as a conditional. They are usually of the form: *If* some premise $P$ is true, *then* some conclusion $C$ is true; $P \to C$.

Suppose that we are to prove the "theorem" concerning integers: *If* $n^2$ is an even integer, *then* $n$ is an even integer. Let $p$ be the proposition: *$n^2$ is an even integer*, and $q$ be the proposition: *$n$ is an even integer*. We are to establish $p \to q$ as a theorem. We could approach the proof directly, establishing that $q$ does indeed follow from the premise, $p$, and this is relatively easy to do.

An alternative method of proof is by means of an indirect approach using the *contrapositive*. The method is known as *proof by contradiction*. Since we are to establish $p \to q$, we can equivalently attempt to establish $\neg q \to \neg p$, the contrapositive.

From the propositions above, $\neg p$ becomes: *$n^2$ is not an even integer* or *$n^2$ is an odd integer*. And $\neg q$ becomes *$n$ is an odd integer*. By using the contrapositive, $\neg q \to \neg p$, we shall establish: *If* $n$ is odd, *then* $n^2$ is odd. In actuality, by assuming the conclusion to be false we hope to arrive at a *contradiction* of the original hypothesis: $n^2$ is even.

The proof is as follows:

**Proof** Assume that $n$ is an odd integer. Then it can be written as $2k + 1$ for $k$ being some integer. Now

$$n^2 = (2k + 1)^2 = 4k^2 + 4k + 1 = 2(2k^2 + 2k) + 1$$

which is some even integer $+1$, thus an odd integer, which contradicts the original hypothesis: $n^2$ is even. Thus the theorem is proved.

Another *example* of proof by contradiction is Euclid's classic proof that there is no largest prime. A direct approach to the proof is fairly advanced and detailed, so we proceed by contradiction. We take as assumptions the usual definitions from number theory: the prime factorization of any integer and the fact that a prime is divisible only by itself and 1. Euclid's proof is thus:

**Proof** Assume, to the contrary, that there is a largest prime $P_n$. Let $P_1, P_2, \ldots, P_n$ be all the primes $\leq P_n$. Form the number $N = 1 + P_1 P_2 \cdots P_n$. $N$ is not divisible by any of the primes $P_i$ (and consequently not by any multiple of a prime). Rather, each division will have a remainder of 1. This means that $N$ is prime. But $N > P_n$, which contradicts the fact that $P_n$ is the largest prime. Consequently, there is no largest prime.

## EXERCISES

1. Write, in English, the converse, the inverse, and the contrapositive of each of the following conditionals. Represent the propositions by the accompanying letter in parentheses.
   (a) All human beings are mortal. [The conditional statement is: "If a being is human ($H$), then the being is mortal ($M$)."]
   (b) If the teacher is wise ($T$), then I will make a good grade ($G$).
   (c) We will all starve ($S$) if the drought does not end ($\neg D$).

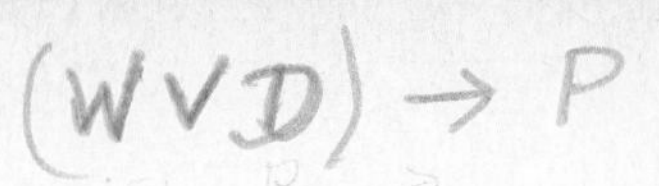


(d) You will be promoted ($P$) if you work hard ($W$) or marry the boss's daughter ($D$).

2. Write the converse, inverse, and contrapositive of each of the following symbolic statements.

(a) $\neg p \rightarrow q$ (b) $(p \vee \neg q) \rightarrow \neg r$ (c) $q \rightarrow \neg(q \rightarrow r)$

3. Prove by contradiction: If the product of two positive integers is odd, then at least one factor is odd.

4. The admonition "spare the rod and spoil the child" can be expressed as the conditional statement: "If you spare the rod, then you will spoil the child." Given that this adage is true, what is it possible to conclude from the following as premises?

(a) The child is spoiled. (b) The child had regular beatings.

(c) The child was never spanked. (d) The child is unspoiled.

## 1.8 Logical Implication and Validity of Arguments

As pointed out earlier, as a logical operator, the conditional need not imply "truth" or causality in that the statements connected need not be related in a "meaningful way." However, the idea of implication or inference is there: *If* something, *then* something else follows.

Statements of this kind generally lead to the notion of whether logical arguments are indeed valid and whether conclusions do follow from premises. This is particularly evident when the proof of mathematical statements and theorems are at stake. Many arguments are presented as acceptable (and are accepted!) without a thought as to their *validity*.

As such arguments are presented, the listener (reader) should decide whether the conclusion reached is *valid;* that is, does the conclusion *necessarily* follow from the hypotheses or premises presented?

To formalize our discussion, we let the premises or hypotheses presented be $P_1, P_2, P_3, \ldots, P_k$, and the given conclusion be $Q$. Our task will be to determine the logical consequence of the premises $P_1, P_2, P_3, \ldots, P_k$. If it turns out that $Q$ does indeed necessarily follow from the premises, we say that the set $P_1, P_2, P_3, \ldots, P_k$ *logically implies* $Q$. Stated symbolically, $(P_1, P_2, P_3, \ldots, P_k) \Rightarrow Q$. In such a case we say that $Q$ is the *logical consequence* of the premises and the argument is valid.

We consider next a method for determining *logical implication*, or the determination of the *validity* of an argument. To be valid, the conclusion must necessarily follow from *all* the premises, taken together. That is, the conclusion, $Q$, must necessarily follow from the *conjunction* of all the premises for all possible truth values. What this says is that the conditional of the conjunction of the premises with the conclusion is true in all cases, or is a *tautology*.

**Definition** Statements $P_1, P_2, \ldots, P_k$ logically imply $Q$ iff the conditional $(P_1 \wedge P_2 \wedge \cdots \wedge P_k) \rightarrow Q$ is a tautology, or $(P_1, P_2, \ldots, P_k) \Rightarrow Q$ iff $(P_1 \wedge P_2 \wedge \cdots \wedge P_k) \rightarrow Q$ is a tautology.

**EXAMPLE 1** Consider the following argument:

> If a person does arithmetic well, his or her checkbook will balance.
> I cannot do arithmetic well. Therefore, my checkbook does not balance.

Our premises are:

> $P_1$: If a person does arithmetic well, his or her checkbook will balance.
> $P_2$: I cannot do arithmetic well.

and the conclusion is

> $Q$: My checkbook does not balance.

If we have as our propositions

> $p$: The person does arithmetic well.
> $q$: The person's checkbook balances.

then we have

$P_1$: $p \to q$
$P_2$: $\neg p$
$Q$: $\neg q$

So our original argument $(P_1 \wedge P_2) \Rightarrow Q$ is

$$[(p \to q) \wedge (\neg p)] \to (\neg q)$$

Is this argument valid? If so, the final conditional has to be a tautology. We construct the truth table shown in Table 1.28. The F circled in the third row indicates that the conditional is not a tautology; thus the argument is *not* valid. The conclusion may be "true," but the *argument is not valid!*

**TABLE 1.28**

| $p$ | $q$ | $[(p \to q)$ ($P_1$) | $\wedge$ | $(\neg p)]$ ($P_2$) | $\to$ | $(\neg q)$ ($Q$) |
|---|---|---|---|---|---|---|
| T | T | T | F | F | T | F |
| T | F | F | F | F | T | T |
| F | T | T | T | T | (F) | F |
| F | F | T | T | T | T | T |

↑ $(P_1 \wedge P_2)$ ↑ $(P_1 \wedge P_2) \to Q$

**EXAMPLE 2** Consider the argument

> All human beings are mortal. Socrates is human. Therefore, Socrates is mortal.

If we use as propositions

$H$: The being is human.
$M$: The being is mortal.
$S$: The being is Socrates.

then we have as premises

$P_1$: $H \to M$
$P_2$: $S \to H$

and the conclusion

$Q$: $S \to M$

The argument, then, is

$$[(H \to M) \wedge (S \to H)] \to (S \to M)$$

**TABLE 1.29**

| $H$ | $M$ | $S$ | $P_1$ $[(H \to M)$ | $\wedge$ | $P_2$ $(S \to H)]$ | $\to$ | $Q$ $(S \to M)$ |
|---|---|---|---|---|---|---|---|
| T | T | T | T | T | T | T | T |
| T | T | F | T | T | T | T | T |
| T | F | T | F | F | T | T | F |
| T | F | F | F | F | T | T | T |
| F | T | T | T | F | F | T | T |
| F | T | F | T | T | T | T | T |
| F | F | T | T | F | F | T | F |
| F | F | F | T | T | T | T | T |

$\uparrow (P_1 \wedge P_2)$   $(P_1 \wedge P_2) \to Q$

We see from Table 1.29 that the conditional is a tautology and the argument is valid.

## EXERCISES

Determine the validity of the following arguments. Use the letters in parentheses as variables for the various propositions.

1. All mathematicians ($M$) are logical ($L$). David ($D$) is logical. Therefore, David is a mathematician. (In this argument the premises are $M \to L$ and $D \to L$. The conclusion is $D \to M$.)

2. If it is Saturday ($S$), then I sleep late ($L$). I am sleeping late. Therefore, it is Saturday.

3. All men who are healthy ($H$) and wealthy ($W$) are smart ($S$). Joe is healthy and poor ($\neg W$). Therefore, Joe is not smart. [The premises are $(H \wedge W) \to S$ and $H \wedge \neg W$.]

4. If John ($J$) or Tom ($T$) wins, then Sally will cry ($S$). Sally does not cry. Therefore, Tom did not win.

5. If I work ($W$), then I will have plenty of money ($M$). If I don't work, then I'll have a good time ($T$). Therefore, I have money or a good time.

6. If you are not patriotic ($\neg P$), then you won't vote ($\neg V$). Hamsters either vote or are happy ($H$). If one is happy, then he is patriotic. Therefore, hamsters are not patriotic.

## 1.9 Other Connectors

Over the pages of this chapter we have studied many operators on logical propositions, each with its own truth table. We might ask: Are there any more? We assert that, yes, there are more, and further ask: How many are there? In other words, how do we go about counting the number of possible distinct operators, that is, truth tables, on two propositions?

To determine this number, we first recall that there are four rows in any truth table on two propositions, and each row can be one of only two values, either a T or an F. Thus there are two distinct choices for a first row; for each of these choices there are two choices for a second row (yielding $2^2 = 4$ possible combinations); for each of these there are two choices for a third row (yielding $2^2 \cdot 2 = 2^3 = 8$); and then again two choices for the fourth row. Thus there are $2^4 = 16$ possible different combinations of T and F (truth tables) on two

**FIGURE 1.3** Counting the number of truth tables for two propositions.

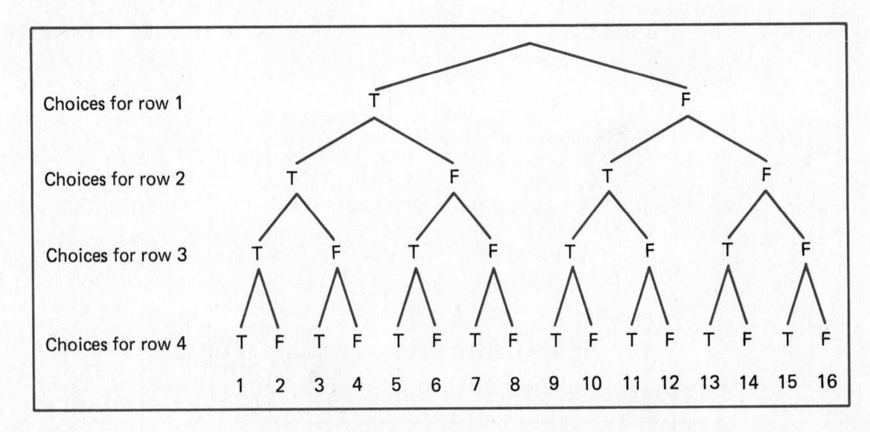

propositions (see the tree diagram of Figure 1.3). These 16 truth tables are exhibited in Table 1.30.

**TABLE 1.30** All Truth Tables on Two Propositions

| $p$ | $q$ | *1* | *2* | *3* | *4* | *5* | *6* | *7* | *8* | *9* | *10* | *11* | *12* | *13* | *14* | *15* | *16* |
|---|---|---|---|---|---|---|---|---|---|---|---|---|---|---|---|---|---|
| T | T | T | T | T | T | T | T | T | T | F | F | F | F | F | F | F | F |
| T | F | T | T | T | T | F | F | F | F | T | T | T | T | F | F | F | F |
| F | T | T | T | F | F | T | T | F | F | T | T | F | F | T | T | F | F |
| F | F | T | F | T | F | T | F | T | F | T | F | T | F | T | F | T | F |

We recognize many of these truth tables from the previous discussion in this chapter:

**(a)** $\wedge$ and $\vee$ are at columns 8 and 2, respectively.

**(b)** $\rightarrow$ and $\leftrightarrow$ are at columns 5 and 7, respectively.

**(c)** $\oplus$ (X-OR) is at column 10. [It should be observed that $(p \oplus q) \Leftrightarrow \neg(p \leftrightarrow q)$.]

**(d)** The converse $(q \rightarrow p)$ is at column 3.

**(e)** Columns 1 and 16 represent the tautology and contradiction, respectively.

**(f)** Proposition $p$ is at column 4 and $\neg p$ is at column 13.

**(g)** Proposition $q$ is at column 6 and $\neg q$ is at column 11.

**(h)** Columns 9, 12, 14, and 15 are the negations of familiar operators, column 9 being $\neg(p \wedge q)$, column 15 being $\neg(p \vee q)$, column 12 being $\neg(p \rightarrow q)$, and column 14 being $\neg(q \rightarrow p)$.

## EXERCISES

1. Find the column in Table 1.30 equivalent to $\neg p \rightarrow q$. To what familiar operator is this equivalent?

2. Find the column in Table 1.30 equivalent to $\neg p \leftrightarrow \neg q$. To what familiar operator is this equivalent?

3. For the conditional $p \rightarrow q$, which column of Table 1.30 shows the inverse, $\neg p \rightarrow \neg q$? The contrapositive, $\neg q \rightarrow \neg p$?

4. (a) By following the counting analysis discussed at the beginning of this section, determine how many possible truth tables there are for *three* variable propositions.
   (b) How many truth tables are there for *four* variable propositions?
   (c) How many truth tables are there for $n$ variable propositions?

**NAND and NOR Operators** There are two operators (truth tables) of Table 1.30 that are particularly worthy of note since they have played an important part in the development of logic design for computers. They are at columns 9 and 15.

Column 9 of Table 1.30 is the truth table for the negation of conjunction, $\neg(p \wedge q)$, and can be read as "not both $p$ and $q$." The two symbols $\neg$ and $\wedge$ can be taken together as a single binary operator on two variables commonly known as the "not and" operator. This has been shortened to become the **NAND** operator. Since, by De Morgan's law, $\neg(p \wedge q) \Leftrightarrow \neg p \vee \neg q$, the NAND operator can also be read as "not $p$ or not $q$."

A very useful fact and one that has been employed in the design of computers and other electronic devices is that all the operators we have studied can be expressed in terms of the NAND operator by applying the laws of De Morgan, involution, and idempotence.

This conversion is illustrated in the following examples, where some of the operators we have studied are converted to expressions that use the NAND operator alone. We will explore these conversions more fully in Chapter 4, where we examine the laws of switching (logic) circuits in detail.

**EXAMPLE 1** Express $p \vee q$ in terms of the NAND operator alone.

$$\begin{aligned} p \vee q &\Leftrightarrow \neg(\neg p \wedge \neg q) && \text{by De Morgan's law} \\ &\Leftrightarrow \underbrace{\neg(\underbrace{\neg(p \wedge p)}_{\text{NAND}} \wedge \underbrace{\neg(q \wedge q)}_{\text{NAND}})}_{\text{NAND}} && \text{by idempotence twice } (p \Leftrightarrow p \wedge p \\ & && \qquad q \Leftrightarrow q \wedge q) \end{aligned}$$

and thus we have $p \vee q$ expressed as the NAND of two NANDs.

**EXAMPLE 2** Express $p \wedge q$ solely in terms of the NAND operator.

$$\begin{aligned} p \wedge q &\Leftrightarrow \neg(\neg(p \wedge q)) && \text{by involution} \\ &\Leftrightarrow \underbrace{\neg(\underbrace{\neg(p \wedge q)}_{\text{NAND}} \wedge \underbrace{\neg(p \wedge q)}_{\text{NAND}})}_{\text{NAND}} && \text{by idempotence} \end{aligned}$$

and here again we have a NAND of two NANDs.

## EXERCISES

5. Express $\neg p$ in terms of the NAND operator alone. (*Hint:* The use of idempotence once will accomplish the transformation.)
6. Express $p \rightarrow q$ in terms of the NAND operator alone. (*Hint:* Use the equivalence at Table 1.11 and Exercise 5.)
7. Express the following statements in terms of the NAND operator alone.
   (a) $p \wedge (\neg p)$ (b) $p \vee (\neg p)$

The negation of disjunction, $\neg(p \vee q)$, is at column 15 of Table 1.30. It can be read as "neither $p$ nor $q$" and has become known as the "not or" or **NOR** operator. It can be rewritten by De Morgan's law as $\neg p \wedge \neg q$, "not $p$ and not $q$."

As with the NAND operator, all operators can be expressed in terms of the NOR operator. Some of these conversions are shown in the following examples.

**EXAMPLE 3** Express $\neg p$ in terms of the NOR operator alone:

$$\neg p \Leftrightarrow \underbrace{\neg(p \vee p)}_{\text{NOR}} \qquad \text{by idempotence } (p \Leftrightarrow p \vee p)$$

and $\neg p$ can be written as a single NOR operator.

**EXAMPLE 4** Express $p \vee q$ in terms of the NOR operator alone.

$$\begin{aligned} p \vee q &\Leftrightarrow \neg(\neg(p \vee q)) && \text{by involution} \\ &\Leftrightarrow \underbrace{\neg(\underbrace{\neg(p \vee q)}_{\text{NOR}} \vee \underbrace{\neg(p \vee q)}_{\text{NOR}})}_{\text{NOR}} && \text{by idempotence} \end{aligned}$$

and $p \vee q$ is the NOR of two NORs.

## EXERCISES

8. Express $p \wedge q$ in terms of the NOR operator alone by using De Morgan's law.
9. Express $p \rightarrow q$ in terms of the NOR operator alone.
10. Express the following statements in terms of the NOR operator alone.
    (a) $p \wedge (\neg p)$ (b) $p \vee (\neg p)$

For the interested reader, we point out that the mathematical representation for the NAND operator is the symbol $|$. Thus $\neg(p \wedge q)$ can be represented as $p|q$. This symbol is known as the Sheffer stroke. The mathematical representation for the NOR operator is the symbol $\downarrow$ (the Peirce arrow). Thus $\neg(p \vee q)$ is represented as $p \downarrow q$. An interesting exercise is to show that

$$\begin{aligned} p \wedge q &\Leftrightarrow (p|q)|(p|q) && \text{(NAND alone)} \\ &\Leftrightarrow (p \downarrow p) \downarrow (q \downarrow q) && \text{(NOR alone)} \end{aligned}$$

and that

$$\begin{aligned} p \vee q &\Leftrightarrow (p|p)|(q|q) && \text{(NAND alone)} \\ &\Leftrightarrow (p \downarrow q) \downarrow (p \downarrow q) && \text{(NOR alone)} \end{aligned}$$

## Summary and Selected References

The subject of logic and logical propositions appears in many textbooks, sometimes as a topic in its own right and in others as a prelude to more advanced subjects. It is the latter approach that has been taken in this chapter. The mathematical model of logical propositions lays the formulation for discrete mathematics as practiced by computer scientists.

There are many excellent references for the subject that take a variety of approaches. For a treatment from a theoretical standpoint, readers may consult Mendelson (1964), Wilder (1965), and Shoenfield (1967).

From a more practical approach applied to computer science, there are a variety of references that include logic and logical propositions as part of their treatment, such as Korfhage (1966, 1974), Mendelson (1970), Lipschutz (1976), Tremblay and Monahar (1975), and Sahni (1981).

## 2.1 Introduction

As a further development of our discrete structures, we introduce the notion of sets in this chapter. Sets are one of the most intuitive and perhaps far-reaching concepts in all mathematics and certainly are not limited to discrete systems. Undoubtedly, the reader is familiar with set notation and some of the associated theory from previous courses.

Our development of sets here will be to focus on the discrete nature of the theory and to develop the necessary portions by using the concepts of logical propositions from Chapter 1.

## 2.2 Representation of Sets and Their Elements

We use the term *set* to mean any collection or grouping of objects that is well defined. The collection itself is considered an entity and it should be possible to identify objects as belonging to the collection or not. For example, the collection of all living Americans is clearly a set since it is easy to determine whether a person belongs to the collection. Similarly, the collection of integers that lie between 0 and 100 inclusive is a set. The collection of books in the Library of Congress is a set, as is the collection of all sentences in this paragraph, or all programs written BASIC that were run in September 1970 (or even 1870!).

If, on the other hand, we consider the collection of all beautiful people, it might be difficult to determine whether or not a person belongs. Thus this collection is not technically a set, nor is the collection of long Pascal programs run yesterday. (Who will define "long" precisely for us?)

The language one uses to describe set theory is easily obtained from the language of propositions and statements from Chapter 1. The fact that it should be possible to determine whether or not an object belongs to a set $A$ is equivalent to the *truth value* of the proposition: "Object $x$ belongs to set $A$." This statement should be clearly true or false; the object is either *in* the set or *out* of (not in) the set. If this proposition were true, we symbolically state $x \in A$, meaning that $x$ is in set $A$; if false, we state $\neg(x \in A)$ or, equivalently, $x \notin A$, that is, $x$ is not in set $A$. The objects of a set are called the *elements* or *members* of the set.

To define some sets we may list the members within braces.

1. The set of vowels of the English alphabet is $\{a, e, i, o, u\}$.
2. The set of even integers between 10 and 20 inclusive is $\{10, 12, 14, 16, 18, 20\}$.
3. The set of solutions to the equation $2x = 14$ is $\{7\}$. A set such as this, which contains only one element, is called a *singleton.*
4. The set of solutions (solution set) of $x^2 - 3x + 2 = 0$ is $\{1, 2\}$. A set with two elements only is called a *doubleton.*
5. The set $\{a, a, b, c, c\}$ should be written $\{a, b, c\}$. In this regard we point out that the elements of a set should be distinct. The term *multiset* is sometimes used when one or more elements is repeated in the listing.

An alternative way to indicate a set is by stating a *property* that is true about all members.

6. Instead of listing the vowels as in set 1, we can indicate the set as $\{x \mid x \text{ is a vowel of the English alphabet}\}$. The "|" means "such that"; thus the expression is to be read: "The set of elements $x$ *such that* $x$ is a vowel of the English alphabet."
7. $\{x \mid x \text{ is an even integer and } 10 \leq x \leq 20\}$ is the same as set 2.
8. $\{x \mid x \text{ is a solution to } x^2 - 3x + 2 = 0\}$ is an alternative to set 4.
9. $\{x \mid x \text{ is a grain of sand on Coney Island}\}$ is a set that is constantly changing but should be well defined at any one given instant.

All the sets above are finite. Set 9 is very large but nonetheless is finite. Examples of infinite sets are:

10. The set of all integers greater than 3, written as $\{x \mid x \text{ is an integer greater than } 3\}$. Also, the members of this set can be listed as $\{4, 5, 6, 7, \ldots\}$.
11. $\{x \mid x \text{ is a real number and } 0 < x < 1\}$. It is impossible to demonstrate a listing of this set.

Most of the sets we will encounter in our study will be finite; thus we will not consider infinite sets further here.

Unless otherwise indicated, sets will be designated by capital letters of the alphabet, such as $A$, $B$, $C$, and so on. The *number* of elements of a finite set $A$ will be designated by $\#(A)$. Thus if set $A = \{x \mid x \text{ is a solution to } x^2 - 3x + 2 = 0\}$, then $\#(A) = 2$. If set $T = \{x \mid x \text{ is a solution to } x^5 - 3x^4 + 5x^2 + 7 = 0\}$, then $\#(T) = 5$. It may be difficult to obtain an actual listing of the elements of this set, but there *are* five members, as we know from college algebra.

Two sets $A$ and $B$ are said to be *equal* ($A = B$) if and only if $A$ and $B$ have the same members. If $A$ and $B$ do not have the same members, we write $A \neq B$ or, equivalently, $\neg(A = B)$. For example, $\{1, 2\} = \{x \mid x \text{ is a solution to } x^2 - 3x + 2 = 0\}$. The order in which the elements are listed is immaterial; thus $\{a, b, c\} = \{c, a, b\}$.

**Venn Diagrams** It is customary to represent sets pictorially. Usually, this is done by means of circles and other geometric figures known as *Venn diagrams.* These diagrams are named after the Englishman John Venn, who used them in his book *Symbolic Logic*, published in 1881, to illustrate principles of logic.

The diagram of a single set $A$ is shown in Figure 2.1. The shaded circle is meant to represent all elements of set $A$. In the diagram the elements of the set can be thought of as points within the circle. Elements not in $A$ are the points outside the circle but within the bounding rectangle. This rectangle, labeled $U$, is known as the universe of discourse, or universal set, or simply as the *universe.* The universe represents all elements that could be under discussion when referring to a set.

One generally assumes a universal set when discussing a particular set. If, for example, we have set $B = \{1, 2, 3\}$ and we want to say something about the elements that are *not* in $B$, what elements do we have? Possibilities include 8, $-19.725$, $\pi^2$, all elephants, white horses, trees, and B-1 bombers. If we limit our discourse to let the universal set be "all positive integers," then we know the elements that are *not* in $B$ constitute the set $\{4, 5, 6, \ldots\}$.

To illustrate further, if we consider set $A = \{x \mid x \text{ is an even integer and } 1 \leq x \leq 10\}$, then $A = \{2, 4, 6, 8, 10\}$. If the universe is envisioned as all the integers between 1 and 10 inclusive, then $U = \{1, 2, 3, 4, 5, 6, 7, 8, 9, 10\}$ and

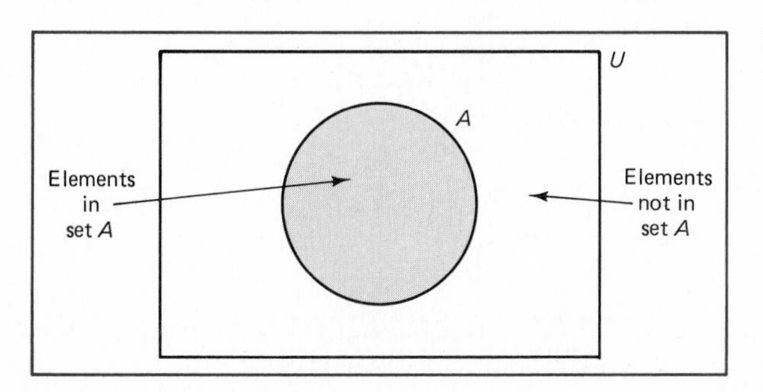


**FIGURE 2.1**

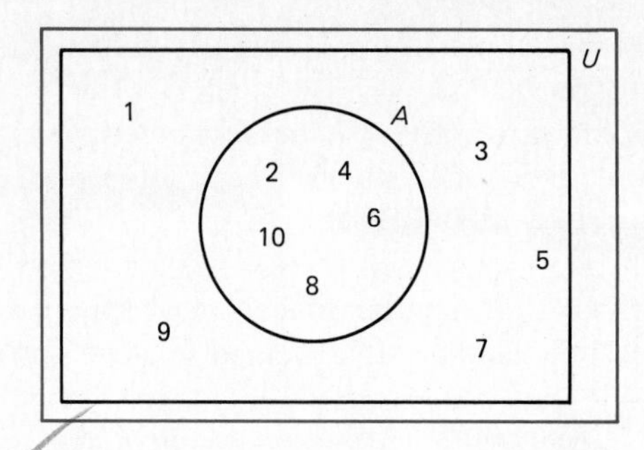


FIGURE 2.2 Set $A = \{2, 4, 6, 8, 10\}$ with $U = \{1, 2, 3, 4, 5, 6, 7, 8, 9, 10\}$.

Complement

the Venn diagram can be represented as shown in Figure 2.2. When the universe is thus limited, the elements that are *not* in set $A$ constitute the set $\{1, 3, 5, 7, 9\}$. So we see that, for instance, $7 \notin A$ or $\neg(7 \in A)$.

This set of elements in the universe that are not in set $A$ is called the *complement* of $A$ and is symbolized by $\bar{A}$. Thus, for our example, $\bar{A} = \{1, 3, 5, 7, 9\}$.

In general, we define the complement by using the symbol for conjunction from Chapter 1:

**Definition** *Complement of a set A*: $\bar{A} = \{x | x \in U \wedge x \notin A\}$

Thus we have the counterpart for negation ($\neg$) from Chapter 1. Using the language of propositions, we have

$$\bar{A} = \{x | x \in U \wedge \neg(x \in A)\}$$

The complement of a set $A$ is shown by the shaded part of Figure 2.3. When we assume that a universe is given, we can write

$$\bar{A} = \{x | x \notin A\}$$

or

$$\bar{A} = \{x | \neg(x \in A)\}$$

FIGURE 2.3 $\bar{A}$ (shaded).

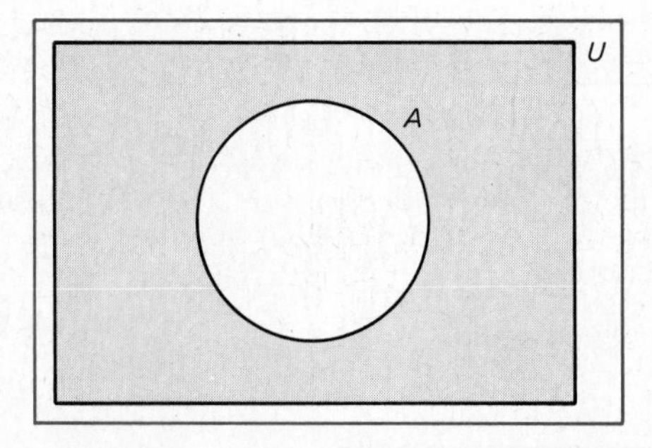

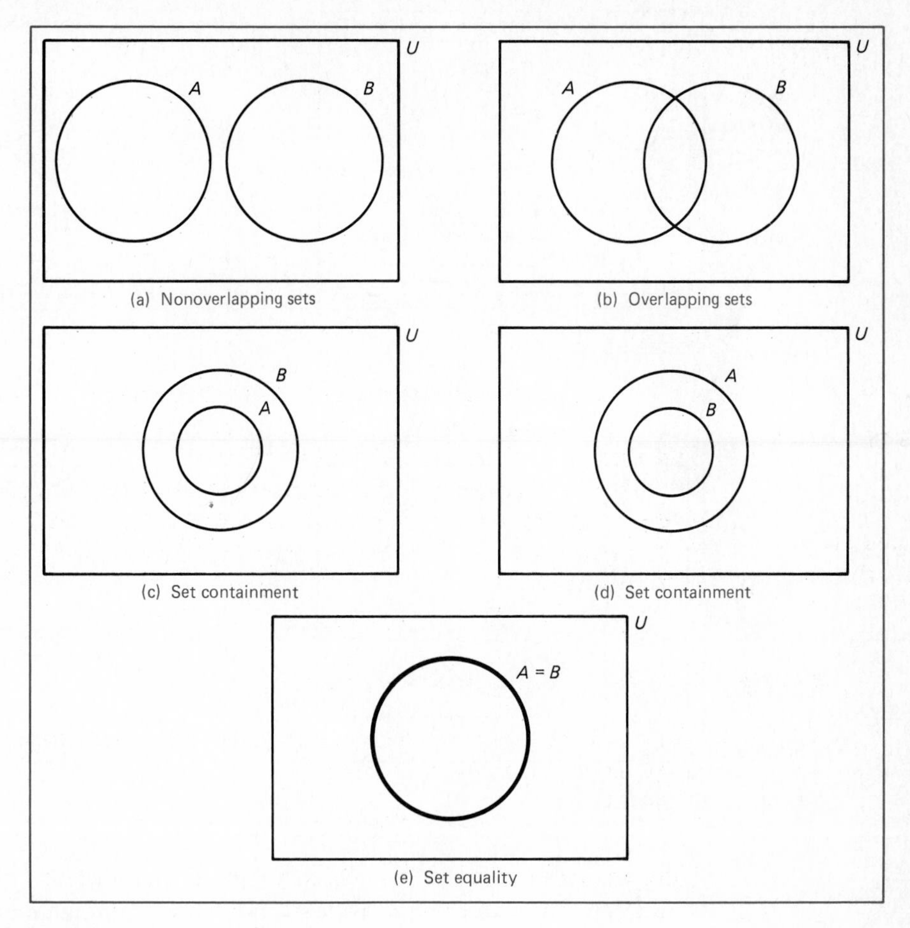


**FIGURE 2.4**

Representation of Two or More Sets with Venn Diagrams: There are a number of ways to represent two sets $A$ and $B$ in Venn diagrams. These are shown in Figure 2.4. If it is clear that the two sets have no common elements, two nonoverlapping circles can be drawn as in Figure 2.4(a). If there is a possibility of some common elements, the overlap can be shown as in Figure 2.4(b). This diagram is the one that is usually depicted for two arbitrary sets. Perhaps one set can be completely contained in the other. This possibility is shown in Figure 2.4(c) and (d). The other option that we have is for the two sets to contain identically the same elements, or $A = B$. This is pictured in Figure 2.4(e). Each of these possibilities will be explained further in the ensuing discussion.

Of course, we need not limit our diagrams to just circles. Any geometric figures properly placed will suffice. Thus the displays in Figure 2.5 can equally well represent two overlapping sets.

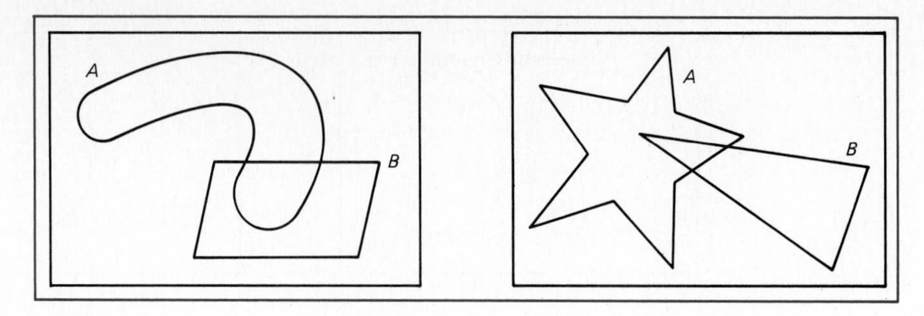


FIGURE 2.5

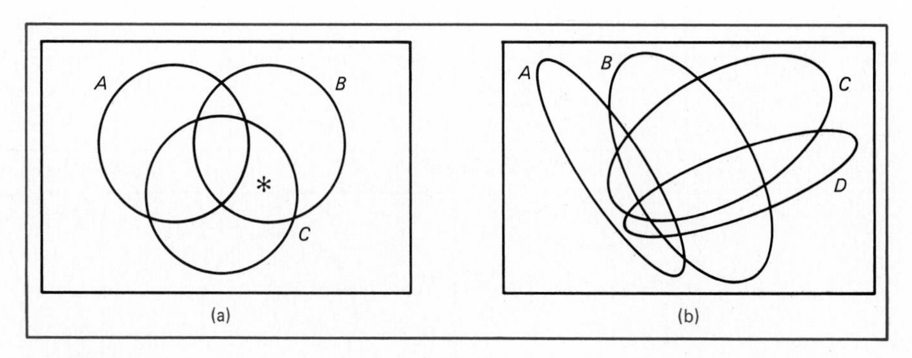


FIGURE 2.6

Venn diagrams can also be used to illustrate more than two sets. Figure 2.6(a) shows three sets $A$, $B$, and $C$. All possibilities of overlap are conveniently represented. The region marked with the asterisk represents those elements in common with sets $B$ and $C$ and at the same time not in set $A$.

It is not possible to illustrate all possible overlaps of four sets with circles. However, by using ellipses as in Figure 2.6(b), it is possible. For the purposes of this text we will limit our discussion to three sets.

## EXERCISES

1. Which of the following collections can be classified as sets?
   (a) All famous books.
   (b) All books longer than 1000 pages.
   (c) The collection of small states.
   (d) All the good teachers at this school.
   (e) All English words beginning with the letter q.
   (f) All even numbers.

(g) All odd numbers divisible by 2.

(h) All interesting computer programs that you have written.

2. If the universal set is $\{x | x$ is a current or past president of the United States$\}$, what is the complement of the following?

(a) All post–World War II presidents whose name begins with the letters N, R, or T.

(b) All female presidents.

(c) All male presidents.

(d) All good presidents.

3. List the elements of the following sets with the universal set being $\{1, 2, 3, 4, \ldots\}$.

(a) $\{x | 10 \leq x < 21\}$

(b) $\{x | x \text{ is odd} \wedge x \leq 12\}$

(c) $\{x | x + 7 = 4\}$

(d) $\{x | x + 7 \neq 4\}$

(e) $\{x | x \text{ is divisible by } 3\}$

(f) $\{x | x \text{ is a perfect square} \wedge x \leq 150\}$

(g) $\{x | x \text{ is a perfect square} \wedge x \text{ does not end with a } 1, 4, 5, 6, \text{ or } 9\}$

(h) $\{x | x \text{ is even or } x > 12\}$

4. List the complements of the sets in Exercise 3.

## 2.3 Subsets and Containment

If, when we are given two sets, $A$ and $B$, it happens that every element of set $A$ is also an element of set $B$, as shown in Figure 2.4(c), we say that set $A$ is a *subset* of set $B$. This is symbolized as $A \subseteq B$. We may also say that $A$ is *contained* in $B$, or equivalently that $B$ contains $A$, or even that $B$ includes $A$.

An alternative symbolism for $A \subseteq B$ is $B \supseteq A$. [Note that the picture for $B \subseteq A$ is shown in Figure 2.4(d).]

The containment symbol, $\subseteq$, has embedded within it the equality sign (similar to $\leq$ for real numbers). This means that $A \subseteq B$ allows for the possibility of $A = B$. If equality does not exist ($A \neq B$), but it is true that $A$ is contained within $B$, we use the symbol $A \subset B$ and say that $A$ is a *proper* subset of $B$. This is referred to as *proper containment* and is also symbolized as $B \supset A$.

**EXAMPLE 1** We have

**(a)** $\{1, 2\} \subseteq \{1, 2, 3, 4, 5\}$

**(b)** $\{2\} \subseteq \{x | x \text{ is a solution of } x^2 - 3x + 2 = 0\}$

**(c)** $\{$Harry Truman, Abraham Lincoln$\} \subseteq \{$all past presidents of the United States$\}$

Also it should be evident that

(d) $\{1, 3\} \nsubseteq \{2, 3, 5, 7, 9\}$
(e) {Alexander Hamilton} $\nsubseteq$ {all past presidents of the United States}

**EXAMPLE 2** It should be pointed out that individual elements themselves of a particular set are *not* subsets of the set. Thus we have

(a) $3 \in \{1, 3, 5\}$, but $3 \nsubseteq \{1, 3, 5\}$, whereas $\{3\} \subseteq \{1, 3, 5\}$.
(b) $\{1\} \notin \{1, 3, 5\}$, but $\{1\} \in \{\{1\}, 1, 3, 5\}$ since $\{1\}$ is an element and $\{1\} \subseteq \{\{1\}, 1, 3, 5\}$ since 1 is an element.

We formally define *set containment* as follows:

**Definition** A set $A$ is a subset of set $B$ ($A \subseteq B$) if and only if every element of set $A$ is also an element of set $B$.

An important idea emerges from this definition. If we are given that $A \subseteq B$, then every element $x$ of set $A$ is necessarily in set $B$ also. The converse situation holds in addition: If all elements of set $A$ are also elements of set $B$, then $A \subseteq B$. This may be stated using the language of the propositional calculus of Chapter 1. We have

$$A \subseteq B \quad \text{iff} \quad (\textit{if } x \in A, \textit{ then } x \in B)$$

or

$$A \subseteq B \quad \text{iff} \quad (x \in A \rightarrow x \in B)$$

If we let

$$p \text{ be the proposition:} \quad x \in A$$

and

$$q \text{ be the proposition:} \quad x \in B$$

then the conditional $p \rightarrow q$ is the logical counterpart of the set-theoretic $A \subseteq B$. This containment can conveniently be represented as $A \rightarrow B$, which has the following meanings when $A$ and $B$ are sets:

1. $A$ is a subset of $B$.
2. If $x \in A$, then $x \in B$.
3. Set $B$ necessarily follows if we have set $A$.
4. Set $A$ is a sufficient condition for set $B$.

This *direct relationship* between set *containment* and the *conditional* statement of logical propositions can be used to establish a number of set properties and theorems. In proving statements about sets, then, we can easily resort to the use of truth tables as a tool of proof.

To illustrate this idea, recall the *transitive* property as introduced in Section 1.5. There we stated that for logical statements $P$, $Q$, and $R$, we have: If $P \Leftrightarrow Q$ and $Q \Leftrightarrow R$, then $P \Leftrightarrow R$. Also (Section 1.8), if $P \Rightarrow Q$ and $Q \Rightarrow R$, then $P \Rightarrow R$. At this point we state that the transitive property holds also for set containment. That is, for sets $A$, $B$, and $C$ we have: If $A \subseteq B$ and $B \subseteq C$, then $A \subseteq C$. We will now prove this by using propositional calculus and truth tables in

□ Theorem 2.1 For sets $A$, $B$, and $C$, if $A \subseteq B$ and $B \subseteq C$, then $A \subseteq C$ (the transitive property of set containment).

**Proof** Recall from the previous discussion that for arbitrary sets, $A$ and $B$, $A \subseteq B$ can be written $A \rightarrow B$. So we want to test the validity of the argument: If ($A \subseteq B$ and $B \subseteq C$), then $A \subseteq C$.

We convert to logical conditionals and ask if from the conjunction of the premises

$$A \rightarrow B\ (A \subseteq B)$$

and

$$B \rightarrow C\ (B \subseteq C)$$

the conclusion $A \rightarrow C\ (A \subseteq C)$ follows. Recalling Section 1.8, we then ask: Does $[(A \rightarrow B) \wedge (B \rightarrow C)] \Rightarrow (A \rightarrow C)$?

Using the truth table for logical implication, we have Table 2.1. The conditional at ⑤ is a tautology; thus the argument is valid, and the transitive property of set containment holds.

**TABLE 2.1**

| $A$ | $B$ | $C$ | $[(A \rightarrow B)$ | $\wedge$ | $(B \rightarrow C)]$ | $\rightarrow$ | $(A \rightarrow C)$ |
|---|---|---|---|---|---|---|---|
| T | T | T | T | T | T | T | T |
| T | T | F | T | F | F | T | F |
| T | F | T | F | F | T | T | T |
| T | F | F | F | F | T | T | F |
| F | T | T | T | T | T | T | T |
| F | T | F | T | F | F | T | T |
| F | F | F | T | T | T | T | T |
| F | F | F | T | T | T | T | T |
| | | | ① | ④ | ② | ⑤ | ③ |

The above is a *general* proof of the transitive property of set containment. A special case can be demonstrated with Venn diagrams, as shown in Figure 2.7. $A \subseteq B$ is shown by Figure 2.7(a), and $B \subseteq C$ is shown by Figure 2.7(b). If we

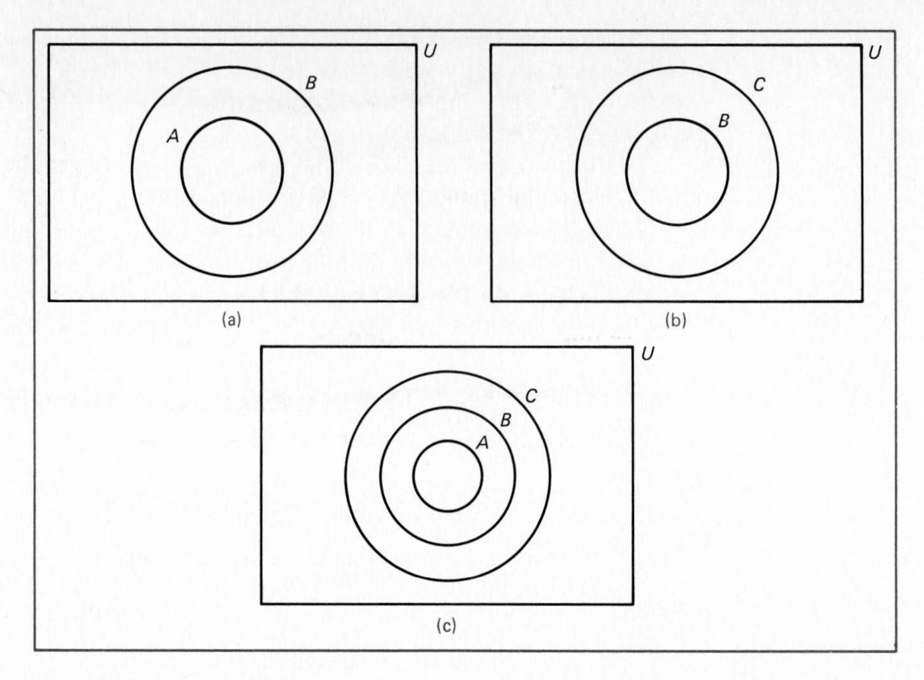


FIGURE 2.7

combine these two diagrams, we have Figure 2.7(c), which clearly shows (for this case) that $A \subseteq C$.

In this light, Venn diagrams are to be used for intuitive purposes and to help illustrate set properties. As is often the case in mathematics, when a figure helps one "see" a problem, the same is true for sets. As is also true, the figure does *not* prove the point. This must be done analytically—for the general case—and in our illustration here, the proof can be carried out by use of truth tables.

Another way to approach the proof of set statements can be illustrated by the following "analytic" method of proving the transitive property of set containment.

The premise $A \subseteq B$ can be restated as

$$\text{For all } x \text{ we have: if } x \in A, \text{ then } x \in B.$$

Or we may say

$$A \subseteq B \quad \text{iff} \quad (\text{for all } x)\,(x \in A \rightarrow x \in B).$$

Similarly

$$B \subseteq C \quad \text{iff} \quad (\text{for all } x)\,(x \in B \rightarrow x \in C).$$

By using conjunction we obtain

$$(A \subseteq B) \wedge (B \subseteq C) \quad \text{iff} \quad (\text{for all } x)\,[(x \in A \rightarrow x \in B) \wedge (x \in B \rightarrow x \in C)].$$

Since the conditional is transitive, we have

$$(A \subseteq B) \wedge (B \subseteq C) \quad \text{iff} \quad (\text{for all } x)\,(x \in A \rightarrow x \in C)$$

which is a restatement of $A \subseteq C$.

Another important property of set containment is contained in the following:

□ Theorem 2.2 For all sets $A$ and $B$ we have $A = B$ iff $A \subseteq B$ and $B \subseteq A$.

**Proof** To prove this theorem, we translate the expression to logical statements:

1. Set property: $A = B$ iff ($A \subseteq B$ and $B \subseteq A$)
2. Propositional statement: $(A \leftrightarrow B) \Leftrightarrow [(A \rightarrow B) \wedge (B \rightarrow A)]$

and note that we are to show logical equivalence. Thus

$$(A \leftrightarrow B) \leftrightarrow [(A \rightarrow B) \wedge (B \rightarrow A)]$$

should be a tautology, which it is shown to be from Table 2.2.

**TABLE 2.2** $A = B$ iff $(A \subseteq B \wedge B \subseteq A)$

| $A$ | $B$ | $(A \leftrightarrow B)$ | $\leftrightarrow$ | $[(A \rightarrow B)$ | $\wedge$ | $(B \rightarrow A)]$ |
|---|---|---|---|---|---|---|
| T | T | T | T | T | T | T |
| T | F | F | T | F | F | T |
| F | T | F | T | T | F | F |
| F | F | T | T | T | T | T |
| | | ① | ⑤ | ② | ④ | ③ |

Compare Table 2.2 with Table 1.15 and see that we are merely reshowing a definition of the biconditional.

This theorem gives a necessary and sufficient condition for set equality. That is, if you are to show that two sets $A$ and $B$ are equal, $A = B$, it must be shown that $A \subseteq B$ *and* $B \subseteq A$. We will use Theorem 2.2 later in the chapter to establish some other set equivalencies.

Null Set ($\varnothing$) Sometimes a set can contain *no* members at all. We call such a set the *empty set* or *null set* and symbolize it with $\varnothing$ or $\{\ \}$. (*Note:* We should *not* use the symbol $\{\varnothing\}$ for the null set, for this is a set that contains exactly one member, namely the null set.)

Examples of the null set are:

The set of BASIC programs executed on August 15, 1870
$\{x \mid x \text{ is an integer and } 2x = 7\}$
The set of symphonic orchestras performing on the planet Jupiter
The set of all elephants looking over your shoulder now

There are two important properties concerning the null set. The first is that the null set is a *subset* of *every* set. We prove this in

□ Theorem 2.3 The null set is a subset of every set ($\varnothing \subseteq A$, for all sets $A$).

**Proof** We prove this by contradiction. Assume, to the contrary, that $\varnothing \nsubseteq A$ for some set $A$. Then there is some element $x \in \varnothing$ that is not in set $A$. This is impossible since there are no elements in $\varnothing$. Therefore, $\varnothing \subseteq A$ for all sets $A$.

**Alternative Proof** (by using truth tables): Let $p$ be the proposition $x \in \varnothing$ and $q$ be the proposition $x \in A$. We ask about the nature of $p \to q$ (or $\varnothing \subseteq A$). Is this true in all cases?

Now $p$ is *always false* since $x \notin \varnothing$ for any $x$; and $q$ may be true or false; hence

| $p$ | $q$ | $p \to q$ |
|---|---|---|
| F | T | T |
| F | F | T |

The truth table is a *tautology;* hence $\varnothing \subseteq A$, for all $A$.

The second important property is that even though there may be several seemingly different examples of the null set, there is indeed only *one*. This is proved in

□ Theorem 2.4 The null set is unique.

**Proof** Suppose that there are two null sets, $\varnothing_1$ and $\varnothing_2$. Since, from Theorem 2.3, we know that the null set is a subset of any set, we have

$$\varnothing_1 \subseteq \varnothing_2 \quad \text{and} \quad \varnothing_2 \subseteq \varnothing_1$$

By Theorem 2.2, it follows that $\varnothing_1 = \varnothing_2$. Hence there is only one null set.

## EXERCISES

1. Prove the following statement about sets $A$, $B$, and $C$:

   If $B \subseteq A$ and $C \subseteq B$, then $C \subseteq A$.

   Use truth tables. Draw Venn diagrams to illustrate.

2. Determine whether the following statement about sets $A$, $B$, and $C$ is true:

   If $A \subseteq B$ and $A \subseteq C$, then $B \subseteq C$.

3. Give examples of sets so that each of the following is true.

   (a) $A \subseteq \varnothing$ (b) $\varnothing \in A$ and $\varnothing \subseteq A$ (c) $A \subseteq B$ and $B \subseteq A$

   (d) $A = B$ and $A \not\subset B$ (e) $A \in B$ and $A \subseteq B$

4. Show the truth table that represents proper containment, $A \subset B$. Recall that $A \subset B$ iff $A \subseteq B$ *and* $A \neq B$.

5. Let the universe of discourse be the set of natural numbers $\{1, 2, 3, 4, \ldots\}$, and define the following sets.

$$A = \{x \mid x \text{ is even}\}$$
$$B = \{x \mid x = 2 \text{ or } x \text{ is odd}\}$$
$$C = \{x \mid x \text{ is a prime number}\}$$

Determine whether each of the following is true.

(a) $B = C$ (b) $B \subseteq C$ (c) $C \subseteq B$ (d) $B \subset C$ (e) $C \neq A$

6. State whether each of the following is true or false.

(a) $\varnothing \in \{a, b, c\}$ (b) $\varnothing \subseteq \{a, b, c\}$ (c) $\{\varnothing\} \in \{a, b, c\}$
(d) $\{\varnothing\} \subseteq \{a, b, c\}$ (e) $\varnothing \in \{\varnothing, \{\varnothing\}\}$ (f) $\varnothing \subseteq \{\varnothing, \{\varnothing\}\}$
(g) $\{\varnothing\} \in \{\varnothing, \{\varnothing\}\}$ (h) $\{\varnothing\} \subseteq \{\varnothing, \{\varnothing\}\}$ (i) $\varnothing \in \{\{\varnothing\}\}$
(j) $\varnothing \subseteq \{\{\varnothing\}\}$ (k) $\{\varnothing\} \in \{\{\varnothing\}\}$ (l) $\{\varnothing\} \subseteq \{\{\varnothing\}\}$
(m) $\varnothing = \{0\}$ (n) $\{\varnothing\} = \{0\}$ (o) $\{\varnothing\} = 0$
(p) $\varnothing \subseteq \{0\}$

## 2.4 Binary Operations on Sets

As was true with propositions and logical statements, there are ways to combine sets in various ways to yield new ones. The combination operations we will consider are *binary* in that they combine *two* sets into a new one.

For illustrative purposes let us suppose that set $A$ consists of all students taking a course in the APL programming language, with set $B$ being those taking a BASIC course. Then when we consider the proposition about some student, $x$: $x \in A$, we mean that "student $x$ is taking an APL course." Similarly, $x \in B$ means that "student $x$ is taking a BASIC course." These are well-defined collections of students and it should be easy to determine whether $x \in A$ and $x \in B$ are indeed true or false. Thus they have truth values.

There are, of course, four possible ways to consider the truth values of these two propositions and thus we have Table 2.3, with $p$ representing the proposition $x \in A$ and $q$ being the proposition $x \in B$. Since we are considering

**TABLE 2.3**

| $p$ | $q$ | |
|---|---|---|
| 1. T ($x \in A$) | T ($x \in B$) | Student $x$ is taking both APL and BASIC |
| 2. T ($x \in A$) | F ($x \notin B$) | Student $x$ is taking APL but not BASIC |
| 3. F ($x \notin A$) | T ($x \in B$) | Student $x$ is not taking APL but taking BASIC |
| 4. F ($x \notin A$) | F ($x \notin B$) | Student $x$ is taking neither APL nor BASIC |

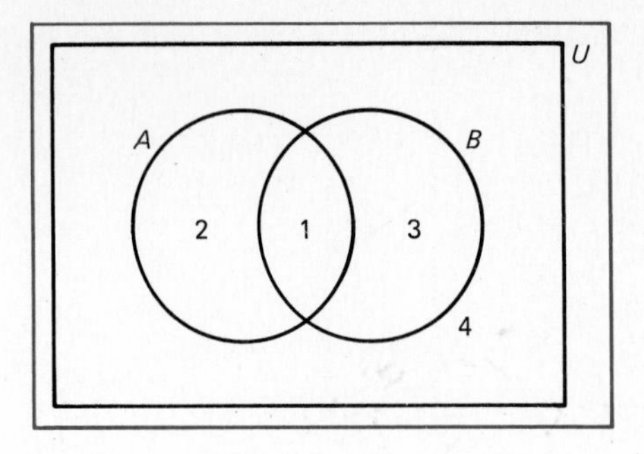

FIGURE 2.8

the fact that these two have a possible overlap, we can also illustrate our situation with the Venn diagram of Figure 2.4(b). We repeat this figure in Figure 2.8 with numbered regions. It is clearly seen that the diagram divides the universe (of students) into four regions. These numbered regions correspond directly to the four rows of the truth table in Table 2.3. The four regions can be described as follows:

| | | |
|---|---|---|
| *Region 1:* | the elements (students) in common with both sets (subjects) | *Line 1:* $x \in A$ and $x \in B$ |
| *Region 2:* | the elements (students) in set $A$ (APL) but not in set $B$ (BASIC) | *Line 2:* $x \in A$ and $x \notin B$ |
| *Region 3:* | the elements (students) not in set $A$ (APL) but in set $B$ (BASIC) | *Line 3:* $x \notin A$ and $x \in B$ |
| *Region 4:* | the elements (students) in neither $A$ (APL) nor $B$ (BASIC) | *Line 4:* $x \notin A$ and $x \notin B$ |

Let us consider the regions of the Venn diagram in Figure 2.8 in detail.

Intersection Region 1 (and line 1 of the truth table) represents the elements that sets $A$ and $B$ have in common. That is, those are the students that are taking *both* APL and BASIC. Stated more formally: This is the set of elements (students) $x$, such that $x \in A$ *and* $x \in B$; or using the conjunction, it is the set $\{x | x \in A \wedge x \in B\}$.

This new set formed from the two original sets is known as the *intersection* of the two sets. This is a binary operation and is symbolized by $A \cap B$. Thus we have the definition of *set intersection:*

$$A \cap B = \{x | x \in A \wedge x \in B\}$$

Figure 2.9 shows set $A$ as the circle with light shading and set $B$ as the moderately shaded circle. The intersection, $A \cap B$, is the darkest shaded region.

Union Another set that we will be interested in forming from the two sets pictured in Figure 2.8 is the set of students that are taking *either* APL *or* BASIC. This set, of course, will include those that are enrolled in both APL and BASIC,

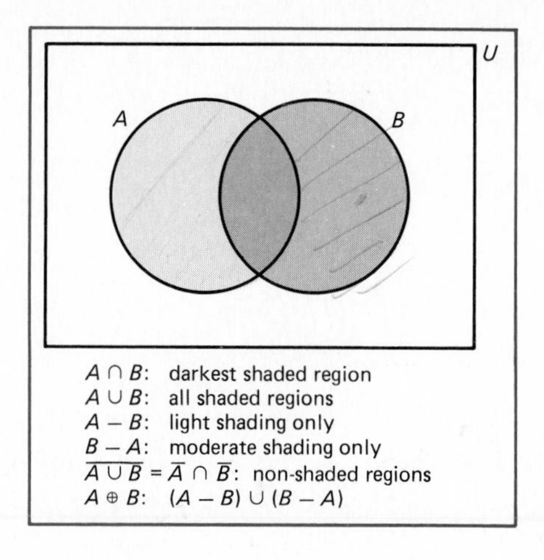


FIGURE 2.9

namely the intersection. Consequently, the "or" in which we are interested is the logical disjunction ($\vee$). This set obviously is made up from elements from regions 1, 2, and 3. The new set we obtain is known as the *union* of sets $A$ and $B$ and is symbolized by $A \cup B$. Thus we have the definition of *set union:*

$$A \cup B = \{x | x \in A \vee x \in B\}$$

The union is represented by all the shaded regions of Figure 2.9. Notice that we have a pictorial representation that $(A \cap B) \subseteq (A \cup B)$.

Relative Complement and Symmetric Difference Region 2 of Figure 2.8 represents those elements that are *in* set $A$ and *not* in set $B$. These constitute the students that are taking APL and not taking BASIC. This set is referred to as the *relative complement* of set $B$ with respect to (relative to) set $A$ and is symbolized by $A - B$ and can also be written as $A \cap \bar{B}$. In Figure 2.9, $A - B$ is shown by the part with light shading only.

Region 3 is $B - A$ or $B \cap \bar{A}$, and is shown by the moderate shading only of Figure 2.9. We have the definitions of relative complement:

$$A - B = A \cap \bar{B} = \{x | x \in A \wedge x \notin B\}$$

$$B - A = B \cap \bar{A} = \{x | x \in B \wedge x \notin A\}$$

The union of these two relative complements, regions 2 and 3 of Figure 2.8, is known as the *symmetric difference* of the two sets. It is symbolized by $A \oplus B$ and is equivalent to the X-OR operator of logical propositions. Thus

$$A \oplus B = (A - B) \cup (B - A) = (A \cap \bar{B}) \cup (B \cap \bar{A})$$

The symmetric difference is represented by the lightly and moderately shaded regions of Figure 2.9.

Region 4 of Figure 2.8 represents those elements that are in neither set $A$ nor set $B$. Of course, this is the set $\overline{A \cup B}$ or $\{x | x \notin A \cup B\}$.

**EXAMPLE 1** Suppose that we are given the following sets:

$$M = \{a, b, c, d, g\}$$

$$N = \{e, f, g, h\}$$

$$P = \{f, h, i\}$$

$$Q = \{b, f, i\}$$

$$R = \{a, b, c\}$$

**(a)** $N \cap P = \{e, f, g, h\} \cap \{f, h, i\} = \{f, h\}$, and $\#(N \cap P) = 2$.

**(b)** $M \cap P = \{a, b, c, d, g\} \cap \{f, h, i\} = \varnothing$, and $\#(M \cap P) = 0$. In the case when the intersection of two sets is empty, we say that the two sets are *disjoint*.

**(c)** $M \cup N = \{a, b, c, d, g\} \cup \{e, f, g, h\} = \{a, b, c, d, e, f, g, h\}$ and $\#(M \cup N) = 8$. Notice that the element $g$, which is repeated, is counted only once in the union.

**(d)** $(P \cap Q) \cup M = (\{f, h, i\} \cap \{b, f, i\}) \cup \{a, b, c, d, g\}$
$= \{f, i\} \cup \{a, b, c, d, g\} = \{a, b, c, d, f, g, i\}$.

**(e)** $N - P = \{e, f, g, h\} - \{f, h, i\} = \{e, g\}$.

**(f)** $M - R = \{a, b, c, a, g\} - \{a, b, c\} = \{d, g\}$.

**(g)** $P \oplus Q = \{f, h, i\} \oplus \{b, f, i\} = \{h, b\}$.

**(h)** $M \oplus P = \{a, b, c, d, g\} \oplus \{f, h, i\} = M \cup P$. When two sets are disjoint, the symmetric difference is the same as the union (since the intersection is empty).

The equivalence of truth tables with Venn diagrams can easily be extended to more than two variables. This concept can be seen for three variables by comparing the truth table in Figure 2.10(a) with the Venn diagram in Figure 2.10(b). The eight numbered regions in the Venn diagram correspond directly

FIGURE 2.10

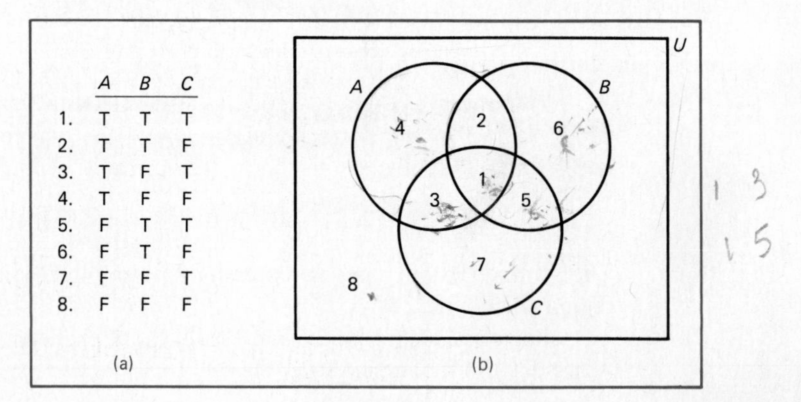

| | A | B | C |
|---|---|---|---|
| 1. | T | T | T |
| 2. | T | T | F |
| 3. | T | F | T |
| 4. | T | F | F |
| 5. | F | T | T |
| 6. | F | T | F |
| 7. | F | F | T |
| 8. | F | F | F |



(a) (b)

with the eight numbers in the truth table. For example, region 5 in the Venn diagram represents the set that is common to sets $B$ and $C$ exclusive of set $A$, or $(B \cap C) - A$. This is where $x \in B$ and $x \in C$ are true and $x \in A$ is false. This is seen to correspond directly to the fifth line in the truth table: $A$ is F, $B$ and $C$ are T.

## EXERCISES

1. For sets

$$A = \{a, b, c, d, e, f, g\}$$
$$B = \{b, c, e\}$$
$$C = \{f, g, h, j\}$$
$$D = \{a, b, c, f\}$$

find each of the following.

(a) $A - (B \cup C)$ (b) $A \cap (B - C)$ (c) $D \oplus (B \cap A)$

(d) $A - \varnothing$ (e) $C \cup (A \oplus B)$ (f) $D - B$

(g) $(A \cup D) - (A \cap D)$ (h) $C \oplus D$

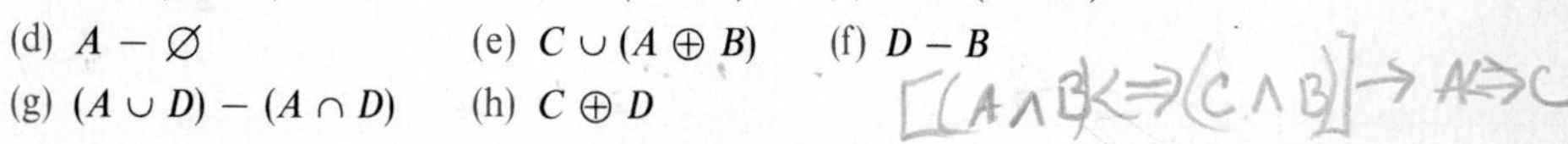

2. Suppose that $A = \{1, 2, 3\}$, $B = \{1, 2\}$, and $C = \{2, 3\}$. If we form $A - B$, we get $\{3\}$; and if we form $C - B$, we get $\{3\}$. Thus $A - B = C - B$, *but* $A \neq C$. Note the dissimilarity of this apparent "subtraction" with subtraction of real numbers. We say that the "cancellation law" does not hold for set difference. Determine whether the cancellation laws hold for $\cap$ and $\cup$. That is, if $A \cap B = C \cap B$, then does $A = C$; and if $A \cup B = C \cup B$, does $A = C$? Find some examples to verify your result. You might even try truth tables.

3. (a) Show by shading a Venn diagram of two sets $A$ and $B$ that $A - (A - B) = A \cap B$.
   (b) Determine whether $A - (B - C) = (A - B) - C$. Your result will show whether relative complement is associative.

4. For arbitrary sets $A$, $B$, and $C$:
   (a) If $A \subseteq B$, what is $A \cup B$? $A \cap B$?
   (b) When does $A - B = B - A$?
   (c) Which regions of Figure 2.10(b) are represented by each of the following?
      (i) $(A \oplus B) - C$
      (ii) $(A \cap C) - (B \cap C)$
      (iii) $C \cup (A \oplus B)$
      (iv) $((A \cup B) - (A \cap B)) - C$
      (v) $A \oplus \bar{A}$

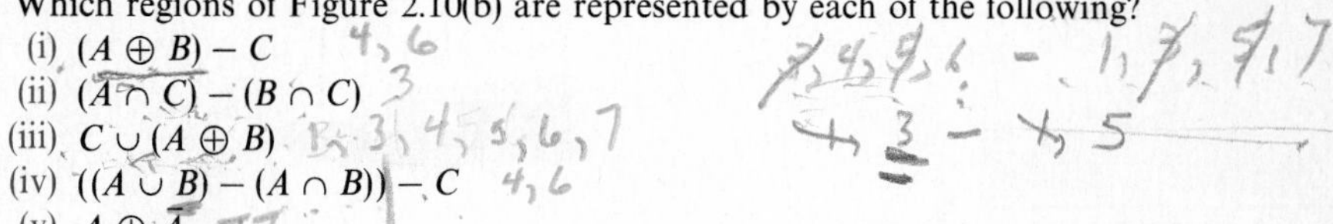

## 2.5 Further Results

Let us at this point state and prove a number of relationships among our binary operations and set containment.

**EXAMPLE 1** For all sets $A$ and $B$, we have $(A \cap B) \subseteq A$ (or alternatively, $(A \cap B) \subseteq B$).

This fact can be observed directly from the Venn diagram of Figure 2.9, in which it is obvious that a set includes its intersection with any other set. But remember, this is only a picture of a special case. Let us be more precise and convert the expression to its counterpart using logical propositions.

The logical statement we have for $(A \cap B) \subseteq A$ is $(A \wedge B) \rightarrow A$ since $\cap$ has the meaning "and" or $\wedge$, and $\subseteq$ has the meaning of the conditional. To ask if $(A \wedge B) \rightarrow A$ is true is to ask if its truth table is a tautology. As we see from Table 2.4, it is a tautology. Therefore, $(A \cap B) \subseteq A$.

**TABLE 2.4**

| $A$ | $B$ | $(A \wedge B)$ | $\rightarrow$ | $A$ |
|---|---|---|---|---|
| T | T | T | T | T |
| T | F | F | T | T |
| F | T | F | T | F |
| F | F | F | T | F |

Another way to argue the case is the following. We want to show that every element $x \in (A \cap B)$ is also an element of $A$, so

If $x \in A \cap B$, then $x \in A \wedge x \in B$, which implies that $x \in A$; thus $(A \cap B) \subseteq A$.

**EXAMPLE 2** For all sets $A$ and $B$, $A \cap B \subseteq A \cup B$. This fact was observed earlier in Figure 2.9. However, by using truth tables (see Table 2.5) we can show that $(A \wedge B) \rightarrow (A \vee B)$ is a tautology.

**TABLE 2.5**

| $A$ | $B$ | $(A \wedge B)$ | $\rightarrow$ | $(A \vee B)$ |
|---|---|---|---|---|
| T | T | T | T | T |
| T | F | F | T | T |
| F | T | F | T | T |
| F | F | F | T | F |

**EXAMPLE 3** For all sets $A$ and $B$, if $A \subseteq B$, then $A \cap B = A$.

If we examine the Venn diagram of Figure 2.4(c), this conditional should be obvious for the figure depicted there. Note, however, that we are to determine whether this logical argument is valid. The argument we test then is

$$(A \rightarrow B) \Rightarrow [(A \wedge B) \leftrightarrow A]$$

and from Table 2.6 we see that the conditional is indeed a tautology.

**TABLE 2.6**

| $A$ | $B$ | $(A \to B)$ | $\to$ | $[(A \wedge B)$ | $\leftrightarrow$ | $A]$ |
|---|---|---|---|---|---|---|
| T | T | T | T | T | T | T |
| T | F | F | T | F | F | T |
| F | T | T | T | F | T | F |
| F | F | T | $\underline{\underline{\text{T}}}$ | F | T | F |

**EXAMPLE 4** The converse of the conditional in Example 3 is also true, which means that we have the important equality (logical equivalence)

$$A \subseteq B \quad \text{iff} \quad A \cap B = A$$

This can be shown to be true by demonstrating that the appropriate biconditional is a tautology (Table 2.7). The importance of this equivalence is that it is sometimes used as a *definition* for set containment. That is, $A$ is defined to be a subset of $B$ iff $A \cap B = A$.

**TABLE 2.7**

| $A$ | $B$ | $(A \to B)$ | $\leftrightarrow$ | $[(A \wedge B)$ | $\leftrightarrow$ | $A]$ |
|---|---|---|---|---|---|---|
| T | T | T | T | T | T | T |
| T | F | F | T | F | F | T |
| F | T | T | T | F | T | F |
| F | F | T | $\underline{\underline{\text{T}}}$ | F | T | F |

**EXAMPLE 5** $A \oplus B = A \cup B$ iff $A \cap B = \varnothing$ [see Example 1(h) of Section 2.4].

Recalling the definition of the symmetric difference and that $\varnothing$ is always false, we have Table 2.8.

**TABLE 2.8**

| | | $\{A \oplus B$ | | | $=$ | $A \cup B\}$ | iff | $\{A \cap B = \varnothing\}$ | | |
|---|---|---|---|---|---|---|---|---|---|---|
| $A$ | $B$ | $\{[(A \wedge \neg B)$ | $\vee$ | $(B \wedge \neg A)]$ | $\leftrightarrow$ | $(A \vee B)\}$ | $\leftrightarrow$ | $[(A \wedge B)$ | $\leftrightarrow$ | $F]$ |
| T | T | F | F | F | F | T | T | T | F | F |
| T | F | T | T | F | T | T | T | F | T | F |
| F | T | F | T | T | T | T | T | F | T | F |
| F | F | F | F | F | T | F | $\underline{\underline{\text{T}}}$ | F | T | F |

**EXAMPLE 6** Show that the following distributive law is *not* true.

$$A - (B \cup C) = (A - B) \cup (A - C)$$

We construct Table 2.9. The final biconditional is not a tautology; hence relative complement does not distribute over union. Does it distribute over intersection? Show your result.

TABLE 2.9

| $A$ | $B$ | $C$ | $[A \wedge \neg(B \vee C)] \leftrightarrow [(A \wedge \neg B) \vee (A \wedge \neg C)]$ | | | | | |
|---|---|---|---|---|---|---|---|---|
| T | T | T | TFF | T | T | F | F | F |
| T | T | F | TFF | T | F | F | T | T |
| T | F | T | TFF | T | F | T | T | F |
| T | F | F | TTT | F | T | T | T | T |
| F | T | T | FFF | T | T | F | F | F |
| F | T | F | FFF | T | T | F | F | F |
| F | F | T | FFF | T | T | F | F | F |
| F | F | F | FFT | F | T | F | F | F |

## EXERCISES

1. Prove that the following statements are true for arbitrary sets $A$, $B$, and $C$.
   (a) $(A \cap B) \subseteq B$
   (b) If $B \subseteq A$ and $C \subseteq A$, then $(B \cup C) \subseteq A$
   (c) If $A \subseteq B$ and $A \subseteq C$, then $A \subseteq (B \cap C)$
   (d) If $A \subseteq B$ and $A \subseteq C$, then $A \subseteq (B \cup C)$
   (e) $A \subseteq (A \cup (A \cap B))$
   (f) $A \subseteq B$ iff $A \cup B = B$
   (g) $A \cup B = \varnothing$ iff $A = \varnothing$ and $B = \varnothing$

2. Determine whether each of the following statements is true for arbitrary sets $A$, $B$, and $C$. If a statement is false, demonstrate a counterexample.
   (a) $A \cap B = \varnothing$ iff $A = \varnothing$ and $B = \varnothing$
   (b) $A = A - \varnothing$
   (c) $A - (B - C) = (A - B) - C$
   (d) $A - (B \cup C) = (A - B) - C$
   (e) $A - (B \cup C) = (A - C) - B$
   (f) $A \oplus B = \varnothing$ iff $A = B$
   (g) $A - (B \cap C) = (A - B) \cap (A - C)$
   (h) $A \subseteq B$ iff $\bar{B} \subseteq \bar{A}$

## 2.6 Properties of Intersection, Union, and Complementation

The properties that we develop in this section pertaining to the set operations of intersection, union, and complementation parallel those of Chapter 1 relating conjunction, disjunction, and negation of logical propositions and statements. The theory of sets as given in this chapter was derived exclusively from the theory of logical statements. Naturally, the same results will hold.

In Chapter 1 we derived a number of logical equivalencies (Section 1.4). They were summarized in Table 1.25. Quite obviously, the notion of logical equivalence between two statements is directly analogous to our concept of set equality between two sets.

Rule 1 of Table 1.25 is that conjunction is commutative. That is, $(p \wedge q) \Leftrightarrow (q \wedge p)$. The analogous statement regarding sets is that set intersection is commutative, or $A \cap B = B \cap A$. This statement, in fact, follows as shown:

$$\begin{aligned} A \cap B &= \{x \mid x \in A \wedge x \in B\} && \text{definition of } \cap \\ &= \{x \mid x \in B \wedge x \in A\} && \text{commutativity of } \wedge \\ &= B \cap A && \text{definition of } \cap \end{aligned}$$

Commutativity of intersection should also be obvious from the Venn diagram of Figure 2.9, where the darkest shaded portion merely represents the elements common to both. Thus the same region is shaded for $A \cap B$ as for $B \cap A$. The truth table for $(A \wedge B) \leftrightarrow (B \wedge A)$ will also bear out this rule.

The union of two sets is commutative also. For

$$\begin{aligned} A \cup B &= \{x \mid x \in A \vee x \in B\} && \text{definition of } \cup \\ &= \{x \mid x \in B \vee x \in A\} && \vee \text{ is commutative} \\ &= B \cup A && \text{definition of } \cup \end{aligned}$$

This identity is evident also from the Venn diagram of Figure 2.9, showing the union of two sets.

Thus we have established our first two rules for set operations:

1. $A \cap B = B \cap A$ intersection is commutative
2. $A \cup B = B \cup A$ union is commutative

Set intersection can be shown to be *associative* by the following.

$$\begin{aligned} A \cap (B \cap C) &= \{x \mid x \in A \wedge (x \in B \wedge x \in C)\} && \text{definition of } \cap \\ &= \{x \mid (x \in A \wedge x \in B) \wedge x \in C\} && \text{associativity of } \wedge \\ &= (A \cap B) \cap C && \text{definition of } \cap \end{aligned}$$

This equality can also be demonstrated by Venn diagrams. We draw two diagrams in Figure 2.11, one to represent $A \cap (B \cap C)$ and the other to represent $(A \cap B) \cap C$. In Figure 2.11(a), $A$ will be indicated by light shading and $B \cap C$ by moderate shading. The darkest area then will represent the intersection of the two. In Figure 2.11(b), we use light shading for $A \cap B$ and moderate for $C$. The darkest area should be the same as in Figure 2.11(a), and indeed it is, thus demonstrating the equality pictorially.

In a similar manner one can show that set union is associative also: $A \cup (B \cup C) = (A \cup B) \cup C$. This follows from the fact that logical disjunction is associative.

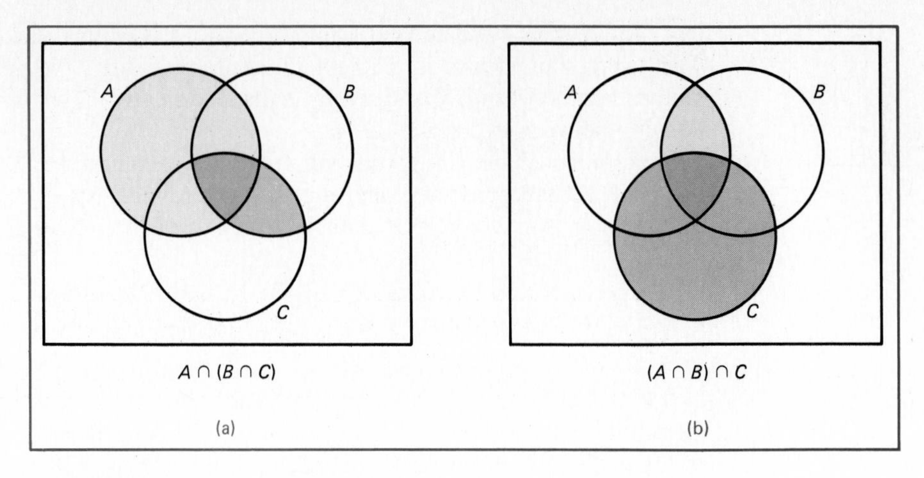


**FIGURE 2.11**

Thus we have our next two rules for set operations.

3. $A \cap (B \cap C) = (A \cap B) \cap C$ intersection is associative
4. $A \cup (B \cup C) = A \cup (B \cup C)$ union is associative

The next equivalence we developed for logical statements was that of *idempotence*. There we showed that $p \Leftrightarrow (p \wedge p)$ and $p \Leftrightarrow (p \vee p)$. The counterpart for sets should be $A = (A \cap A)$ and $A = (A \cup A)$. These can be established from

$$
\begin{aligned}
A \cap A &= \{x \mid x \in A \wedge x \in A\} && \text{definition of } \cap \\
&= \{x \mid x \in A\} && \text{since } \wedge \text{ is idempotent} \\
&= A && \text{set definition}
\end{aligned}
$$

and

$$
\begin{aligned}
A \cup A &= \{x \mid x \in A \vee x \in A\} && \text{definition of } \cup \\
&= \{x \mid x \in A\} && \text{since } \vee \text{ is idempotent} \\
&= A && \text{set definition}
\end{aligned}
$$

Thus we have

5. $A = A \cap A$ idempotence of intersection
6. $A = A \cup A$ idempotence of union

The absorption laws can be easily demonstrated by Venn diagrams. They are rules (7) and (8) in Table 1.25: $p \Leftrightarrow p \wedge (p \vee q)$ and $p \Leftrightarrow p \vee (p \wedge q)$. The analogous statements regarding sets would be $A = A \cap (A \cup B)$ and $A = A \cup (A \cap B)$. These laws are demonstrated in Figure 2.12, in which the darker region in Figure 2.12(a) is set $A$, which is the union (all shaded portions) shown in Figure 2.12(b).

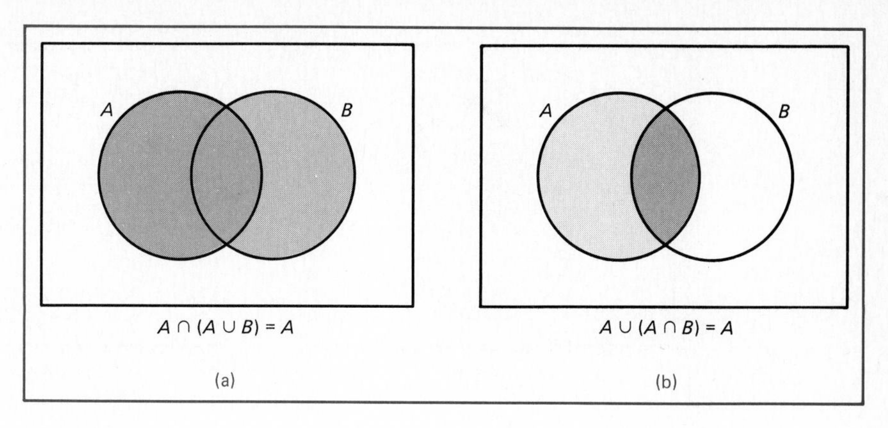


FIGURE 2.12

These laws can also be established from

$$
\begin{aligned}
A \cap (A \cup B) &= \{x \mid x \in A \wedge (x \in A \vee x \in B)\} && \text{definition of } \cap \text{ and } \cup \\
&= \{x \mid x \in A\} && \text{absorption law of logical propositions} \\
&= A && \text{set definition}
\end{aligned}
$$

The other absorption law can be shown in a similar manner. Thus we have

7. $A = A \cap (A \cup B)$ absorption
8. $A = A \cup (A \cap B)$ absorption

Distribution of intersection over union,

$$A \cap (B \cup C) = (A \cap B) \cup (A \cap C)$$

and of union over intersection,

$$A \cup (B \cap C) = (A \cup B) \cap (A \cup C)$$

can be established similarly from the laws of logical propositions. We will demonstrate the first:

$$
\begin{aligned}
A \cap (B \cup C) &= \{x \mid x \in A \wedge (x \in B \vee x \in C)\} && \text{definition of } \cap \text{ and } \cup \\
&= \{x \mid (x \in A \wedge x \in B) \vee (x \in A \wedge x \in C)\} && \text{distributive property of } \wedge \text{ over } \vee \\
&= \{x \mid x \in (A \cap B) \vee x \in (A \cap C)\} && \text{definition of } \cap \\
&= (A \cap B) \cup (A \cap C) && \text{set definition}
\end{aligned}
$$

This law can also be shown by using Venn diagrams. We draw two diagrams. One will represent $A \cap (B \cup C)$. The darkest region will be the

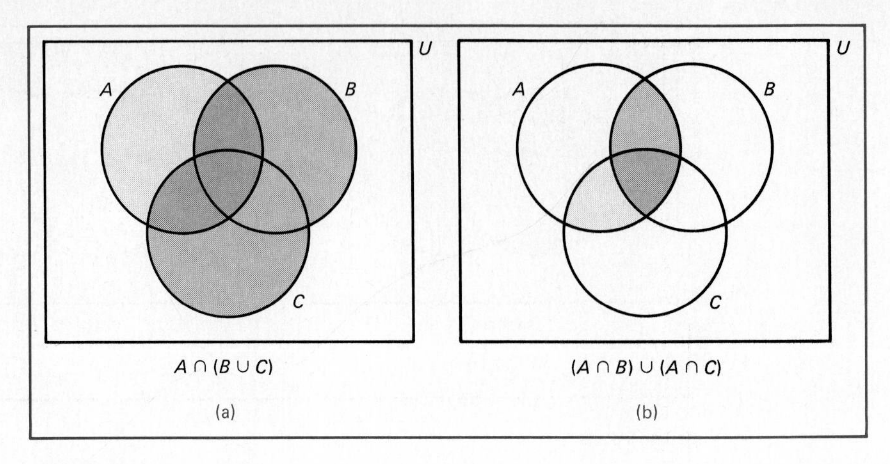


**FIGURE 2.13**

desired result as shown in Figure 2.13(a). $(A \cap B) \cup (A \cap C)$ is shown in Figure 2.13(b). All of the shaded area will represent the result. As can be seen, the two final results are identical.

The other distributive law can be displayed similarly. These results give our next two rules of set operations.

9. $A \cap (B \cup C) = (A \cap B) \cup (A \cap C)$ distributive law
10. $A \cup (B \cap C) = (A \cup B) \cap (A \cup C)$ distributive law

De Morgan's laws hold for set operations also. These rules are

11. $\overline{A \cap B} = \bar{A} \cup \bar{B}$ De Morgan's law
12. $\overline{A \cup B} = \bar{A} \cap \bar{B}$ De Morgan's law

The first of these laws (11) follows from

| | |
|---|---|
| $\overline{A \cap B} = \{x \mid \neg(x \in A \wedge x \in B)\}$ | definition of complement |
| $= \{x \mid \neg(x \in A) \vee \neg(x \in B)\}$ | DeMorgan's law of logical propositions |
| $= \{x \mid x \notin A \vee x \notin B\}$ | definition of set membership |
| $= \bar{A} \cup \bar{B}$ | set definition |

This can be shown from the Venn diagrams in Figure 2.14, where the shaded areas of parts (a) and (b) represent the same region. The other De Morgan law (12) can be shown similarly.

Rule (13), the law of double negation, has its set counterpart:

13. $\bar{\bar{A}} = A$ law of double negation (involution)

Laws (14) of Table 1.25 have their set counterparts also. In these laws pertaining to logical propositions we operated with conjunction and disjunction

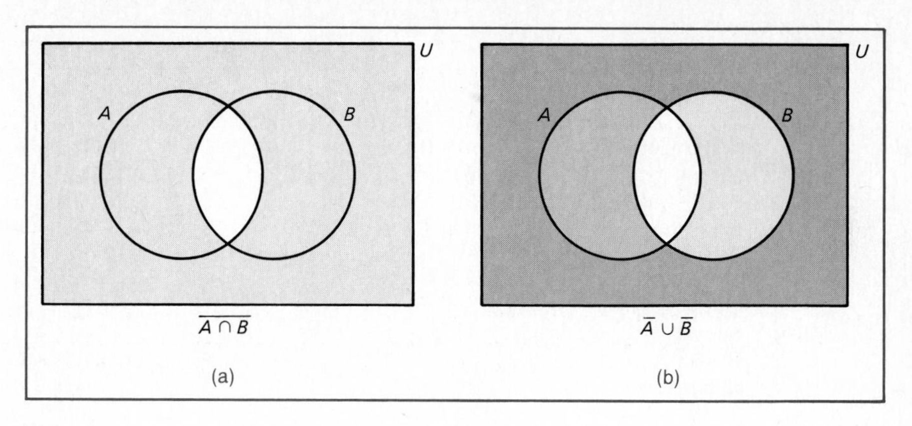


**FIGURE 2.14**

on the tautology ($T$) and the contradiction ($F$). The corresponding elements for sets will be the universe ($U$) for the logical tautology and the null set ($\varnothing$) for the contradiction.

Law (14a) of propositions stated that

$$p \wedge T \Leftrightarrow p$$

Our set counterpart for this identity law will be

$$A \cap U = A$$

This is evident when one considers the diagram in Figure 2.15, in which the darker region is set $A$.

Law (14b), $p \vee F \Leftrightarrow p$, has its counterpart, $A \cup \varnothing = A$. This can be seen from

$$\begin{aligned} A \cup \varnothing &= \{x \mid x \in A \vee x \in \varnothing\} && \text{definition of } \cup \\ &= \{x \mid x \in A \vee F\} && \text{definition of } \varnothing \\ &= \{x \mid x \in A\} && \text{law (14b), Table 1.25} \\ &= A && \text{set definition} \end{aligned}$$

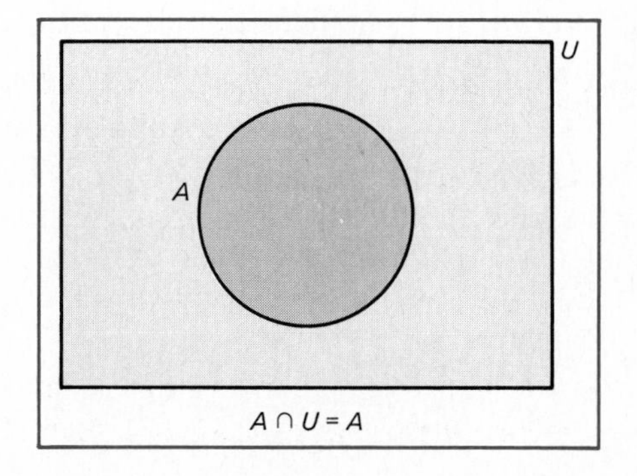


**FIGURE 2.15**

**TABLE 2.10**

| | Rule | Intersection $\cap$ | Union $\cup$ |
|---|---|---|---|
| (1)–(2) | Commutative | $A \cap B = B \cap A$ | $A \cup B = B \cup A$ |
| (3)–(4) | Associative | $A \cap (B \cap C) = (A \cap B) \cap C$ | $A \cup (B \cup C) = (A \cup B) \cup C$ |
| (5)–(6) | Idempotence | $A = A \cap A$ | $A = A \cup A$ |
| (7)–(8) | Absorption | $A = A \cap (A \cup B)$ | $A = A \cup (A \cap B)$ |
| (9)–(10) | Distribution | $A \cap (B \cup C) = (A \cap B) \cup (A \cap C)$ | $A \cup (B \cap C) = (A \cup B) \cap (A \cup C)$ |
| (11)–(12) | De Morgan | $\overline{A \cap B} = \bar{A} \cup \bar{B}$ | $\overline{A \cup B} = \bar{A} \cap \bar{B}$ |
| (13) | Double negation | $A = \bar{\bar{A}}$ | |
| (14) (a)–(b) | Identity | $A \cap U = A$ | $A \cup \varnothing = A$ |
| (c)–(d) | Boundness | $A \cap \varnothing = \varnothing$ | $A \cup U = U$ |
| (e)–(f) | Complement | $A \cap \bar{A} = \varnothing$ | $A \cup \bar{A} = U$ |

The other laws of (14) can be obtained similarly, giving

**14.** (a) $A \cap U = A$ (b) $A \cup \varnothing = A$ — identity

(c) $A \cap \varnothing = \varnothing$ (d) $A \cup U = U$ — boundness

(e) $A \cap \bar{A} = \varnothing$ (f) $A \cup \bar{A} = U$ — complement

The previous laws established for sets can be illustrated as shown in Table 2.10. The entries parallel the rules shown in Table 1.25 for logical equivalencies.

The same property of *duality as* that evidenced with logical propositions is seen to be true with regard to sets. That is, the *dual* of a set statement is obtained by replacing all $\cup$'s with $\cap$'s, all $\cap$'s with $\cup$'s, and each occurrence of the universal set, $U$, with the null set, $\varnothing$, and each $\varnothing$ with $U$.

Let us establish some results of set theory by using the rules of Table 2.10.

**EXAMPLE 1** $A - (A - B) = A \cap B$ [*see Exercise 3(a) of Section 2.4*].

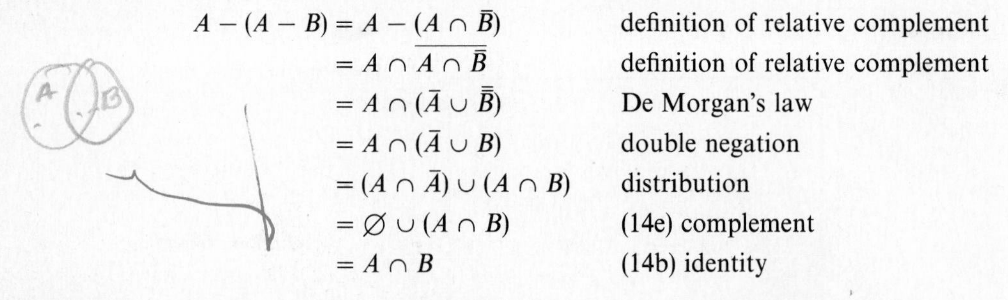

| | |
|---|---|
| $A - (A - B) = A - (A \cap \bar{B})$ | definition of relative complement |
| $= A \cap \overline{A \cap \bar{B}}$ | definition of relative complement |
| $= A \cap (\bar{A} \cup \bar{\bar{B}})$ | De Morgan's law |
| $= A \cap (\bar{A} \cup B)$ | double negation |
| $= (A \cap \bar{A}) \cup (A \cap B)$ | distribution |
| $= \varnothing \cup (A \cap B)$ | (14e) complement |
| $= A \cap B$ | (14b) identity |

**EXAMPLE 2** $A - (B \cap C) = (A - B) \cup (A - C)$.

$$A - (B \cap C) = A \cap (\overline{B \cap C}) \quad \text{definition of relative complement}$$
$$= A \cap (\bar{B} \cup \bar{C}) \quad \text{De Morgan's law}$$
$$= (A \cap \bar{B}) \cup (A \cap \bar{C}) \quad \text{distribution}$$
$$= (A - B) \cup (A - C) \quad \text{definition of relative complement}$$

**EXAMPLE 3** $A \cup (A \cap B \cap C) \cup (\bar{C} \cap A) = A$.

$$A \cup (A \cap B \cap C) \cup (\bar{C} \cap A) = [A \cup (A \cap B \cap C)] \cup (\bar{C} \cap A) \quad \text{associative}$$
$$= [A \cup [A \cap (B \cap C)]] \cup (\bar{C} \cap A) \quad \text{associative}$$
$$= A \cup (\bar{C} \cap A) \quad \text{absorption}$$
$$= A \cup (A \cap \bar{C}) \quad \text{commutative}$$
$$= A \quad \text{absorption}$$

**EXAMPLE 4** $A \oplus \bar{A} = U$.

$$A \oplus \bar{A} = (A \cap \bar{\bar{A}}) \cup (\bar{A} \cap \bar{A}) \quad \text{definition of symmetric difference}$$
$$= (A \cap A) \cup (\bar{A} \cap \bar{A}) \quad \text{involution}$$
$$= A \cup \bar{A} \quad \text{idempotence (twice)}$$
$$= U \quad \text{complement}$$

## EXERCISES

1. Prove that the following statements are true about arbitrary sets $A$, $B$, and $C$ by using results of Table 2.10.
   (a) $A \cap (\bar{A} \cup B) = A \cap B$ (b) $A \cap B = \overline{\bar{A} \cup \bar{B}}$
   (c) $A \cup B = \overline{\bar{A} \cap \bar{B}}$ (d) $(A \cap B) \cup (A \cap \bar{B}) = A$
   (e) $A \cup ((\overline{B \cup A}) \cap B) = U$ (f) $A \oplus \varnothing = A$
   (g) $A \oplus A = \varnothing$ (h) $A - B = A - (A \cap B)$

2. (a) Show by shading a Venn diagram of three sets that relative complement is not associative. That is, $(A - B) - C \neq A - (B - C)$. (See Exercise 3(b) of Section 2.4)
   (b) Show by means of Table 2.10 that it is true, nevertheless, that
   (i) $(A - B) - C = (A - C) - B$
   (ii) $(A - B) - C = (A - C) - (B - C)$
   (iii) $(A - B) - C = A - (B \cup C)$
   (c) Determine whether symmetric difference (X-OR) is associative. That is, does $(A \oplus B) \oplus C = A \oplus (B \oplus C)$?

3. We showed in Example 1 that $A \cap B = A - (A - B)$. That is, the intersection of two sets can be accomplished by completing a relative complement twice. Finding elements in one set that are not in another can be programmed easily enough by a simple search. Write a procedure in your favorite programming language to

determine the relative complement of two sets (lists or arrays) sent as arguments. Then write a program to call this procedure *twice* to determine the intersection of two sets (lists or arrays).

## 2.7 Number of Elements in Finite Sets

If we try to determine the number of elements in the union of two sets, $A$ and $B$, by merely adding together the numbers in each set, the intersection will be counted twice, once for set $A$ and again when we add in set $B$. Since we are to count the elements in the intersection only once, we have to subtract out the number in the intersection once. This leads to the formula for determining the number of elements in the union of sets $A$ and $B$:

$$\#(A \cup B) = \#(A) + \#(B) - \#(A \cap B)$$

**EXAMPLE 1** Suppose, in a group of 200 art and biology students, we know that 150 are taking art; 87 are taking biology. We wish to find the number of students taking both art and biology. If we let set $A$ represent those students taking art and set $B$ be those taking biology, then $\#(A) = 150$, $\#(B) = 87$, and the total number of students then is $\#(A \cup B) = 200$.

Then from the formula above we have

$$\#(A \cup B) = \#(A) + \#(B) - \#(A \cap B)$$
$$200 = 150 + 87 - \#(A \cap B)$$
$$\#(A \cap B) = 150 + 87 - 200 = 37$$

The number of students taking art and not biology—the set $A \cap \bar{B}$—will be $\#(A) - \#(A \cap B) = 150 - 37 = 113$. Similarly, the number taking biology and not art—the set $B \cap \bar{A}$—will be $\#(B) - \#(A \cap B) = 50$.

This situation can be illustrated by the following Venn diagram with the numbers representing the number of students in each region. From Figure 2.16, we see also that $\#(A \cup B) = \#(A - B) + \#(B - A) + \#(A \cap B)$.

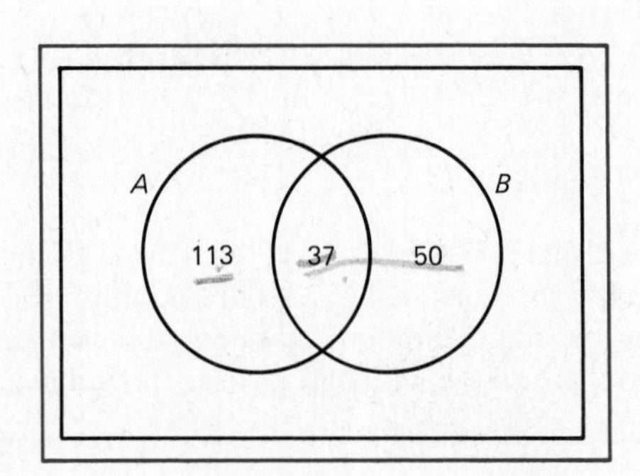


FIGURE 2.16

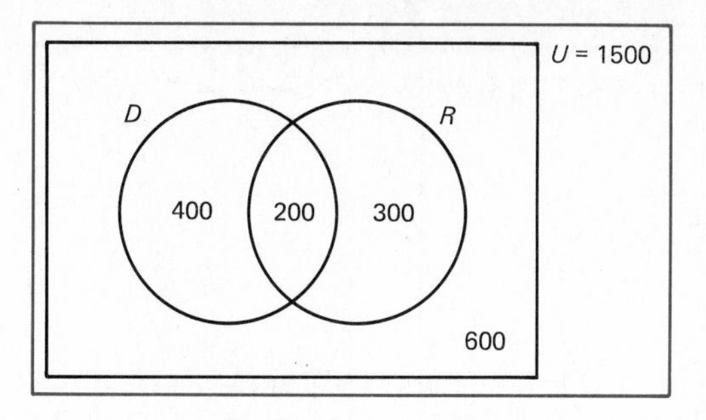


FIGURE 2.17

**EXAMPLE 2** In an election precinct consisting of 1500 registered voters, suppose we know that 400 voters pulled the Democratic party lever (voted straight Democratic), 300 voted straight Republican, and 200 split their votes between Democrats and Republicans. How many of the registered voters did not vote?

If sets $D$ and $R$ represent the sets of people who voted Democratic and Republican, respectively, we have $\#(D - R) = 400$, $\#(R - D) = 300$, and $\#(D \cap R) = 200$. Then $\#(D \cup R) = 400 + 300 + 200 = 900$. Thus the number that did not vote, $\#\overline{(D \cup R)} = 1500 - 900 = 600$. This is depicted in Figure 2.17.

The formula above can be extended to three or more sets. For three sets, $A$, $B$, and $C$, we have

$$\#(A \cup B \cup C) = \#(A) + \#(B) + \#(C) - \#(A \cap B) - \#(A \cap C) - \#(B \cap C) + \#(A \cap B \cap C)$$

**EXAMPLE 3** A group of 100 students was surveyed to find the extent to which there was knowledge of three programming languages: FORTRAN, BASIC, and Pascal. The report found that 45 of the students knew FORTRAN, 35 knew BASIC, and 25 knew Pascal. Furthermore, 10 knew both FORTRAN and BASIC, 15 knew both FORTRAN and Pascal, 5 knew both BASIC and Pascal, and 5 knew all three. What can be gleaned from these findings?

Let sets $F$, $B$, and $P$ represent the students that knew FORTRAN, BASIC, and Pascal, respectively. We may wish to find the total number who knew one or more of the three languages. Thus we want $\#(F \cup B \cup P)$. This can be formed by

$$\begin{aligned}\#(F \cup B \cup P) &= \#(F) + \#(B) + \#(P) - \#(F \cap B) - \#(F \cap P) - \\ &\quad \#(B \cap P) + \#(F \cap B \cap P) \\ &= 45 + 35 + 25 - 10 - 15 - 5 + 5 \\ &= 80\end{aligned}$$

Then 80 of the 100 students knew at least one of the three languages; 20 did not know any of the three.

This example can be demonstrated very nicely by using a Venn diagram of Figure 2.18(a) in which we have three sets drawn. From the statement of the

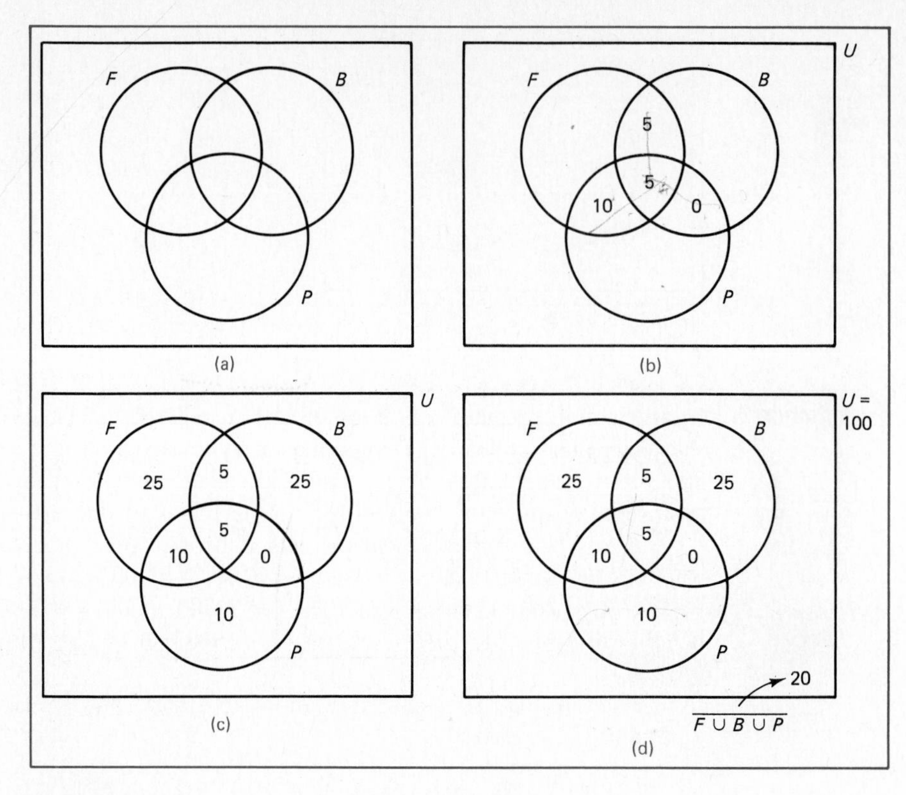


**FIGURE 2.18**

problem we have

$$U = 100$$
$$\#(F) = 45$$
$$\#(B) = 35$$
$$\#(P) = 25$$
$$\#(F \cap B) = 10$$
$$\#(F \cap P) = 15$$
$$\#(B \cap P) = 5$$
$$\#(F \cap B \cap P) = 5$$

As we work from "inside out," we note that $\#(F \cap B \cap P) = 5$ and thus the number who knew FORTRAN and BASIC but *not* Pascal is $\#(F \cap B) - \#(F \cap B \cap P) = 10 - 5 = 5$; those who knew FORTRAN and Pascal but *not* BASIC is $15 - 5 = 10$; and those who knew BASIC and Pascal but not FORTRAN is $5 - 5 = 0$. We fill in these numbers as shown in Figure 2.18(b).

To find those who knew FORTRAN *only* we compute $\#(F) - \#$(those in $F$ common to $B$ and $P) = 45 - 10 - 5 - 5 = 25$. The number who knew BASIC only will be $35 - 5 - 5 - 0 = 25$; and the number who knew Pascal only will be $25 - 5 - 10 - 0 = 10$. Thus we have Figure 2.18(c).

If we add the seven regions shown, we find that there were 80 students who knew one or more of the three languages. This leaves 20 in $\#(\overline{F \cup B \cup P})$ who knew none of the three. Our final diagram will be Figure 2.18(d).

## EXERCISES

1. Show that if two sets, $A$ and $B$, are disjoint, then $\#(A \cup B) = \#(A) + \#(B)$.

2. Suppose in a programming team it was found that 12 programmers knew FORTRAN and 8 knew COBOL. How many different programmers are there if
   (a) No one knew two programming languages?
   (b) Five of the programmers knew both languages?
   (c) Six knew FORTRAN only?
   (d) No one knew COBOL only?

3. At University Y the athletic Hall of Fame inducts prior members of both the basketball and football teams. Forty-eight of the inductees were former basketball players and 33 were football players. If the total number of inductees were 75, how many played both sports?

4. (a) Suppose that 80% of all families own a television set and 50% own a stereo system. If 30% own both, what percentage of all families own at least one of the entertainment devices?
   (b) Would it be possible for only 20% to own both a TV set and a stereo system?

5. Of 100 students surveyed it was found that some had gone to summer school and some had taken Saturday classes in order to graduate early. Of the 100, 25 had gone to summer school, 47 had neither gone to summer school nor taken Saturday classes, and 10 had gone both to summer school and had taken Saturday classes. How many had gone to summer school only? How many had taken Saturday classes only?

6. An advertising firm surveyed 110 headache sufferers and reported on the usage of three brands of pain-relieving drugs, brand A, brand B, and brand C. The report stated that 50 of the people had used brand A, 40 had used brand B, and 30 had used brand C. Furthermore, 25 had used both brands A and B, 20 had used brands B and C, and 12 had used brands A and C. Five had used all three. How many people had used brand A alone? Brand B alone? Brand C alone? Brands A or C but not brand B? None of the three brands?

7. A government committee reported that after questioning 100 people addicted to three drugs, marijuana, heroin, or cocaine, the following usages were in evidence. Of the 100, 80 had used marijuana, 55 had used heroin, and 75 had used cocaine. The users of both marijuana and heroin numbered 40, of both marijuana and cocaine there were 50, and 45 had used both heroin and cocaine. There were 25 who had used all three drugs. Are the committee's findings consistent?

## 2.8 Power Sets

Consider the set $A = \{0, 1\}$. Let us list the subsets of $A$. First there is $A$ itself, $\{0, 1\}$, for $A \subseteq A$. Then there are the singleton sets:

$$\{0\} \subseteq A$$

$$\{1\} \subseteq A$$

Then, since $\varnothing$ is a subset of every set, we have $\varnothing \subseteq A$. Thus there are four distinct subsets of a two-element set.

If we consider the three-element set $B = \{a, b, c\}$ we have as its subsets:

The set itself: $\{a, b, c\}$
The doubletons, the sets with two elements: $\{a, b\}$, $\{a, c\}$, $\{b, c\}$
The singletons, the one-element sets: $\{a\}$, $\{b\}$, $\{c\}$
The empty set: $\varnothing$

Thus we see there are eight distinct subsets of a set with three elements.

This concept leads to the following:

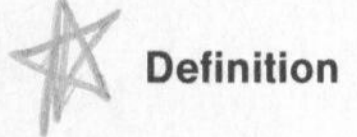

**Definition** The set of all subsets of a given set $A$ is known as the *power set* of $A$ and is denoted by $P(A)$.

A more formal definition of power set is $P(A) = \{X \mid X \subseteq A\}$.

Another notation that is used for the power set of a set $A$ is $2^A$. This stems from the fact that the number of elements in $P(A)$ is $2^{\#(A)}$. This is stated in the following:

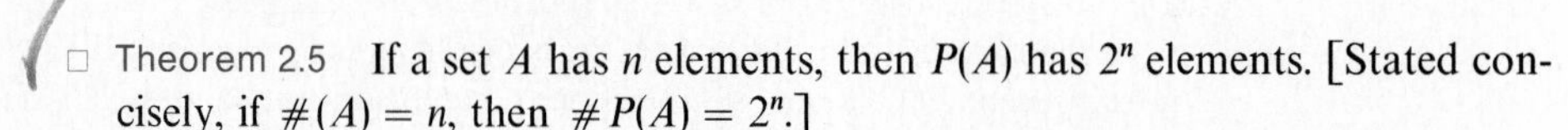

□ Theorem 2.5 If a set $A$ has $n$ elements, then $P(A)$ has $2^n$ elements. [Stated concisely, if $\#(A) = n$, then $\# P(A) = 2^n$.]

**Proof** When we construct a subset of set $A$ having $n$ elements, we decide for each of the elements of $A$ whether to include it. Thus each subset of $A$ is created by making a binary ("yes or no") choice for each element. The number of subsets, then, are doubled for each element in $A$. This gives us

$$\underbrace{2 \cdot 2 \cdots 2}_{n \text{ times}} = 2^n \text{ elements}$$

**EXAMPLE 1** If we consider a set with one element, say $A = \{a\}$, then $P(A)$ has two subsets [$\#(A) = 1$ and $2^1 = 2$], one with the single element, $\{a\}$, and the other without the element, that is, $\varnothing$. We have

$$P(\{a\}) = \{\{a\}, \varnothing\}$$

**EXAMPLE 2** Let $A = \{0, 1\}$; then we have the following choices:

$$\{\text{yes, yes}\} = \{0, 1\} = A$$
$$\{\text{yes, no}\} = \{0\}$$
$$\{\text{no, yes}\} = \{1\}$$
$$\{\text{no, no}\} = \varnothing$$

Thus we have

$$P(A) = \{\varnothing, \{0\}, \{1\}, \{0, 1\}\}$$

**EXAMPLE 3** Let $A = \varnothing$; thus $\# A = 0$. Then $\# P(A) = 2^0 = 1$ and $P(A) = \{\varnothing\}$, a set with one element.

## EXERCISES

1. List the elements in the power set of each of the following.
   (a) $\{\varnothing\}$ (b) $\{\sqrt{3}\}$ (c) $\{a, b, c, d\}$ (d) $\{a, b, \{a\}\}$ (e) $\{\{a, b, c\}\}$
2. Does $P(A \cap B) = P(A) \cap P(B)$?
3. Does $P(A \cup B) = P(A) \cup P(B)$?
4. If $A \subseteq B$, is $P(A) \subseteq P(B)$?
5. Answer true or false to each statement about an arbitrary set $A$.
   (a) $A \subseteq P(A)$ (b) $\varnothing \in P(A)$ (c) $\varnothing \subseteq P(A)$ (d) $A \in P(A)$
6. If $A \cap B = \varnothing$, discuss $P(A) \cap P(B)$.

## 2.9 Product Sets (Cartesian Products) and Binary Relations

In mathematics as well as in everyday life, it is useful to form pairs of related objects from sets. It is desirable to discuss properties of objects relative to others either from the same set or perhaps from another. It is not possible to make such statements simply in terms of set membership. Such statements are called relational statements, or just *relations*. When *two* objects are thus being compared, the statements are *binary relations*.

The following are illustrations of such statements from everyday life. (It is easy to see from the first two why these statements are called relations.)

1. John is the husband of Elizabeth.
2. Cain is the brother of Abel.

3. Bill is older than Henry.
4. Goliath is taller than David.
5. Milk is produced by cows.
6. North Carolina borders (is adjacent to) Virginia.
7. Sacramento is the capital of California.
8. Algebra I is a prerequisite for Algebra II.

The following are illustrations of statements of relation from mathematics.

9. Ten is greater than 3.
10. Fourteen equals 7 times 2.
11. Thirty is a multiple of 5.
12. $\{a\}$ is a subset of $\{a, b, c\}$.
13. $m$ divides $n$.
14. Triangle $ABC$ is congruent to triangle $DEF$.
15. Line $x$ is parallel to line $y$.

An underlying order is evident from some of the pairs above, that is, from the relations "is taller than," "is older than," "is greater than," "is a subset of." However, an implied ill-defined order is evident in "is the husband of," "is produced by," and "is the capital of." In these examples, the order is implied in that certainly "Elizabeth is the husband of John" is not what any speaker would intend.

Others of the example relations show no implied order whatsoever. "Abel is the brother of Cain" certainly means the same thing as "Cain is the brother of Abel." And we know from geometry that if "line $x$ is parallel to line $y$," then "line $y$ is parallel to line $x$."

Binary relations are always statements not about just pairs of elements, but about *ordered pairs*. When the order of a pair is not important, one usually just uses the set notation, for instance $\{10, 3\}$. This is the same set as $\{3, 10\}$. When order is important the notation used is $\langle 10, 3\rangle$. This pair is different from $\langle 3, 10\rangle$. For example, $\langle 10, 3\rangle$ is in the relation "is greater than," and $\langle 3, 10\rangle$ is not.

Let us consider two arbitrary sets, $A$ and $B$. The set of *all* ordered pairs $\langle a, b\rangle$ with $a \in A$ and $b \in B$ is called the *product* set or *Cartesian product* (from the French mathematician René Descartes) of $A$ and $B$. This is designated by $A \times B$ (the *cross product*) and by definition

$$A \times B = \{\langle a, b\rangle | a \in A \text{ and } b \in B\}$$

When considering the Cartesian product of some set $A$ with itself, we have $A \times A$, which can be denoted as $A^2$.

**EXAMPLE 1** Let $A = \{1, 2, 3, 4\}$ and $B = \{a, b, c\}$. Then

$$A \times B = \{\langle 1, a\rangle, \langle 1, b\rangle, \langle 1, c\rangle, \langle 2, a\rangle, \langle 2, b\rangle, \langle 2, c\rangle, \langle 3, a\rangle, \langle 3, b\rangle, \langle 3, c\rangle, \langle 4, a\rangle, \langle 4, b\rangle, \langle 4, c\rangle\}$$

Note that this set is *not the same as* $B \times A$.

$$B \times A = \{\langle a, 1\rangle, \langle a, 2\rangle, \langle a, 3\rangle, \langle a, 4\rangle, \langle b, 1\rangle, \langle b, 2\rangle, \langle b, 3\rangle, \langle b, 4\rangle, \langle c, 1\rangle, \langle c, 2\rangle, \langle c, 3\rangle, \langle c, 4\rangle\}$$

Cartesian products can be represented in graphical fashion for purposes of visualization. The first set mentioned in the product, say in Example 1, $A \times B$, will be enumerated along a vertical axis and the second along a horizontal axis. All the points generated then will be the Cartesian product (see Table 2.11).

**TABLE 2.11** $A \times B$ Where $A = \{1, 2, 3, 4\}$ and $B = \{a, b, c\}$

| $A$ \ $B$ | $a$ | $b$ | $c$ |
|---|---|---|---|
| 1 | × | × | × |
| 2 | × | × | ⊗ ← the ordered pair $\langle 2, c\rangle$ |
| 3 | × | × | × |
| 4 | × | × | × |

Let us consider a set $T = \{2, 3, 4, 5, 6\}$ and the cross product $T \times T$ or $T^2$ as pictured in Table 2.12.

**TABLE 2.12** $T \times T$

| $T$ \ $T$ | 2 | 3 | 4 | 5 | 6 |
|---|---|---|---|---|---|
| 2 | × | × | × | × | × |
| 3 | × | × | × | × | × |
| 4 | × | × | × | × | × |
| 5 | × | × | × | × | × |
| 6 | × | × | × | × | × |

One is usually interested in only a subset of a cross product, a set of ordered pairs that have some mutual meaning or relationship. Suppose that in this case we are interested only in those ordered pairs $\langle x, y\rangle$ that have the relation: $x$ "is a multiple of" $y$. Thus we can represent this relation as a *subset* of the entire cross product, as shown in Table 2.13, where the ordered pairs are $\{\langle 2, 2\rangle, \langle 3, 3\rangle, \langle 4, 2\rangle, \langle 4, 4\rangle, \langle 5, 5\rangle, \langle 6, 2\rangle, \langle 6, 3\rangle, \langle 6, 6\rangle\}$.

**TABLE 2.13** Relation "Is a Multiple of" on $T = \{2, 3, 4, 5, 6\}$

| $T$ | 2 | 3 | 4 | 5 | 6 |
|---|---|---|---|---|---|
| 2 | × | | | | |
| 3 | | × | | | |
| 4 | × | | × | | |
| 5 | | | | × | |
| 6 | × | × | | | × |

Let us consider a set $S$ of certain states in the United States, $S = \{$Missouri, Illinois, Arkansas, Indiana, Kentucky$\}$. A map of these states is shown in Figure 2.19(a) with a diagram of the relation "is adjacent to" shown in Figure 2.19(b).

**Definition** Let $A$ and $B$ be sets. A *binary relation*, $R$, from $A$ to $B$ is a subset of $A \times B$. If $\langle a, b\rangle \in A \times B$ is in the relation $R$, we say that $\langle a, b\rangle \in R$ or, equivalently, $a\, R\, b$. If $\langle a, b\rangle \notin R$, we denote this by $a \not R\, b$.

**FIGURE 2.19**

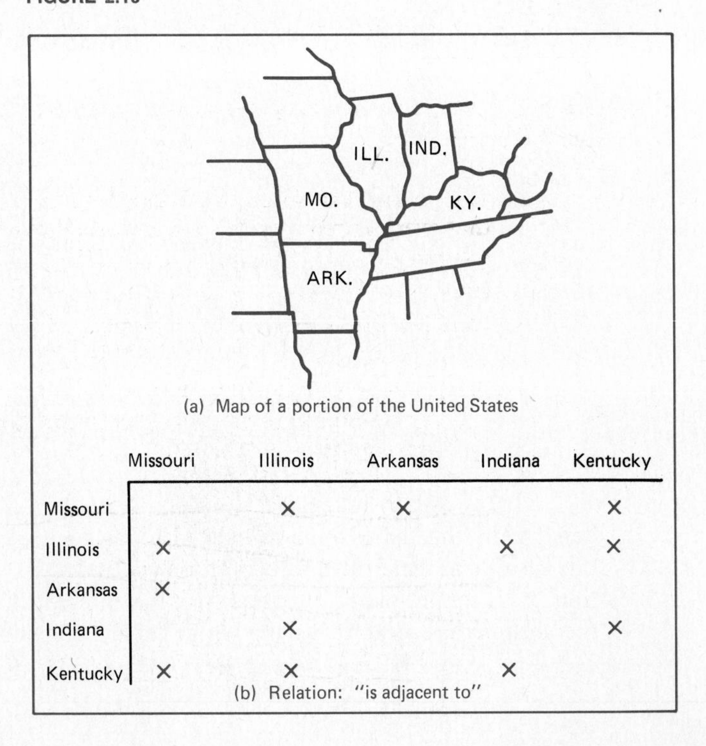

(a) Map of a portion of the United States

| | Missouri | Illinois | Arkansas | Indiana | Kentucky |
|---|---|---|---|---|---|
| Missouri | | × | × | | × |
| Illinois | × | | | × | × |
| Arkansas | × | | | | |
| Indiana | | × | | | × |
| Kentucky | × | × | | × | |

(b) Relation: "is adjacent to"

If $R$ is a relation from set $A$ to itself, we say that $R$ is a relation *on* $A$. In the example of the relation "is a multiple of" defined on set $T = \{2, 3, 4, 5, 6\}$ (Table 2.13), let us denote the relation as $M$. Thus it is clear that, for instance, the ordered pair $\langle 4, 2 \rangle \in M$ or $4\ M\ 2$. Also, it is clear that $\langle 5, 3 \rangle \notin M$ or $5 \not M 3$.

In the relation "is adjacent to" of Figure 2.19, if we denote the relation by $B$, we have: Kentucky $B$ Illinois, but Kentucky $\not B$ Arkansas.

An important relation is one that we have been studying in this chapter. This is the relation relating two sets by set containment or inclusion ($\subseteq$). If, when we are given two sets, $A$ and $B$, we state that $A \subseteq B$, we have the relation: $A$ "is contained in" $B$.

**EXAMPLE 2** If we are given the set $A = \{a, b, c\}$, every member of the power set, $P(A)$, is related to set $A$ by set containment. This is illustrated in Table 2.14.

**TABLE 2.14** Relation "Containment" among Members of the Power Set of $\{a, b, c\}$

| | $\varnothing$ | $\{a\}$ | $\{b\}$ | $\{c\}$ | $\{a, b\}$ | $\{a, c\}$ | $\{b, c\}$ | $\{a, b, c\}$ |
|---|---|---|---|---|---|---|---|---|
| $\varnothing$ | × | × | × | × | × | × | × | × |
| $\{a\}$ | | × | | | × | × | | × |
| $\{b\}$ | | | × | | × | | × | × |
| $\{c\}$ | | | | × | | × | × | × |
| $\{a, b\}$ | | | | | × | | | × |
| $\{a, c\}$ | | | | | | × | | × |
| $\{b, c\}$ | | | | | | | × | × |
| $\{a, b, c\}$ | | | | | | | | × |

Special Relations Some relations have properties of special significance. In the following let $R$ be a relation on a set $A$.

1. $R$ is said to be *reflexive* if for every $a \in A$, we have $a\ R\ a$. This means that every element in the universe of discourse, set $A$, is related to itself through $R$. A notable example of this property is set containment, where we have shown that for every set $A$, we have $A \subseteq A$. Some other examples of reflexive relations are:
   (a) Is equal to.
   (b) Is parallel to, for lines in a plane. (Lines are parallel if they have the same slope, or if they are vertical, with no slope).
   (c) Is the same color as.
   (d) Is less than or equal to (on the set of real numbers).
   (e) Is greater than or equal to, for real numbers.
   In the diagrams pictured previously, ×'s will be at *all* points on the main diagonal if the relation is reflexive.
2. A relation $R$ is said to be *irreflexive* if *no* element of $A$ is related to itself; that is, for all $a \in A$ we have $a \not R a$ or $\langle a, a \rangle \notin R$. Examples are:
   (a) Proper containment ($\subset$) among sets.
   (b) Is greater than.

(c) Is taller than.
(d) Is the sister of.
(e) Is the father of.

3. A relation is *symmetric* if when we have that $\langle a, b\rangle \in R$, then we also have $\langle b, a\rangle \in R$. Using the language of logical equivalence we have: $a\ R\ b \rightarrow b\ R\ a$. Examples of some symmetric relations are:
   (a) Is perpendicular to (with respect to lines in a plane).
   (b) Has a border in common with.
   (c) Is the same height as.
   (d) Is a sibling of. (*Note:* we cannot state that "is a brother of" is symmetric. Although it may be true that "David is a brother of Katy," it is certainly not true that "Katy is a brother of David.")
   (e) Is equal to.

   In the case of the diagrams pictured earlier, for every $\times$ not on the main diagonal there will be its corresponding $\times$ reflected on the other side of the main diagonal.
4. The *antisymmetric* property of relations states that: If for some $a$ and $b$ we have $a\ R\ b$ and also $b\ R\ a$, then $a = b$. Using the language of logical propositions, we have

$$((a\ R\ b) \wedge (b\ R\ a)) \rightarrow (a = b)$$

   Examples are:
   (a) Is a subset of. (For example, if $A \subseteq B$ and $B \subseteq A$, then $A = B$.)
   (b) Is less than or equal to.
   (c) Is greater than or equal to.
5. A relation is *asymmetric* if the symmetric relation is not true in any case. That is, if $a\ R\ b$, then $b \not R\ a$ for all $a$ and $b$. Examples of asymmetric relations are:
   (a) Is the father of.
   (b) Is greater than.
6. The *transitive* relation is one with which the reader should be familiar since it has been mentioned several times earlier in the text. This property states that: If $a\ R\ b$ and $b\ R\ c$, then $a\ R\ c$, for all $a, b, c \in A$. We have

$$((a\ R\ b) \wedge (b\ R\ c)) \rightarrow (a\ R\ c).$$

   This property states that whenever two elements are linked together through a third element, they are linked directly. Relations that are transitive are:
   (a) Is a subset (proper also) of. For sets $A$, $B$, $C$, if $A \subseteq B$ and $B \subseteq C$, then $A \subseteq C$.
   (b) Is equal to. If $a = b$ and $b = c$, then $a = c$.
   (c) Is a multiple of. If $m$ is a multiple of $n$ and $n$ is a multiple of $p$, then $m$ is a multiple of $p$, for integers $m$, $n$, and $p$.

(d) Is a divisor of. This reverses the linkages of the "multiple of" relation. If $p$ is a divisor of $n$ and $n$ is a divisor of $m$, then $p$ is a divisor of $m$.
(e) Is due north of (unless you are already at the North Pole).
(f) Is an ancestor (or descendant) of.
(g) Is less than (or equal to).

**EXAMPLE 3** *(Illustrations of Relations Having Some of the Special Properties)*

**(a)** Consider the "universal relation" on a set $\{a, b, c\}$, shown in Table 2.15. This relation is reflexive, symmetric, and transitive. It is not irreflexive, asymmetric, or antisymmetric.

**TABLE 2.15**

| $U$ | $a$ | $b$ | $c$ |
|---|---|---|---|
| $a$ | × | × | × |
| $b$ | × | × | × |
| $c$ | × | × | × |

**(b)** The "diagonal" relation (also known as the identity relation) (Table 2.16.) is reflexive (any time that all diagonal elements are in the relation, the relation is reflexive), symmetric (the representation is symmetric about the diagonal), transitive (trivially so, because $\langle x, x\rangle$ is in the relation). It is also antisymmetric, for any time that $\langle x, x\rangle$ and $\langle x, x\rangle$ is in the relation, we have $x = x$. It is not irreflexive nor asymmetric.

**TABLE 2.16**

| $D$ | $a$ | $b$ | $c$ |
|---|---|---|---|
| $a$ | × | | |
| $b$ | | × | |
| $c$ | | | × |

**(c)** The relation $R$ (Table 2.17) is not reflexive ($\langle c, c\rangle \notin R$). Neither is it irreflexive (because two diagonal elements do appear). It is obviously not symmetric nor is it asymmetric ($\langle b, b\rangle \in R$ and its reverse pairing $\langle b, b\rangle \in R$). It is not transitive ($\langle a, c\rangle \in R$, $\langle c, b\rangle \in R$ but $\langle a, b\rangle \notin R$). Additionally, it is not antisymmetric ($\langle b, c\rangle \in R$, $\langle c, b\rangle \in R$, and $b \neq c$).

**TABLE 2.17**

| $R$ | $a$ | $b$ | $c$ |
|---|---|---|---|
| $a$ | × | | × |
| $b$ | | × | × |
| $c$ | | × | |

**(d)** The relation $S$ (Table 2.18) is obviously neither reflexive, irreflexive, symmetric, nor asymmetric. It is, however, transitive (nowhere does this property fail; for instance, $\langle c, c\rangle \in S$, $\langle c, a\rangle \in S$ and $\langle c, a\rangle \in S$) and it is antisymmetric (nowhere does this property fail to hold either).

**TABLE 2.18**

| $S$ | $a$ | $b$ | $c$ |
|---|---|---|---|
| $a$ | × | | |
| $b$ | × | | |
| $c$ | × | | × |

**(e)** The relation $T$ (Table 2.19) is reflexive, symmetric, and transitive. None of the other properties hold.

**TABLE 2.19**

| $T$ | $a$ | $b$ | $c$ |
|---|---|---|---|
| $a$ | × | | |
| $b$ | | × | × |
| $c$ | | × | × |

## EXERCISES

**1.** (a) Draw a picture of the relation $R$ on set $A = \{1, 2, 3, 4\}$ where $R = \{\langle 1, 1\rangle, \langle 1, 2\rangle, \langle 2, 3\rangle, \langle 2, 4\rangle, \langle 3, 1\rangle, \langle 3, 3\rangle, \langle 4, 4\rangle\}$.

(b) Determine which of the special properties discussed in this section are true relative to relation $R$.

**2.** Show the ordered pairs and give a picture of the relation "is divisible by" on the set $B = \{5, 6, 7, 8, 9, 10, 11, 12, 13, 14, 15\}$. Which of the special properties of this section are true of this relation?

**3.** Given the set $M = \{1, 2, 3\}$. Determine which of the special properties of this section are true of the following relations: $R$, $S$, $T$, $V$, and $W$.

(a)

| $R$ | 1 | 2 | 3 |
|---|---|---|---|
| 1 | | | × |
| 2 | | × | |
| 3 | | | |

(b)

| $S$ | 1 | 2 | 3 |
|---|---|---|---|
| 1 | × | × | × |
| 2 | × | × | × |
| 3 | | | |

(c)

| $T$ | 1 | 2 | 3 |
|---|---|---|---|
| 1 | × | | × |
| 2 | | × | |
| 3 | × | | × |

(d)

| $V$ | 1 | 2 | 3 |
|---|---|---|---|
| 1 | | | × |
| 2 | | × | |
| 3 | × | | |

(e)

| $W$ | 1 | 2 | 3 |
|---|---|---|---|
| 1 | | × | |
| 2 | × | | × |
| 3 | | × | |

4. Give an example of a relation that is
   (a) Reflexive but is not symmetric or transitive.
   (b) Symmetric but is not reflexive or transitive.
   (c) Transitive but is not reflexive or symmetric.
   (d) Reflexive and transitive and not symmetric.
   (e) Antisymmetric and irreflexive.
   (f) Asymmetric and transitive.
5. The empty relation is one that contains no ordered pairs. Which of the relations of this section are true with regard to the empty relation?
6. Is the cross product between two sets commutative? Among three sets, associative?
7. Show that if a relation is asymmetric, then it is irreflexive.

## 2.10 Partial Orders

In isolation, properties of relations, in general, do not carry much information about the structure of a set. However, they may be grouped together in such ways to develop useful classifications. One of these groupings that has had considerable impact within computer science is known as a *partial order*. This is a relation that is *reflexive, antisymmetric,* and *transitive*. Many of the discrete systems that we will examine in this book are partial orders and their connection with the discipline will be explored as they are covered. For now, let us examine some "well-known" partial orders.

For purpose of illustration, consider the set of integers, $S = \{2, 3, 4, 5, \ldots\}$, and the relation "divides exactly," denoted by $D$, on the set $S$. Thus, for integers $a$ and $b$, the meaning of $a\ D\ b$ is "$a$ divides $b$ exactly." For instance, we have $2\ D\ 6$, $10\ D\ 50$, $3\ \not{D}\ 19$, and so on.

To see that "divides exactly" is a partial order, we need to show that the relation, $D$, is reflexive, antisymmetric, and transitive. Now, it is clear that $a\ D\ a$, for any $a \in S$; thus *$D$ is reflexive*. $D$ is also *transitive* since, for $a$, $b$, and $c \in S$ we

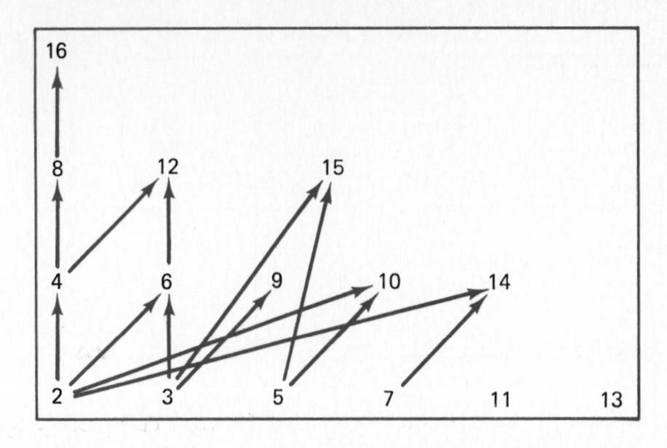


**FIGURE 2.20** Partial Ordering "Divides" on $\{2, 3, \ldots, 16\}$.

have: If $a\ D\ b$ and $b\ D\ c$, then obviously $a\ D\ c$. To note *antisymmetry* we have only to observe that for $a, b \in S$, if $a\ D\ b$ and $b\ D\ a$, then $a$ must be the same element as $b$, or $a = b$. Thus $D$ is a partial order on the set $S$.

It is important to note that some pairs of elements of $S$ are not related at all by the relation, $D$. Such pairs are said to be *noncomparable*. For instance, for the pair $\{3, 10\}$, we have $3\ \not{D}\ 10$ and $10\ \not{D}\ 3$. The ordering is not complete; it is partial: thus the name "partial order." Any set that is thus ordered partially by some relation is called a partially ordered set, called *poset* for short.

Diagrams of partial orders are very helpful to visualize the ordering involved. The divisibility ordering can be shown very nicely as in Figure 2.20, where the arrows specify the order of pairs of elements. In the figure we show the integers from 2 to 16.

When one has a project to perform that can be broken down into a number of individual tasks such that some tasks can be done only after some others have been completed, we say that the tasks are ordered. When the completion of a task, say $T_1$, is required before task $T_2$ can be begun, we have the ordered pair $\langle T_1, T_2 \rangle$, with the relation between the two being "must be completed before" or "is a prerequisite for."

Consider the project of scrambling eggs. This project can be broken down in an overly simplified manner into the following tasks: (1) crack the eggs ($C$); (2) mix the eggs with milk and cheese ($M$); (3) heat grease in a skillet ($H$); (4) pour the egg mixture into the skillet with hot grease ($P$); and (5) stir ($S$).

Obviously, task $C$ must be completed before task $M$ but not necessarily before task $H$ is completed. But all three of these have to be completed before task $P$, which itself is a prerequisite for task $S$. We have the obvious pairings: $\langle C, M \rangle$, $\langle M, P \rangle$, $\langle H, P \rangle$, and $\langle P, S \rangle$. The order we have is *partial* in that there is *no* prescribed order between $C$ and $H$ or between $M$ and $H$.

The diagram of this project can be illustrated as in Figure 2.21, in which the arrows indicate the order in which the individual tasks are to be carried out. Transitivity of the prerequisite relation is obvious from the diagram. For example, $\langle C, M \rangle$ and $\langle M, P \rangle$ are in the relation and this obviously implies that $\langle C, P \rangle$ must be present. Observe that the relation is antisymmetric also. There are no tasks $T_1$ and $T_2$ such that $\langle T_1, T_2 \rangle$ and $\langle T_2, T_1 \rangle$ are in the relation. If it were the case that these two exist, then we would have to have $T_1 = T_2$.

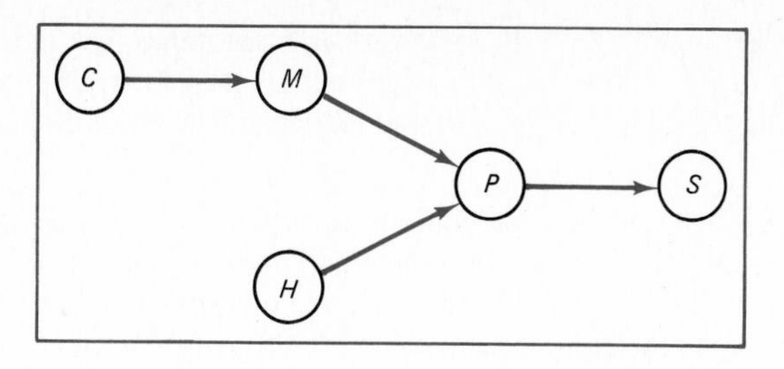


**FIGURE 2.21** Partially Ordered Project.

Thus antisymmetry is trivially true. (The hypotheses are false, so the conditional is true.)

This is an example of what is known as an *incompletely specified* partial order in that the reflexive property fails to hold. Nevertheless, it is still considered a partial order in that there is no prescribed order between certain pairs of tasks of the project.

## EXERCISES

1. Determine which of the following relations on the set $S = \{1, 2, 3\}$ are partial orders; incompletely specified partial orders.

(a)

| | 1 | 2 | 3 |
|---|---|---|---|
| 1 | × | × | |
| 2 | × | × | |
| 3 | | | × |

(b)

| | 1 | 2 | 3 |
|---|---|---|---|
| 1 | | | |
| 2 | | × | × |
| 3 | | × | × |

(c)

| | 1 | 2 | 3 |
|---|---|---|---|
| 1 | × | | × |
| 2 | | × | |
| 3 | × | | × |

(d)

| | 1 | 2 | 3 |
|---|---|---|---|
| 1 | × | | |
| 2 | | × | × |
| 3 | | | × |

2. Show that $\subseteq$ on all sets defines a partial order.

3. Show that proper containment $\subset$ on all sets defines an incompletely specified partial order.

4. Find the fallacy in the following reasoning that every symmetric and transitive relation $R$ is also reflexive:

   Since $R$ is symmetric, if $a\ R\ b$, then $b\ R\ a$. From these two, transitivity implies that $a\ R\ a$. Therefore, $R$ is reflexive.

5. A linear order is an ordering relation, $L$, on a set $S$ such that for every pair of distinct elements $a$ and $b$ of $S$, we have either $a\ L\ b$ or $b\ L\ a$. Explain how this ordering differs from a partial order.

6. Show that "is less than," $<$, on the set of integers is a linear order (see Exercise 5).

7. Assume that we have a tennis tournament among four players $\{A, B, C, D\}$, and the outcomes of the form "wins over" (denoted by $W$) form the following relation:

$$W = \{\langle A, B\rangle, \langle B, D\rangle, \langle B, C\rangle, \langle A, C\rangle, \langle A, D\rangle\}$$

Draw a diagram of $W$. Is $W$ a partial order? A linear order?

8. Consider the project of playing a record on a turntable. The individual tasks can be broken down to: (1) turn the system on; (2) take the record out of the jacket; (3) place the record on the spindle; and (4) engage the needle. Is this project a partial order?

9. An *equivalence relation* is a relation that is reflexive, symmetric, and transitive. For example, people with the same last names are indistinguishable from each other (or equivalent) if all one has to rely on is a listing of last names. If we let "has the same last name as" be the relation denoted by $L$, then obviously person $x$ is related by $x \; L \; x$. Hence the relation is reflexive. Also for persons $x$ and $y$, if $x \; L \; y$, then $y \; L \; x$. Hence the relation is symmetric. And if we introduce a third person, $z$, we have that if $x \; L \; y$ and $y \; L \; z$, then $x \; L \; z$. Thus the relation is transitive.

   Determine which of the following are equivalence relations:
   (a) "Has the same number of pages as" on the set of all books.
   (b) "Has more pages than" on the set of all books.
   (c) "Is born in the same year as" on the set of all people.
   (d) "Is similar to" on the set of all triangles in a plane.
   (e) "Is perpendicular to" on the set of lines in a plane.
   (f) "Is greater than or equal to" on the set of real numbers.
   (g) "Is properly contained in," $\subset$, on sets.
   (h) "Is a friend of" on the set of all people.

## Summary and Selected References

The subject of sets and their operations is so fundamental to the study of mathematics that some treatment is given in almost every mathematics book and familiarity with the topic abounds.

Theoretical treatments of sets can be found in Stoll (1963, 1974), Halmos (1960), and Monk (1969). Sets as a fundamental model of applied discrete structures are treated in Stanat and McAllister (1977), Mendelson (1970), Lin and Lin (1981), Tremblay and Monahar (1975), Levy (1980), and Preparata and Yeh (1973). The books by Fisher (1977) and Kapps and Bergman (1975) give especially nice treatments of relations and posets.

# Boolean Algebra

# 3

## 3.1 Introduction: Mathematical Systems

What we have seen from Chapters 1 and 2 is an example of two seemingly different discrete concepts having the same underlying formal structure. On the one hand, we studied propositions and statements and the operations by which we joined them to form new statements. In the other case we examined the structure of collections of objects known as sets and the operations for combining them to form new sets. We developed the structure of sets from that of logical propositions, so it appeared that set theory, in fact, depended on the theory of logic for its existence. Such is not the case. We could have just as easily started from a definition of sets and its related theory and developed the theory of logical propositions. The fact that these two theories have an *identical* structure is the important point.

Let us see how the two structures are similar.

1. In each case we defined elements of interest to us. On the one hand we discussed propositions, symbolized as variables $p$, $q$, $r$, and so on; on the other hand, we symbolized our variable sets as $A$, $B$, $C$, and so on. These were the entities of concern to us, the ones we learned how to combine and manipulate. These were the objects of our system.
2. We then defined ways to combine our objects to get new ones.
   (a) We defined binary operations on our objects, our "elements of interest." For propositions these were conjunction ($\wedge$), disjunction ($\vee$), conditional ($\rightarrow$), biconditional ($\leftrightarrow$) and X-OR ($\oplus$). The

corresponding binary operators for sets were intersection (∩), union (∪), set containment (⊆), set equivalence (=), and symmetric difference (⊕).

(b) Each system had a unary operator: negation (¬) for propositions and complement (‾) for sets. These also had equivalent meanings.

3. Each system had two important special elements that served unique purposes. For propositions we had the *tautology* ($T$) and the *contradiction* ($F$). For sets, the corresponding special elements were, respectively, the *universal set* ($U$) and the *null set* (∅). When elements have important special properties in the system, they are called "distinguished."
4. Each system had a set of axioms, or postulates, to govern how the elements are combined by the operators. These are the "rules of the game," the laws that tell us how the system operates. From these rules, or axioms, we developed certain theorems or additional properties of the system. Of course, each of the two systems had its own special flavor (validity of arguments, number of elements in sets, etc.), but the underlying theory of the structure was identical.

What has been described above is known as a *mathematical system.* In general, a mathematical system consists of:

1. Elements of interest; those objects to be studied and discussed. This set of elements is sometimes known as the "carrier set" in that it "carries" the elements in which we are interested.
2. A set of operations or operators that is defined on the carrier set. These operators combine and manipulate the elements to form new ones in the carrier set. These operators usually are binary and unary.
3. Important, special "distinguished" elements of the carrier set that act in ways unique to the system.
4. A set of axioms that describe the rules by which the operators act on the elements and from which the theory of the system can be established.

For many years now, you have studied the mathematical system of real numbers. What you have learned fits precisely with the foregoing description of a mathematical system.

1. The elements of interest (carrier set) are the real numbers: the integers, the rationals, the irrationals, and so on.
2. The binary operations on numbers were defined to be addition (+) and multiplication (×). From these two you learned how to perform subtraction, division, and exponentiation. You studied the ways that numbers could be combined two at a time to form new numbers by

using these operators. There are also unary operators on numbers: the unary negation: $-5$, $-a$, $-5/9$, and so on, and the unary positive: $+3$, $+19/8$, and so on.

3. The system has its own special numbers that stand out uniquely. The number *zero* (0) is distinguished in that for any real number $a$, $a \times 0 = 0$ and $a + 0 = a$. Similarly, the number *one* (1) is unique, in that for any real number, $a$, $a \times 1 = a$. These numbers act as identity elements.
4. You learned many axioms. For instance, you learned, among others, that addition and multiplication are commutative: $a + b = b + a$ and $ab = ba$; and that they are both associative: $a + (b + c) = (a + b) + c$ and $a(bc) = (ab)c$. From these you proved theorems and established the theory of arithmetic. For example, you learned that if $xy = 0$, then $x = 0$ or $y = 0$. As another theorem, you learned that every number can be uniquely represented as a product of primes.

Thus you learned a mathematical system. The technical name for this system of arithmetic is a *field*. There are many systems and each has its own name. The overall study of this classification and theory of mathematical systems is known, in general, as algebra. Thus when you studied high school and college algebra, you were learning the rules of the system of real numbers.

But let us come back to the systems of study in this book. The underlying structure evident in the study of propositions and sets is known as Boolean algebra. This algebra is named in honor of a British mathematician, George Boole (1813–1864), who first examined, in some depth, the laws of logic. The important application of the laws is that they form the basis for the design of logic circuits of electronic computers.

To capture the full flavor of this system, we will adopt some new symbolism so that we can be completely general in our discussion.

1. For the elements of interest (the carrier set) we will use lowercase letters from the end of the alphabet: $w$, $x$, $y$, $z$. These will represent our variable elements. Recall that for propositions we use $p$, $q$, $r$, and so on, and for sets, $A$, $B$, $C$, and so on.
2. For our operations we will adopt the conventions that are prevalent in the literature. These are not always standard.

   For the operation that corresponds to conjunction ($\wedge$) of propositions and intersection ($\cap$) of sets, we will use the raised dot ($\cdot$). Sometimes we will use juxtaposition of the variables. Thus $xy$ will have the same meaning as $x \cdot y$. This operator is referred to as the *meet* of $x$ and $y$ and is read as "$x$ and $y$."

   We will use $+$ for the operation corresponding to disjunction ($\vee$) and union ($\cup$). $x + y$ is referred to as the *join* of $x$ and $y$ and can be read as "$x$ or $y$."

   For the unary complement of an element $x$, we use $\bar{x}$.

3. Our distinguished element representing the tautology ($T$) or the universe ($U$) will be "1." It will be known as the *unit* element. For the contradiction ($F$) and null set ($\varnothing$) we will use "0," which will be known as the *zero* element. It is clear that these two elements shall be distinct; $0 \neq 1$.
4. The correspondence among the systems is shown in Table 3.1.

**TABLE 3.1**

| | *Propositions* | *Sets* | *Boolean Algebra* |
|---|---|---|---|
| Variable Elements | $p, q, r$ | $A, B, C$ | $x, y, z$ |
| Operations | | | |
| binary | $\wedge$ | $\cap$ | $\cdot$ (or juxtaposition) |
| | $\vee$ | $\cup$ | $+$ |
| unary | $\neg$ | $-$ | $-$ |
| Special distinguished | $T$ (tautology) | $U$ (universe) | 1 |
| elements | $F$ (contradiction) | $\varnothing$ (null set) | 0 |
| Sample expressions | $p \wedge q$ | $A \cap B$ | $x \cdot y$ or $xy$ |
| | $p \vee q \vee r$ | $A \cup B \cup C$ | $x + y + z$ |
| | $\neg(p \vee q)$ | $\overline{A \cup B}$ | $\overline{(x + y)}$ |
| | $\neg p \wedge \neg q$ | $\bar{A} \cap \bar{B}$ | $\bar{x}\bar{y}$ |
| Sample rules | $p \Leftrightarrow p \wedge p$ | $A = A \cap A$ | $x = xx$ |
| | $p \Leftrightarrow p \vee p$ | $A = A \cup A$ | $x = x + x$ |
| | $p \Leftrightarrow p \wedge (p \vee q)$ | $A = A \cap (A \cup B)$ | $x = x(x + y)$ |
| | $p \Leftrightarrow p \vee (p \wedge q)$ | $A = A \cup (A \cap B)$ | $x = x + xy$ |
| | $p \wedge \neg p \Leftrightarrow F$ | $A \cap \bar{A} = \phi$ | $x\bar{x} = 0$ |
| | $p \vee \neg p \Leftrightarrow T$ | $A \cup \bar{A} = U$ | $x + \bar{x} = 1$ |

## EXERCISES

1. Write the following logical expressions as expressions in Boolean algebra by letting $x$, $y$, and $z$ represent $p$, $q$, and $r$, respectively.
   (a) $(p \wedge \neg q) \vee \neg(q \wedge \neg r \wedge \neg p)$
   (b) $(q \vee T) \wedge (p \vee F) \wedge \neg(p \wedge q \wedge T)$
   (c) $(p \wedge q \wedge r) \vee (\neg p \wedge \neg q \wedge \neg r) \vee (p \wedge \neg q \wedge \neg r)$
   (d) $(p \wedge T) \vee (\neg p \wedge F) \vee (q \wedge \neg T)$
2. Write the following set expressions as expressions in Boolean algebra by letting $x$, $y$, and $z$ represent $A$, $B$, and $C$, respectively.
   (a) $(A \cap \bar{B}) \cup (\overline{B \cap \bar{C} \cap \bar{A}})$
   (b) $(B \cup U) \cap (A \cup \varnothing) \cap (\overline{A \cap B \cap U})$
   (c) $(A \cap B \cap C) \cup (\bar{A} \cap \bar{B} \cap \bar{C}) \cup (A \cap \bar{B} \cap \bar{C}) \cup U$
   (d) $(A \cap U) \cup (\bar{A} \cap \varnothing) \cup (B \cap \bar{U})$

## 3.2 Axioms and Laws for a Boolean Algebra

Now, obviously, all the results proven for propositions and for sets as summarized in Tables 1.25 and 2.10 hold for a Boolean algebra. However, one of the requirements for a set of axioms of a mathematical system is that they be *independent*. Basically, this means that it should not be possible to derive some of them (as theorems) from the others. Consequently, we can limit our set of axioms to be those that cannot be derived from the others. There are several approaches to take, but one that is standard is that the rules given in Table 3.2 be taken as *axioms*.

We will then show that all the other rules we have discussed in Chapters 1 and 2 can be derived from them.

**TABLE 3.2** Axioms of a Boolean Algebra

| *Commutative laws* | *Identity laws* |
|---|---|
| 1a. $x \cdot y = y \cdot x$ (or $xy = yx$) | 3a. $x \cdot 1 = x$ |
| 1b. $x + y = y + x$ | 3b. $x + 0 = x$ |
| *Distributive laws* | *Complement laws* |
| 2a. $x(y + z) = xy + xz$ | 4a. $x\bar{x} = 0$ |
| 2b. $x + yz = (x + y)(x + z)$ | 4b. $x + \bar{x} = 1$ |

DUALITY: Observe that in each of the pairs of axioms each statement is the *dual* of the other. Recall that the dual of a statement is one in which every instance of each binary operator is replaced with the other binary operator, and the special elements are exchanged: 0 for 1 and 1 for 0. This principle of duality is very important for the development of the theory. We saw it at work in our discussion both with propositions and with our theory of sets.

□ Principle of Duality If a statement or equivalence is derivable from axioms 1a–4b, then its dual is also derivable from the axioms.

IDEMPOTENCE: To show how the principle of duality works and to also show how the other rules of the system can be derived, consider the laws of *idempotence:*

$$x = x \cdot x \quad \text{and} \quad x = x + x$$

Note that these two equalities are duals of each other. We need not accept them as axioms, for they can be derived from axioms 1a–4b, as shown in the following:

□ Theorem 3.1 (Idempotence) For $x \in B$: (a) $x = x \cdot x$ and (b) $x = x + x$.

**Proof** (a)

$$\begin{aligned} x \cdot x &= x \cdot x + 0 && \text{by axiom 3b} \\ &= x \cdot x + x \cdot \bar{x} && \text{by axiom 4a} \\ &= x \cdot (x + \bar{x}) && \text{by axiom 2a} \\ &= x \cdot 1 && \text{by axiom 4b} \\ &= x && \text{by axiom 3a} \end{aligned}$$

(b)

$$\begin{aligned} x + x &= (x + x) \cdot 1 && \text{by axiom 3a} \\ &= (x + x) \cdot (x + \bar{x}) && \text{by axiom 4b} \\ &= x + x \cdot \bar{x} && \text{by axiom 2b} \\ &= x + 0 && \text{by axiom 4a} \\ &= x && \text{by axiom 3b} \end{aligned}$$

If we examine these derivations in detail, we see that at each step, the parallel axioms used are duals. Idempotence is proved for both meet and join. Furthermore, only one needs proving, for since one follows from the axioms, its dual must also follow by the principle of duality.

Let us derive some of the other rules established in Chapters 1 and 2.

**BOUNDNESS (DOMINANCE):** The pairs of rules for boundness

$$p \wedge F \Leftrightarrow F, \quad p \vee T \Leftrightarrow T \qquad \text{for propositions}$$

and

$$A \cap \varnothing = \varnothing, \quad A \cup U = U \qquad \text{for sets}$$

are, of course, duals. The Boolean algebra counterparts are $x \cdot 0 = 0$ and $x + 1 = 1$. We will establish one of these, and the other will follow from duality.

□ **Theorem 3.2** $x + 1 = 1$.

**Proof**

$$\begin{aligned} x + 1 &= (x + 1) \cdot 1 && \text{by axiom 3a} \\ &= (x + 1) \cdot (x + \bar{x}) && \text{by axiom 4b} \\ &= (x + \bar{x}) \cdot (x + 1) && \text{by axiom 1a} \\ &= x + (\bar{x} \cdot 1) && \text{by axiom 2b} \\ &= x + \bar{x} && \text{by axiom 3a} \\ &= 1 && \text{by axiom 4b} \end{aligned}$$

□ **Theorem 3.3** $x \cdot 0 = 0$.

**Proof** Follows from Theorem 3.2 by the principle of duality.

Theorems 3.2 and 3.3 establish the rules of boundness.

**UNIQUENESS OF COMPLEMENT:** An important consequence of our structure that has not been previously mentioned is that each element $x$ have *one and only one* complement, $\bar{x}$; and it has the property that $x \cdot \bar{x} = 0$ and $x + \bar{x} = 1$. We will show this fact in Theorem 3.4, where we will *assume* that element $x$ has *two* complements, $z$ and $w$, and then establish that they are really the same element.

**Theorem 3.4 (Uniqueness of Complement)** Every element $x$ of a Boolean algebra has one and only one complement, $\bar{x}$, such that $x \cdot \bar{x} = 0$ and $x + \bar{x} = 1$.

**Proof** Let us assume that for element $x$, there are two elements $z$ and $w$ such that

$$xz = 0 \quad \text{and} \quad x + z = 1$$

$$xw = 0 \quad \text{and} \quad x + w = 1$$

Thus we assume two complements for $x$.

We will show by the following parallel steps that these "two" complements equal each other by showing that they are equal to the same expression.

| $z$ | | $w$ |
|---|---|---|
| $z = z + 0$ | axiom 3b | $w = w + 0$ |
| $= z + xw$ | since $xw = 0$ and $xz = 0$ | $= w + xz$ |
| $= (z + x)(z + w)$ | axiom 2b | $= (w + x)(w + z)$ |
| $= (x + z)(z + w)$ | axiom 1b | $= (x + w)(w + z)$ |
| $= 1 \cdot (z + w)$ | since $x + z = 1$ and $x + w = 1$ | $= 1 \cdot (w + z)$ |
| $= (z + w) \cdot 1$ | axiom 1a | $= (w + z) \cdot 1$ |
| $= z + w$ | axiom 3a | $= w + z$ |

Therefore, $z = z + w = w + z = w$. Thus these "two" complements are really only *one;* and we shall call the *one* complement, $\bar{x}$.

Theorem 3.4 may at first glance may seem somewhat abstract. But if one relates it to the theory of sets of Chapter 2, the result becomes

> For $A$ being an arbitrary set, if $A \cap B = \varnothing$ and $A \cup B = U$, then $B = \bar{A}$.

This result can easily be verified by using a Venn diagram.

INVOLUTION: We now use the uniqueness of complement concept to prove the law of *double negation*, or involution: that is, $\bar{\bar{x}} = x$.

**Theorem 3.5 (Involution)** For $x$ being an element of a Boolean algebra, we have $\bar{\bar{x}} = x$.

**Proof** By axioms 1a and 4a we have $x \cdot \bar{x} = \bar{x} \cdot x = 0$.
By axioms 1b and 4b we have $x + \bar{x} = \bar{x} + x = 1$.

Consequently, we have by Theorem 3.4 that $x$ is the unique complement of $\bar{x}$, or $x = \bar{\bar{x}}$.

ABSORPTION: The *absorption* laws are straightforward and have been proven in Chapters 1 and 2; thus we state them without proof.

□ Theorem 3.6 (Absorption Laws) For $x$ and $y$ being arbitrary elements of a Boolean algebra, we have

(a) $x = x \cdot (x + y)$ and
(b) $x = x + (x \cdot y)$.

ASSOCIATIVITY: The laws of associativity, $x + (y + z) = (x + y) + z$ and $x(yz) = (xy)z$, look deceptively easy to establish. The fact is, however, that they are a bit difficult to prove. An outline of the proof is explored in Exercises 10 and 11 and the interested reader is invited to explore the details.

For now we state that associativity holds as

□ Theorem 3.7 (Associativity) For $x$, $y$, and $z$ being arbitrary elements of a Boolean algebra, we have

(a) $x(yz) = (xy)z$ and
(b) $x + (y + z) = (x + y) + z$

DE MORGAN'S LAWS

□ Theorem 3.8 (De Morgan's Laws) For $x$ and $y$ being arbitrary elements of a Boolean algebra, we have

(a) $\overline{xy} = \bar{x} + \bar{y}$ and
(b) $\overline{x + y} = \bar{x}\bar{y}$

**Proof** We will outline the proof of part (b) and ask the reader to fill in the pertinent details (Exercise 6). Part (a) follows from duality (Exercise 7). Observe in part (b) that we are to show that the complement of $x + y$ is $\bar{x}\bar{y}$. If this is indeed the case, then the meet of the two will be 0: $[(x + y) \cdot \bar{x}\bar{y} = 0]$ and the join will be 1: $[(x + y) + \bar{x}\bar{y} = 1]$. Furthermore, by Theorem 3.4, $\bar{x}\bar{y}$ will be the unique complement.

To this end, we show that $(x + y) \cdot (\bar{x}\bar{y}) = 0$:

$$(x + y)\bar{x}\bar{y} = \bar{x}\bar{y}(x + y) = \bar{x}\bar{y}x + \bar{x}\bar{y}y = 0 \cdot \bar{y} + \bar{x} \cdot 0 = 0$$

We also show that $(x + y) + \bar{x}\bar{y} = 1$:

$$(x + y) + \bar{x}\bar{y} = (x + y + \bar{x})(x + y + \bar{y}) = (1 + y)(1 + x) = 1$$

Thus De Morgan's laws hold for a Boolean algebra.

The axioms given and theorems proven of a Boolean algebra can be displayed as shown in Table 3.3. Note that these correspond directly with Tables 1.25 and 2.10.

These laws can, of course, be used to establish other properties (theorems) of a Boolean algebra. The rules cited are from Table 3.3

**TABLE 3.3**

| | *Rule* | *Meet* | *Join* | *Theorem or axiom* |
|---|---|---|---|---|
| (1)–(2) | Commutative | $xy = yx$ | $x + y = y + x$ | Axiom 1 |
| (3)–(4) | Associative | $x(yz) = (xy)z$ | $x + (y + z) = (x + y) + z$ | Theorem 3.7 |
| (5)–(6) | Idempotence | $x = x \cdot x$ | $x = x + x$ | Theorem 3.1 |
| (7)–(8) | Absorption | $x = x \cdot (x + y)$ | $x = x + (x \cdot y)$ | Theorem 3.6 |
| (9)–(10) | Distribution | $x(y + z) = xy + xz$ | $x + yz = (x + y)(x + z)$ | Axiom 2 |
| (11)–(12) | De Morgan | $\overline{(xy)} = \bar{x} + \bar{y}$ | $\overline{(x + y)} = \bar{x}\bar{y}$ | Theorem 3.8 |
| (13) | Double negation (involution) | $x = \bar{\bar{x}}$ | | Theorem 3.5 |
| (14) (a)–(b) | Identity | $x \cdot 1 = x$ | $x + 0 = x$ | Axiom 3 |
| (c)–(d) | Boundness | $x \cdot 0 = 0$ | $x + 1 = 1$ | Theorems 3.2 and 3.3 |
| (e)–(f) | Complement | $x\bar{x} = 0$ | $x + \bar{x} = 1$ | Axiom 4 |

**EXAMPLE 1** $\bar{0} = 1$.

$$\begin{aligned} \bar{0} &= \bar{0} + 0 && \text{rule (14b) (identity)} \\ &= 0 + \bar{0} && \text{rule (2) (commutativity)} \\ &= 1 && \text{rule (14f) (complement)} \end{aligned}$$

**EXAMPLE 2** $(x + y)(x + z)\overline{(\bar{x}y)} = x$.

$$\begin{aligned} (x + y)(x + z)\overline{(\bar{x}y)} &= (x + yz)\overline{(\bar{x}y)} && \text{rule (10)} \\ &= (x + yz)(x + \bar{y}) && \text{rules (11) and (13)} \\ &= x + (yz)\bar{y} && \text{rule (10)} \\ &= x + \bar{y}(yz) && \text{rule (1)} \\ &= x + (\bar{y}y)z && \text{rule (3)} \\ &= x + (y\bar{y})z && \text{rule (1)} \\ &= x + 0 \cdot z && \text{rule (14e)} \\ &= x + z \cdot 0 && \text{rule (1)} \\ &= x + 0 && \text{rule (14c)} \\ &= x && \text{rule (14b)} \end{aligned}$$

**EXAMPLE 3** If $x\bar{y} = 0$, then $xy = x$.

$$\begin{aligned} xy &= xy + 0 && \text{rule (14b)} \\ &= xy + x\bar{y} && \text{hypothesis } (x\bar{y} = 0) \\ &= x(y + \bar{y}) && \text{rule (9)} \\ &= x \cdot 1 && \text{rule (14f)} \\ &= x && \text{rule (14a)} \end{aligned}$$

## EXERCISES

1. Prove the following statements by using rules in Table 3.3.
   (a) $\bar{1} = 0$ (dual of Example 1)
   (b) $x(\bar{x} + y) = xy$
   (c) $x + \bar{x}y = x + y$ (dual of statement (b))
   (d) $x(\overline{x\bar{y}}) = xy$ (see Example 1, Section 2.6)
   (e) $xy(\overline{xz}) = xy\bar{z}$
   (f) $x\bar{y} + \bar{x}y = (x + y)(\overline{xy})$
   (g) $(\overline{\bar{x}y}) + x\bar{y} = x + \bar{y}$
2. Prove Theorem 3.3. Do not use the fact that it follows from Theorem 3.2 by duality.
3. Show that the cancellation law of meet does not hold for a Boolean algebra. That is, show that we cannot state that $y = z$ if $xy = xz$. You may exhibit a counterexample from either set theory or the theory of logical propositions to prove this.
4. Show that the cancellation law of join does not hold for a Boolean algebra. That is, show that $x + y = x + z$ does not imply that $y = z$ (see Exercise 3).
5. Prove the absorption laws of Theorem 3.6.
6. Fill in the details of the proof of De Morgan's law given in Theorem 3.8(b).
7. Prove Theorem 3.8(a).
8. For two elements, $x$ and $y$, of a Boolean algebra, the X-OR operator, $\oplus$, is defined by $x \oplus y = x\bar{y} + \bar{x}y$. This is the same as the X-OR connector of logical propositions and the symmetric difference of sets. Show that $\oplus$ has the following properties:
   (a) $x \oplus 0 = x$
   *Proof:* $x \oplus 0 = x\bar{0} + \bar{x}0 = x \cdot 1 + 0 = x$
   (b) $x \oplus x = 0$ (c) $x \oplus \bar{x} = 1$
   (d) $x \oplus y = y \oplus x$ (e) $x \oplus 1 = \bar{x}$
   (f) $x(y \oplus z) = xy \oplus xz$ (g) $x + (x \oplus y) = x + y$
   (h) $(x \oplus y) \oplus y = x$ (i) $(x \oplus y) \oplus x = y$
9. In Exercise 2(c) of Section 2.6 you were asked to determine whether the X-OR

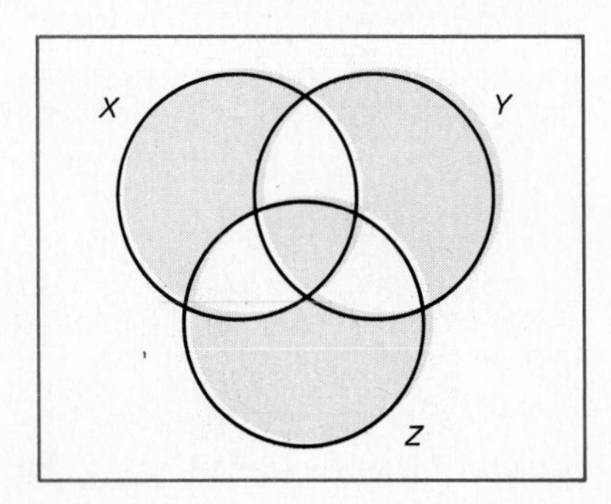

operator (of Exercise 8) is associative. That is, does

$$(x \oplus y) \oplus z = x \oplus (y \oplus z)?$$

(a) Show that this is true by demonstrating that each of the expressions above yields the shaded portions of the Venn diagram shown on page 92.

(b) Show from the definition given in Exercise 8 that

$$x \oplus y \oplus z = xyz + x\bar{y}\bar{z} + \bar{x}y\bar{z} + \bar{x}\bar{y}z$$

**10.** *Modified cancellation law:* Exercises 3 and 4 show that the cancellation laws do *not* hold for Boolean algebras. There is a modified cancellation law that does hold, however. This is stated by:

If $xy = xz$ *and* $\bar{x}y = \bar{x}z$, then $y = z$.

(a) Fill in the details of the following proof, showing that these conditions do actually yield $y = z$.

$$y = y \cdot 1 = y(x + \bar{x}) = yx + y\bar{x} = xz + \bar{x}z = z(x + \bar{x}) = z \cdot 1 = z$$

Thus $y = z$.

(b) Prove the dual statement:

If $x + y = x + z$ and $\bar{x} + y = \bar{x} + z$, then $y = z$.

**11.** *Associativity:* We use the modified cancellation law of Exercise 10 to prove that $x + (y + z) = (x + y) + z$.

We will show that

(1) $[x + (y + z)] \cdot x = [(x + y) + z] \cdot x$, and
(2) $[x + (y + z)] \cdot \bar{x} = [(x + y) + z] \cdot \bar{x}$

and from the modified cancellation law, we have

$$x + (y + z) = (x + y) + z.$$

Now

$$[x + (y + z)] \cdot x = x \qquad \text{by absorption}$$

and

$$[(x + y) + z] \cdot x = x(x + y) + xz = x + xz = x \qquad \text{by absorption}$$

Thus statement (1) is established. Also,

$$[x + (y + z)] \cdot \bar{x} = \bar{x}x + \bar{x}(y + z) = \bar{x}(y + z)$$

and

$$[(x + y) + z] \cdot \bar{x} = \bar{x}(x + y) + \bar{x}z = \bar{x}x + \bar{x}y + \bar{x}z = \bar{x}(y + z)$$

Thus statement (2) is established; and the associative law $x + (y + z) = (x + y) + z$ is established.

(a) Fill in the details of the proof.

(b) Prove the other associative law: $x(yz) = (xy)z$.

## 3.3 Examples of Boolean Algebras

In this section we show several examples of Boolean algebras.

**EXAMPLE 1** As our first example of a Boolean algebra let the carrier set, $B$, be the set of all integral divisors of 30: $B = \{1, 2, 3, 5, 6, 10, 15, 30\}$.

We need to define what the two binary operators, *meet* and *join*, will be. For this purpose, let the meet of two elements of $B$ be defined to be their *greatest common divisor* (gcd). The gcd of two integers is defined to be the largest exact divisor of the two integers. Thus for any two elements $x, y \in B$, the operation $x \cdot y$ will be interpreted as gcd $(x, y)$.

meet

Then the *meet* of 5 and 10, or, $5 \cdot 10 = \text{gcd}(5, 10) = 5$; and $2 \cdot 5 = \text{gcd}(2, 5) = 1$.

For the *join* of two elements we will use their *least common multiple* (lcm), the smallest integer that the two integers divide exactly. Thus for our system, the *join* of 3 and 6, or, $3 + 6 = \text{lcm}(3, 6) = 6$, and $2 + 15 = \text{lcm}(2, 15) = 30$.

Let the unary operator, or complement of $x \in B$, be defined to be the quotient of 30 divided by $x$, $30/x$, so

$$\overline{1} = 30/1 = 30 \qquad \overline{30} = 30/30 = 1$$
$$\overline{2} = 30/2 = 15 \qquad \overline{15} = 30/15 = 2$$
$$\overline{3} = 30/3 = 10 \qquad \overline{10} = 30/10 = 3$$
$$\overline{5} = 30/5 = 6 \qquad \overline{6} = 30/6 = 5$$

Thus it is seen that each element of $B$ has a unique complement. The axioms 1a–4b are satisfied by the system by the following:

- *Axiom 1—commutativity:* Without showing every case, it is obvious from the definition of gcd that this axiom holds. For example,

$$\text{gcd}(3, 10) = \text{gcd}(10, 3) = 1$$
$$\text{gcd}(5, 15) = \text{gcd}(15, 5) = 5$$

Also, from the definition of lcm, commutativity holds:

$$\text{lcm}(5, 6) = \text{lcm}(6, 5) = 30$$
$$\text{lcm}(10, 2) = \text{lcm}(2, 10) = 10$$

- *Axiom 2—distribution:* Again, some examples will make this law evident:

$$\begin{aligned} 5 \cdot (2 + 15) &= \text{meet}(5, \text{join}(2, 15)) \\ &= \text{gcd}(5, \text{lcm}(2, 15)) \\ &= \text{gcd}(5, 30) \\ &= 5 \end{aligned}$$

and

$$\begin{aligned} 5 \cdot 2 + 5 \cdot 15 &= \text{join}\ (\text{meet}\ (5, 2), \text{meet}\ (5, 15)) \\ &= \text{lcm}\ (\text{gcd}\ (5, 2), \text{gcd}\ (5, 15)) \\ &= \text{lcm}\ (1, 5) \\ &= 5 \end{aligned}$$

Also,

$$\begin{aligned} 10 + 15 \cdot 3 &= \text{lcm}\ (10, \text{gcd}\ (15, 3)) \\ &= \text{lcm}\ (10, 3) \\ &= 30 \end{aligned}$$

and

$$\begin{aligned} (10 + 15) \cdot (10 + 3) &= \text{gcd}\ (\text{lcm}\ (10, 15), \text{lcm}\ (10, 3)) \\ &= \text{gcd}\ (30, 30) \\ &= 30 \end{aligned}$$

- *Axiom 3:* There are unit and zero elements, namely, 30 and 1, respectively:

for any $x \in B$, we have gcd $(x, 30) = x$; i.e., $x \cdot 30 = x$ so 30 is the unit.
for any $x \in B$, we have lcm $(x, 1) = x$; i.e., $x + 1 = x$ so 1 is the zero.

- *Axiom 4:* We need to show that the complement laws hold for the system. That is, we need to show:

| | |
|---|---|
| gcd $(x, \bar{x}) = 1$ | (the zero element) |
| lcm $(x, \bar{x}) = 30$ | (the unit element) |
| gcd (1, 30) = 1 | lcm (1, 30) = 30 |
| gcd (2, 15) = 1 | lcm (2, 15) = 30 |
| gcd (3, 10) = 1 | lcm (3, 10) = 30 |
| gcd (5, 6) = 1 | lcm (5, 6) = 30 |

Thus we have shown that all the axioms of a Boolean algebra hold. Our system is a Boolean algebra.

Boolean algebras lend themselves very nicely to diagrams. In Figure 3.1 we illustrate the Boolean algebra of the divisors of 30 under gcd and lcm. The unit element (30) is placed on top and the zero element (1) on the bottom. Each of the elements is pictured as a point and is reachable (in the sense of following the arrows upward) from another element only if it is a multiple of the lower element. Figure 3.1 is known as a *Hasse diagram.*

An underlying order is evident in Figure 3.1. For example, the element 1 is "less than" the element 3; 3 is "less than" 15; and so on. But what about the elements 2 and 15? Can we say that one is "less than" the other? They are

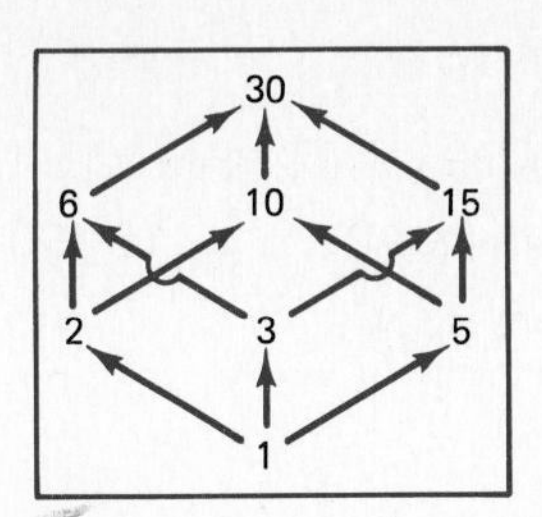


**FIGURE 3.1** Boolean algebra of the divisors of 30.

not even related by being a divisor or multiple. The element 1 is "less than" all the other elements, and all the elements are "less than" 30. This is an example of a *partial order*, as introduced in Chapter 2. Show that the properties of reflexivity, antisymmetry, and transitivity are satisfied.

**EXAMPLE 2** As a second example of a Boolean algebra, consider the power set of a given set. Recall from Chapter 2 that the power set of some set $A$ is the set of all subsets of $A$. For example, if $A = \{0, 1\}$, then the power set of $A$, $P(A)$ is

$$P(A) = \{\varnothing, \{0\}, \{1\}, \{0, 1\}\}$$

Thus there are four elements in the carrier set.

This power set, $P(A)$, forms a Boolean algebra under the usual set operations of intersection and union. The unary complement will be the set-theoretic complement relative to the original set. Thus

$$\bar{\varnothing} = \{0, 1\} \qquad \overline{\{0\}} = \{1\} \qquad \overline{\{1\}} = \{0\} \qquad \overline{\{0, 1\}} = \varnothing$$

It should not be difficult to show that all the axioms of a Boolean algebra are satisfied. The Hasse diagram is shown in Figure 3.2.

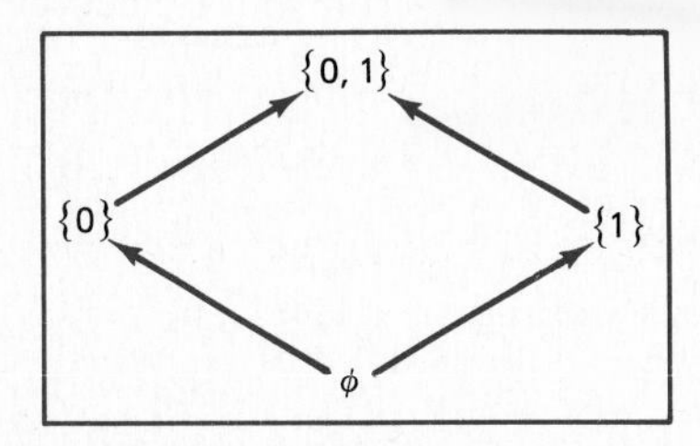


**FIGURE 3.2** Boolean algebra of the power set of {0, 1}.

**EXAMPLE 3** As our last example, let us look at the Boolean algebra with the fewest number of elements in its carrier set. We have noted earlier that there have to be at least two elements, the zero element and the unit element, for they have to be distinct. So let us consider such a set for the carrier set. We let $B = \{0, 1\}$. Do there need to be other elements?

If the axioms are satisfied by using only these two elements, then they should be sufficient. How can we define our operations of $\cdot$ and $+$ and that of complement so that we obtain a Boolean algebra?

Now obviously since there are only two elements, each is the complement of the other, $\bar{0} = 1$ and $\bar{1} = 0$.

The *complement* laws need to be satisfied:

Axiom 4a states that $x \cdot \bar{x} = 0$, so we should have

$$0 \cdot 1 = 0 \tag{i}$$

$$1 \cdot 0 = 0 \tag{ii}$$

and

Axiom 4b states that $x + \bar{x} = 1$, so we should have

$$0 + 1 = 1 \tag{iii}$$

$$1 + 0 = 1 \tag{iv}$$

and

To obtain other properties of $\cdot$ and $+$, consider the *identity* laws. Since 1 is the unit element, axiom 3a, $x \cdot 1 = x$, yields the following:

$$0 \cdot 1 = 0 \tag{i}$$

$$1 \cdot 1 = 1 \tag{v}$$

and since 0 is the zero element, axiom 3b, $x + 0 = x$, gives us

$$0 + 0 = 0 \tag{vi}$$

$$1 + 0 = 1 \tag{iv}$$

The results above can be displayed very nicely in "join" and "meet" tables (Table 3.4).

**TABLE 3.4**

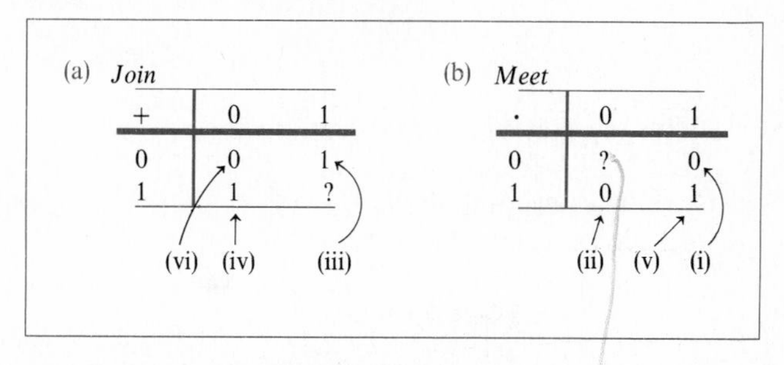

(a) *Join*

| + | 0 | 1 |
|---|---|---|
| 0 | 0 | 1 |
| 1 | 1 | ? |

(vi) (iv) (iii)

(b) *Meet*

| · | 0 | 1 |
|---|---|---|
| 0 | ? | 0 |
| 1 | 0 | 1 |

(ii) (v) (i)

The two items left out of the table, $1 + 1$ and $0 \cdot 0$, can be supplied from the boundness laws in Theorems 3.2 and 3.3.

From Theorem 3.3 we have that $x \cdot 0 = 0$, so we have

$$0 \cdot 0 = 0 \tag{vii}$$

and

$$1 \cdot 0 = 0 \tag{ii}$$

Also from Theorem 3.2, $x + 1 = 1$, so we have

$$0 + 1 = 1 \qquad \text{(iii)}$$

and

$$1 + 1 = 1 \qquad \text{(viii)}$$

So now we have complete tables for our operations (Table 3.5).

**TABLE 3.5**

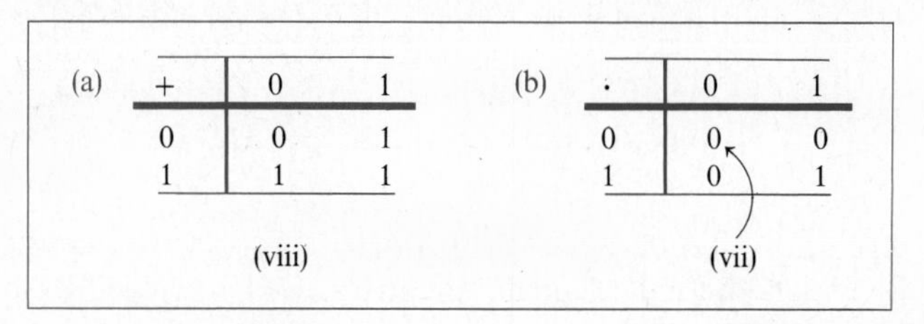

The commutativity axioms are evident from the tables. The identity and complement laws have been established. So now we have to verify that distribution (axioms 2a and 2b) holds.

Two cases will be shown here. The reader can supply the others.

1. $0 \cdot (1 + 0)$ should equal $0 \cdot 1 + 0 \cdot 0$.

   The left side: $0 \cdot (1 + 0) = 0 \cdot 1 = 0$

   The right side: $0 \cdot 1 + 0 \cdot 0 = 0 + 0 = 0$

2. $1 + 1 \cdot 0$ should equal $(1 + 1) \cdot (1 + 0)$.

   The left side: $1 + 1 \cdot 0 = 1 + 0 = 1$

   The right side: $(1 + 1) \cdot (1 + 0) = 1 \cdot 1 = 1$

Thus the system is a Boolean algebra.

We can portray Table 3.5(a) and (b) in another form, as shown in Table 3.6.

**TABLE 3.6**

| | A | B | Operations ∨ + | ∧ · |
|---|---|---|---|---|
| | 1 | 1 | 1 | 1 |
| Combinations | 1 | 0 | 1 | 0 |
| of elements | 0 | 1 | 1 | 0 |
| | 0 | 0 | 0 | 0 |

If we compare this table to the truth tables of disjunction and conjunction of Chapter 1, we see that there is a direct correspondence with

1 being T
0 being F
$+$ being $\vee$ (disjunction)
$\cdot$ being $\wedge$ (conjunction)

This should not be surprising since the $\{0, 1\}$ Boolean algebra can be traced directly to the laws of logical propositions. What is interesting, though, is that the $\{0, 1\}$ Boolean algebra has direct and practical application to the heart of computers, as will be discussed in Chapter 4.

## EXERCISES

1. Show that the integral divisors of 70, that is, the set $\{1, 2, 5, 7, 10, 14, 35, 70\}$, form a Boolean algebra by using the operations of Example 1. Draw the Hasse diagram.
2. (a) Show that the integral divisors of 24 do *not* form a Boolean algebra by using the operations of Example 1. Where does the theory fall down? [*Hint:* What is $\bar{6}$? What are gcd $(6, \bar{6})$ and lcm $(6, \bar{6})$?]
   (b) Draw the Hasse diagram.
3. (a) Do the integral divisors of 60 form a Boolean algebra by using the operations of Example 1?
   (b) Draw the Hasse diagram.
4. Verify that the power set of $M = \{0, 1, 2\}$ is a Boolean algebra by using the operations as defined in Example 2. Draw the Hasse diagram. (This exercise will be referenced in the next section.)
5. Does the power set of any set form a Boolean algebra by using the operations of Example 2? (See Section 3.6.)
6. Draw the Hasse diagram for Example 3.

## 3.4 Partial Orders on Boolean Algebras

Every Boolean algebra has defined on it a partial order relation. This relation was explained in Section 2.10 and reference made to it in Example 1 of Section 3.3. The relation as defined is very similar to the usual arithmetic relation, $\leq$, "is less than or equal to," and also to the subset relation, $\subseteq$. You should recall that a partial order relation satisfies the properties of (1) reflexivity,

(2) antisymmetry, and (3) transitivity. Also any set that is partially ordered under some relation is known as a *poset.*

The symbol that will be used for the partial ordering relation for Boolean algebras will be "$\leq$." There is some risk involved in using this symbol, since inevitably one will interpret it as "is less than or equal to." We use it only to indicate that it will act somewhat in the manner as "less than or equal to" acts on real numbers. A closer parallel is the symbol "$\subseteq$" of sets.

The meaning given to "$x \leq y$" for $x$ and $y$ being elements of a Boolean algebra is that "*$x$ precedes $y$.*" Alternatively, we say "$y$ succeeds $x$."

Let us define the partial order, $\leq$, as follows:

**Definition** Let $x$ and $y$ be arbitrary elements of a Boolean algebra. A partial order relation, $\leq$, is defined by stating:

$$x \leq y \quad \text{if and only if} \quad x \cdot y = x$$

(One may think of this less abstractly by recalling an equivalent statement regarding sets: $A \subseteq B$ iff $A \cap B = A$. This was proven in Example 4 of Section 2.5.)

It is readily seen that $\leq$ is a partial ordering relation:

1. *$\leq$ is reflexive:* This is obvious since $x \cdot x = x$ from idempotency.
2. *$\leq$ is antisymmetric:* Here we have to show that

$$(x \leq y \wedge y \leq x) \rightarrow x = y$$

Since $x \leq y$ implies that $x = x \cdot y$ and $y \leq x$ implies that $y = y \cdot x$, then $x = x \cdot y = y \cdot x = y$; thus $\leq$ is antisymmetric.

3. *$\leq$ is transitive:* To show transitivity we have to show that

$$(x \leq y \wedge y \leq z) \rightarrow x \leq z$$

Since $x \leq y$ we have $x = xy$, and since $y \leq z$ we have $y = yz$. By combining these two results, we have

$$x = xy = x(yz) = (xy)z = xz$$

Thus $x \leq z$.

There is an alternative definition that is sometimes used for $\leq$. It can be stated as: $x \leq y$ iff $x + y = y$. Note that this definition corresponds to the set-theoretic property: $A \subseteq B$ iff $A \cup B = B$ [see Exercise 1(f) of Section 2.5].

To show that this is an equivalent statement for defining $\leq$, we prove the following theorem.

□ Theorem 3.9 $x \leq y$ iff $x + y = y$.

**Proof** Since this is an "if and only if" proof we have to prove it "going both ways."

1. Assume that $x \leq y$. Then $x = xy$ by definition. So

$$
\begin{aligned}
x + y &= xy + y && \text{substitution} \\
&= y + yx && \text{commutativity (twice)} \\
&= y && \text{absorption}
\end{aligned}
$$

2. Conversely, assume that $x + y = y$. We are to show that $x = xy$. This portion of the proof is left as an exercise. (See Exercise 1.)

Let us examine this partial ordering relation, $\leq$, with regard to an example given previously. Consider Exercise 4 of Section 3.3, where we are given as the carrier set, the power set of $\{0, 1, 2\}$. Here $B = \{\varnothing, \{0\}, \{1\}, \{2\}, \{0, 1\}, \{0, 2\}, \{1, 2\}, \{0, 1, 2\}\}$. The binary operations defined are $\cap$ and $\cup$ for $\cdot$ and $+$. The unary complement is the complement relative to $\{0, 1, 2\}$; thus

$$
\begin{aligned}
\overline{\varnothing} &= \{0, 1, 2\} & \overline{\{0, 1, 2\}} &= \varnothing \\
\overline{\{0\}} &= \{1, 2\} & \overline{\{1, 2\}} &= \{0\} \\
\overline{\{1\}} &= \{0, 2\} & \overline{\{0, 2\}} &= \{1\} \\
\overline{\{2\}} &= \{0, 1\} & \overline{\{0, 1\}} &= \{2\}
\end{aligned}
$$

What does $\leq$ mean in this example? Let us take as examples, $\{2\}$ and $\{0, 2\}$. Then

$$\{2\} \leq \{0, 2\} \quad \text{iff} \quad \{2\} \cap \{0, 2\} = \{2\}$$

The right-hand side is true; therefore, $\{2\} \leq \{0, 2\}$.

Consider $\{1\}$ and $\{0, 2\}$. $\{1\} \cap \{0, 2\} = \varnothing \neq 1$, so we cannot say that $\{1\} \leq \{0, 2\}$. This case gives the essence of the *partial* order. Some of the elements are ordered relative to each other, others are not.

When two elements *are* ordered under $\leq$, then we say the elements are comparable and one *precedes* the other. When two elements are not ordered they are said to be *noncomparable*.

As shown in Example 1 of Section 3.3, the partial order of a Boolean algebra gives rise to a diagram. The "unit" element is placed on the top and the "zero" element is on the bottom. The elements are then placed so that they are reachable (by arrows) only if they are *comparable* by the relation. Thus for our current example we have the Hasse diagram shown in Figure 3.3. Note that this

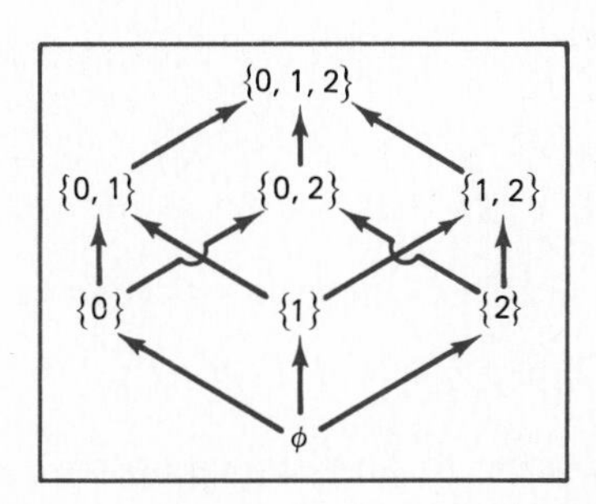


**FIGURE 3.3**

diagram is the same as the one in Figure 3.1. This similarity will be true of any eight-element Boolean algebra.

## EXERCISES

1. Prove Theorem 3.9(b); that is, if $x + y = y$, then $x = xy$.
2. For the Boolean algebra of Example 2 of Section 3.3, determine which pairs of elements are ordered by $\leq$.
3. Draw the Hasse diagram for the power set of the set $\{a, b, c, d\}$. Which pairs of elements are ordered by $\leq$? Which pairs of elements are noncomparable?

## 3.5 Further Properties of Partial Orders on Boolean Algebras

A number of further properties can be established regarding the partial order relation of a Boolean algebra. An interesting observation to be made is that the partial order, $\leq$, corresponds directly to containment, $\subseteq$, of sets.

**EXAMPLE 1** For elements $x$ and $y$ of a Boolean algebra we have $xy \leq x$ (or $xy \leq y$). In set-theoretic terms, $A \cap B \subseteq A$. This is evident from a Venn diagram of two sets, and was proven as Example 1 of Section 2.5.

To show that $xy \leq x$, we have to demonstrate that $xy = (xy)x$ by the definition of a partial order. Indeed, this is true from

$$(xy)x = x(xy) = (xx)y = xy$$

Thus $xy \leq x$.

By using the alternative definition of partial orders, we could also demonstrate that $xy + x = x$. The absorption law verifies this equality.

**EXAMPLE 2** Show that if $x \leq y$ and $x \leq z$, then $x \leq (y \cdot z)$.

Since $x \leq y$ we have $x = xy$, and since $x \leq z$ we have $x = xz$. Thus, by substitution,

$$x = xy = (xz)y = x(zy) = x(yz)$$

Therefore, $x \leq yz$.

Alternatively, $x \leq y$ means $x + y = y$ and $x \leq z$ means $x + z = z$. By substitution and distribution,

$$yz = (x + y)(x + z) = x + yz$$

Therefore, $x \leq yz$ (since $x + yz = yz$). Of course, in set-theoretic terms, we have

$$[(A \subseteq B) \wedge (A \subseteq C)] \rightarrow A \subseteq (B \cap C)$$

(Draw the appropriate Venn diagram to verify this property.)

**EXAMPLE 3** Show that if $x \leq y$, then $x\bar{y} = 0$. (Set theoretically, we have: If $A \subseteq B$, then $A \cap \bar{B} = \varnothing$.)

Since $x \leq y$, then $x + y = y$, and by substitution for $y$ we get

$$x\bar{y} = x\overline{(x + y)} = x\bar{x}\bar{y} = 0$$

## EXERCISES

Prove the following properties of a Boolean algebra. Assume that $x$, $y$, and $z$ are arbitrary elements of some carrier set.

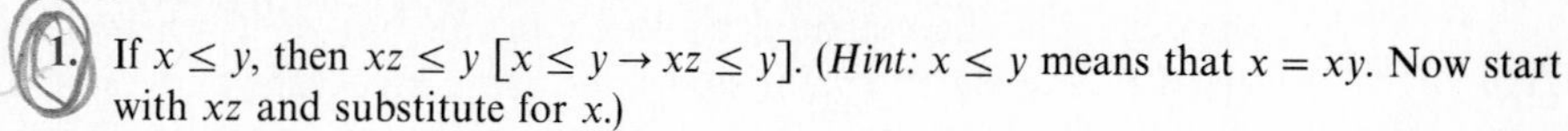

1. If $x \leq y$, then $xz \leq y$ $[x \leq y \rightarrow xz \leq y]$. (*Hint:* $x \leq y$ means that $x = xy$. Now start with $xz$ and substitute for $x$.)
2. $x \leq y \rightarrow \bar{y} \leq \bar{x}$.

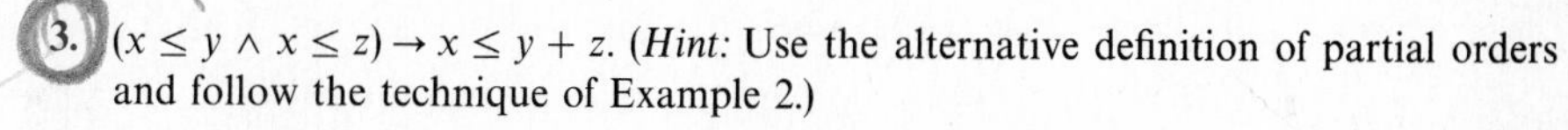

3. $(x \leq y \wedge x \leq z) \rightarrow x \leq y + z$. (*Hint:* Use the alternative definition of partial orders and follow the technique of Example 2.)
4. $x \leq y \rightarrow x \leq y + z$.

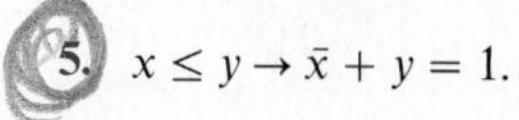

5. $x \leq y \rightarrow \bar{x} + y = 1$.
6. $x\bar{y} = 0 \rightarrow x \leq y$. This is the converse of Example 3. (*Hint:* Start with $x = x \cdot 1$ and use the complement law. See also Example 3 of Section 3.2.)
7. $x \leq \bar{y} \leftrightarrow xy = 0$. (See Exercise 6 and Example 3.)
8. $xy \leq x + y$. (Use the techniques of Example 1.)
9. $(x \oplus y = y) \rightarrow x \leq y$. (*Hint:* Show that $x + y = y$. This can be accomplished by observing that

$$x = x(y + \bar{y}) = xy + x\bar{y}$$
$$y = y(x + \bar{x}) = xy + \bar{x}y$$

Now, form the join of $x$ and $y$ and make the appropriate substitution.)

## 3.6 Number of Elements in a Boolean Algebra

From the examples given previously in this chapter, one might imagine that there is considerable flexibility in constructing a Boolean algebra. It might be thought that the carrier set may contain any arbitrary number of elements. This is not the case. We show some of the possibilities and impossibilities below.

To start with, we know that the carrier set $B$ must contain *at least two elements*, a unit element and a zero element. This was explored in Example 3 of Section 3.3.

Can there be a three-element Boolean algebra? If so, the carrier set $B$ must contain distinct zero and unit elements. Call these elements 0 and 1; and let the third element of $B$ be designated by $a$. So the carrier set $B = \{0, a, 1\}$. Now, we know from Example 1 and Exercise 1(a) of Section 3.2 that $\bar{0} = 1$ and $\bar{1} = 0$. Furthermore, these complements are unique (Theorem 3.4). What is $\bar{a}$? From above we know that $\bar{a} \neq 0$ and $\bar{a} \neq 1$. The only possibility left is for $a$ to be self-complementary, or $\bar{a} = a$. Axiom 4a, then, yields $a \cdot \bar{a} = a \cdot a = 0$. This is clearly impossible since we know from idempotency that $a \cdot a = a$. Thus there is *no* three-element Boolean algebra.

A four-element Boolean algebra was exhibited in Example 2 of Section 3.3; and an eight-element one was shown in Example 1.

It turns out that every finite Boolean algebra has *exactly* $2^n$ elements for some $n > 0$. A proof of this fact is beyond our scope at present but we further state here that each Boolean algebra of $2^n$ elements is unique; that is, they can be shown to have the same structure and same Hasse diagram. Recall that the number of elements of the power set of a set of $n$ elements is $2^n$. This is no coincidence. There is an intimate relationship between Boolean algebras and the subsets of a given universal set in that they are structurally the same.

## EXERCISES

1. Show by example that no five-element Boolean algebra can exist.
2. Show that a Boolean algebra cannot have an odd number of elements.

## Summary and Selected References

Boolean algebra is by far the most important algebraic structure for practicing computer scientists and those with a more theoretical bent as well. The nature of application will become apparent in the next chapter. Early treatments of the subject and the relationship to logic can be found in Hohn (1966) and Arnold (1962). A theoretical coverage is presented in Halmos (1963).

More recent books that incorporate Boolean algebra as an integral part of their treatment include Birkhoff and Bartee (1970), Prather (1976), Fisher (1977), Korfhage (1966, 1974), Gersting (1982), Kapps and Bergman (1975), and Berztiss (1975).

# The Algebra of Switching Circuits

# 4

## 4.1 Introduction

The study of Boolean algebra is very appealing as an abstract study of a mathematical system. Its theory is important as a topic in its own right. The fact that its properties and theories underlie those of logical propositions and those of sets is an example of the power of mathematics: once theoretical properties have been established in one model of a system, the same principles apply in the other.

But aside from this appeal, the importance of Boolean algebra lies in its practicality to electronics and computer science. The theory itself concerns the control of the operation of *discrete* devices. These are devices that can be thought of in an "on-off" manner. We are all familiar with such devices: a light switch, the pushbutton control of an elevator, and the on-off switch of a radio. The devices that control these mechanisms are called *digital.*

There are other controls that act in a continuous manner. You are familiar with such controls as the volume control on a radio, an accelerator pedal on a car, and rheostat (dimmer) light switches. These are *analog* controls and are described by the mathematical theories of analysis and calculus, which are outside the scope of this book.

Our study will be focused on digital and discrete devices. The digital controls of interest to us are those that allow electricity to flow in electric circuits. We will not be concerned with the continuous flow of electricity, as such, but rather with the devices that, indeed, let it flow. Their application is at the heart of computers, and the design of such controls is of interest to those constructing computing devices. We will leave the actual design of these circuits to more advanced courses and for the present concern ourselves with how the algebra we have been studying is applicable.

## 4.2 Logic Gates and Circuits

An electronic switch is a device in an electric circuit which allows current to flow or not to flow. The device will be considered as an on-off device, thus discrete. When the switch is in the *on* position we say there is a *closed* circuit and current is allowed to flow. This is indicated in Figure 4.1(a) where the switch, $x$, is closed and current can pass from position 1 to position 2. When the switch is in the *off* position, the circuit is *open*, as shown in Figure 4.1(b), and no current can flow through switch $x$ from position 1 to position 2.

There are two fundamental ways of interconnecting two switches. These are connections in *series* or in *parallel.*

Connection in Series: The AND Gate Two switches $x$ and $y$ connected in *series* are shown in Figure 4.2(a). It is evident that current can flow from position 1 to position 2 only when both $x$ and $y$ are closed (on). If either one is open (off), current cannot flow. Thus we have current if it is true that *both* $x$ and $y$ are closed. This situation can be shown in Figure 4.2(b). If we let the flow of current be symbolized by 1 or T for "on," and 0 or F for "off," our table in Figure 4.2(b) can be represented as in Figure 4.2(c). It is seen that we have the truth table for the logical *conjunction* where switches $x$ and $y$ replace propositions $p$ and $q$. Note also that when using 1 and 0 we have the Boolean table for "meet."

When switches are viewed as inputs to a circuit, to be controlled by some external means, the terminology used is taken from the algebra of logic. The interconnections are known as *logic gates.* Thus when two switches are connected in *series*, or in *conjunction*, we have what is known as an *AND gate.* Its diagram is shown in Figure 4.2(d).

Connection in Parallel: The OR Gate Figure 4.3(a) shows two switches, $x$ and $y$, connected in *parallel.* Here it can be seen that current can flow from position 1 to position 2 if either one of $x$ or $y$ is closed or if both are closed. This situation can be seen in Figure 4.3(b).

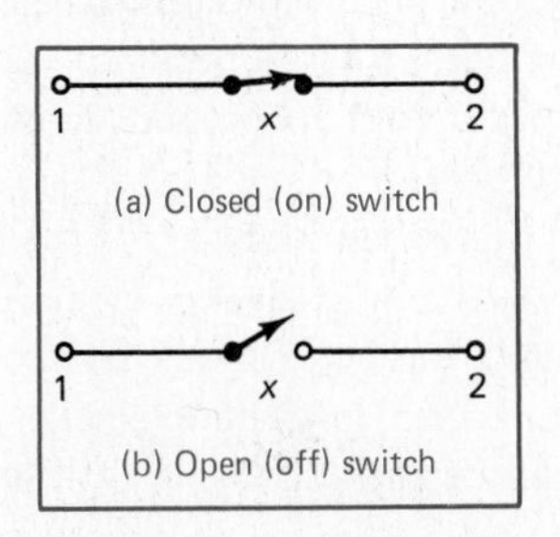


FIGURE 4.1

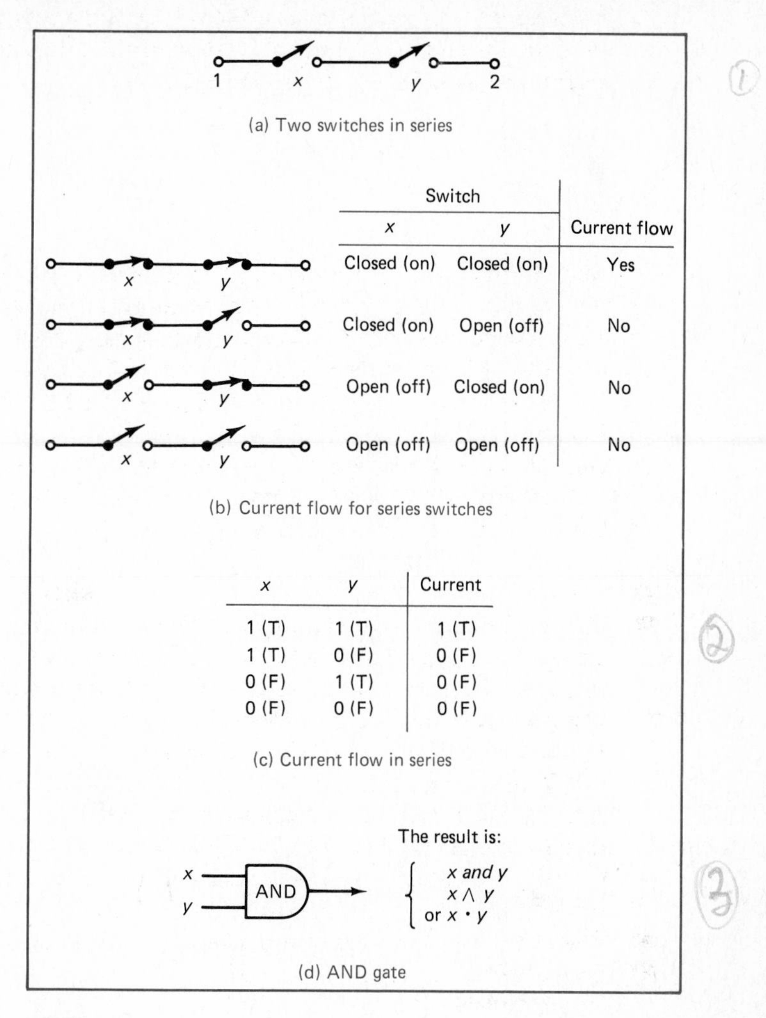


(a) Two switches in series

| Switch |  |  |
|---|---|---|
| x | y | Current flow |
| Closed (on) | Closed (on) | Yes |
| Closed (on) | Open (off) | No |
| Open (off) | Closed (on) | No |
| Open (off) | Open (off) | No |

(b) Current flow for series switches

| x | y | Current |
|---|---|---|
| 1 (T) | 1 (T) | 1 (T) |
| 1 (T) | 0 (F) | 0 (F) |
| 0 (F) | 1 (T) | 0 (F) |
| 0 (F) | 0 (F) | 0 (F) |

(c) Current flow in series



(d) AND gate

**FIGURE 4.2**

If we let the flow of current be symbolized by 1 (or T) for "on" and 0 (or F) for "off," we see in Figure 4.3(c) the correspondence of parallel circuits with the logical *disjunction*. We also have the Boolean table for join. The logic gate thus formed is the OR gate. It is denoted in Figure 4.3(d).

AND gates and OR gates are models of *binary* connectors and hence should be limited to two inputs. However, some authors choose to let there be more than two inputs to these gates. This is acceptable theoretically, due to the property of associativity of meet and join. Thus the AND gate of Figure 4.4(a) can be interpreted as the diagram of either Figure 4.4(b) or (c). An analogous interpretation is applicable for OR gates also.

Inverters: NAND and NOR Gates Switching circuits also have the counterpart of the unary complement. For any switch $x$, we can include a switch, called

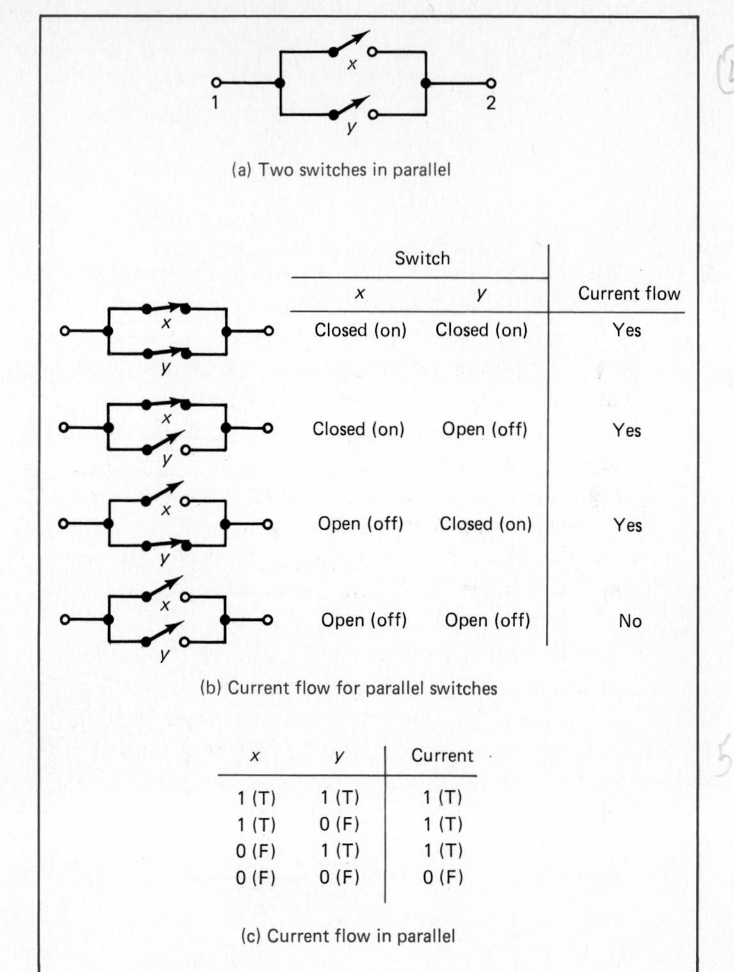


(a) Two switches in parallel

| Switch x | Switch y | Current flow |
|---|---|---|
| Closed (on) | Closed (on) | Yes |
| Closed (on) | Open (off) | Yes |
| Open (off) | Closed (on) | Yes |
| Open (off) | Open (off) | No |

(b) Current flow for parallel switches

| x | y | Current |
|---|---|---|
| 1 (T) | 1 (T) | 1 (T) |
| 1 (T) | 0 (F) | 1 (T) |
| 0 (F) | 1 (T) | 1 (T) |
| 0 (F) | 0 (F) | 0 (F) |

(c) Current flow in parallel



The result is:

$$\begin{cases} x \text{ or } y \\ x \vee y \\ \text{or } x + y \end{cases}$$

(d) OR gate

**FIGURE 4.3**

**FIGURE 4.4**

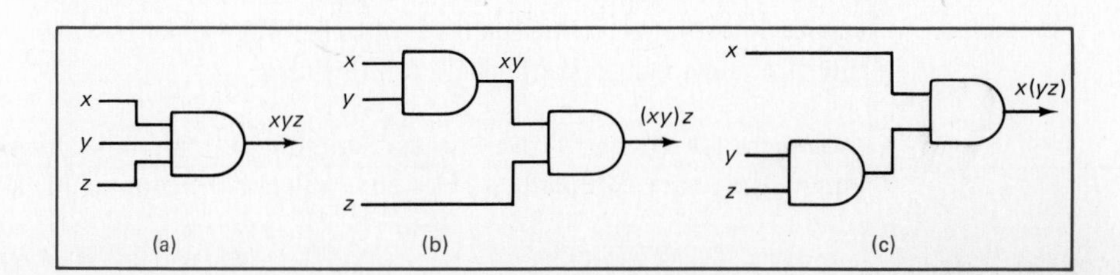


(a) (b) (c)

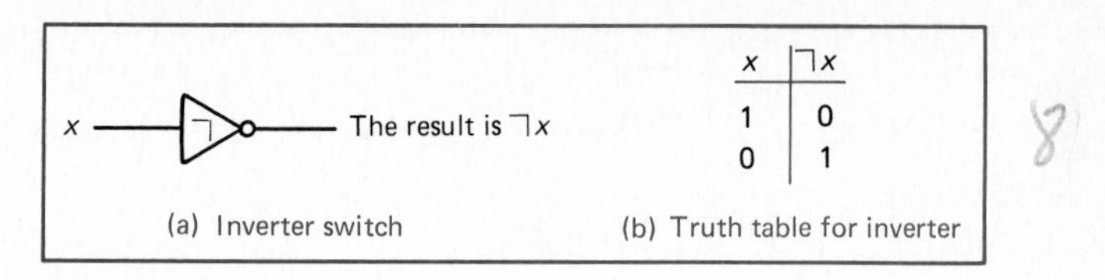


(a) Inverter switch (b) Truth table for inverter

**FIGURE 4.5**

the *inverter*, for $\neg x$, or $\bar{x}$. This is symbolized in Figure 4.5(a). The inverter has the truth table shown in Figure 4.5(b).

The inverter switch can also be used to negate the result of any other gate. For instance, to negate an AND gate, one may use the diagram of Figure 4.6(a). Figure 4.6(b) is also commonly used for this purpose to denote the "not and" as a single gate. This special gate symbol is called the *NAND gate* and its truth table is found at Figure 4.6(c). Recall the special emphasis made regarding the NAND operator of logical propositions of Section 1.9.

When the OR gate is negated by an inverter, as in Figure 4.7(a), one may use the single "not or" gate designation as shown in Figure 4.7(b). This is the symbol for the *NOR gate*. Its truth table is at Figure 4.7(c).

Combinations of Gates Circuits can be made up in a variety of ways depending on the diagram in question. These logic circuits, made up of several logic gates, may have several inputs. Each one is fed into a logic gate and then various combinations are taken to form the final product. For example, suppose that a design called for the result: $xy + \bar{z}$, that is, the disjunction of the complement of $z$ with the conjunction of $x$ and $y$. We would have the

**FIGURE 4.6**

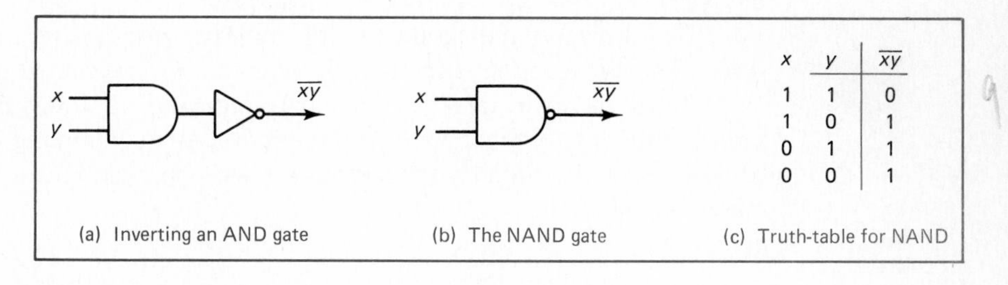


| x | y | $\overline{xy}$ |
|---|---|---|
| 1 | 1 | 0 |
| 1 | 0 | 1 |
| 0 | 1 | 1 |
| 0 | 0 | 1 |

(a) Inverting an AND gate (b) The NAND gate (c) Truth-table for NAND

**FIGURE 4.7**

x y $\overline{x+y}$ x y $\overline{x+y}$

| x | y | $\overline{x+y}$ |
|---|---|---|
| 1 | 1 | 0 |
| 1 | 0 | 0 |
| 0 | 1 | 0 |
| 0 | 0 | 1 |

(a) Inverting an OR gate (b) The NOR gate (c) Truth-table for NOR

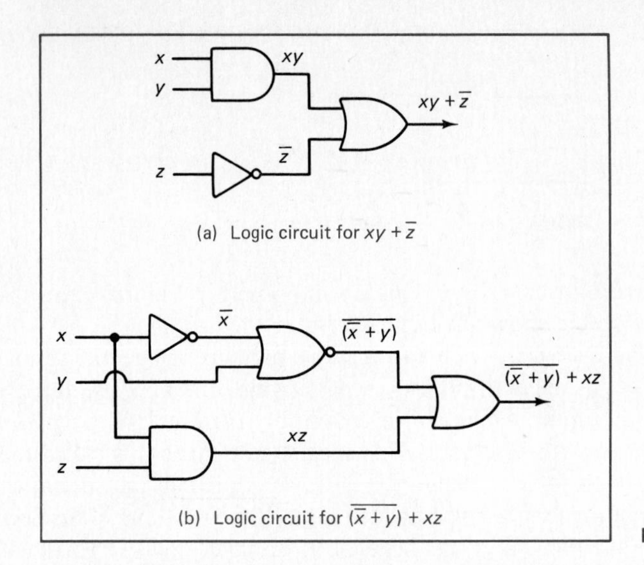


(a) Logic circuit for $xy + \bar{z}$

(b) Logic circuit for $\overline{(\bar{x} + y)} + xz$

**FIGURE 4.8** Logic circuits.

circuit shown in Figure 4.8(a). Figure 4.8(b) shows another way in which AND gates and OR gates can be used in combination with inverters.

**EXAMPLE 1** In a committee of three, each member controls one of three switches $x$, $y$, and $z$. Issues are to be decided by a simple majority and each member votes by closing his switch for "yes" and opening his switch for "no." A question, then, will be decided "yes" if one of the following is true: $x$ and $y$ are closed, $x$ and $z$ are closed, $y$ and $z$ are closed, or all three of $x$, $y$, and $z$ are closed. This is equivalent to the statement: $xy + xz + yz + xyz$. You should show that this is equivalent to $xy + xz + yz$ by using absorption.

The design of the circuit for this majority decision is shown in Figure 4.9(a). The three inputs into the OR gate can be interpreted as either $xy + (xz + yz)$ or as $(xy + xz) + yz$ and can be indicated as shown in the figure as either of these two. Since $xy + xz + yz$ can be written, among other ways, as $x(y + z) + yz$, the circuit for this majority decision making can be designed as Figure 4.9(b).

**EXAMPLE 2** Let us design a circuit that allows current to pass if and only if each of two switches are both closed or both open. If the switches are labeled $x$ and $y$, this is equivalent to the logical biconditional connective: $x \leftrightarrow y$. Recall that the biconditional is equivalent to $(x \wedge y) \vee (\neg x \wedge \neg y)$. Expressed in terms of Boolean expressions, we have $xy + \bar{x}\bar{y}$. A corresponding logic circuit is shown in Figure 4.10(a). By using De Morgan's law on the second conjunction, $\bar{x}\bar{y}$, we obtain the expression $\overline{(x + y)}$. The circuit, shown in Figure 4.10(b), then, is equivalent to the one in Figure 4.10(a).

There are other ways to represent the biconditional also. These will be explored in Exercises 2 and 3.

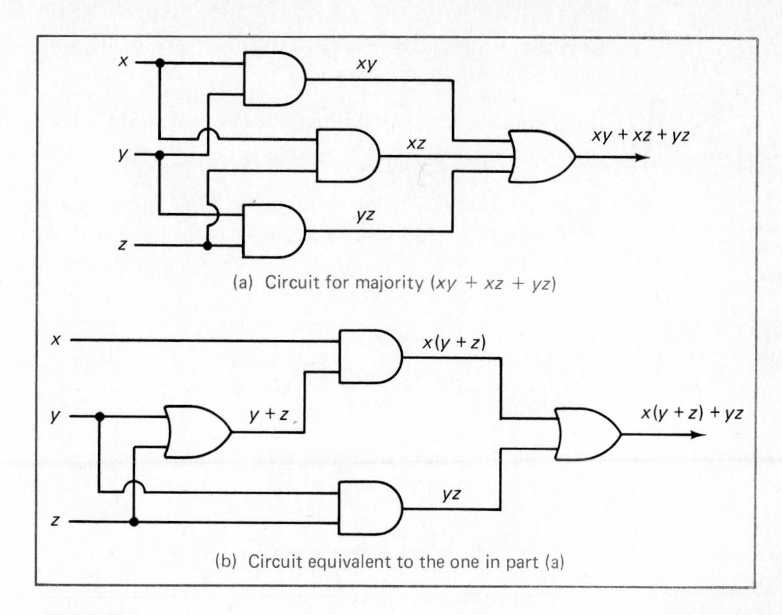

(a) Circuit for majority $(xy + xz + yz)$

(b) Circuit equivalent to the one in part (a)

**FIGURE 4.9**

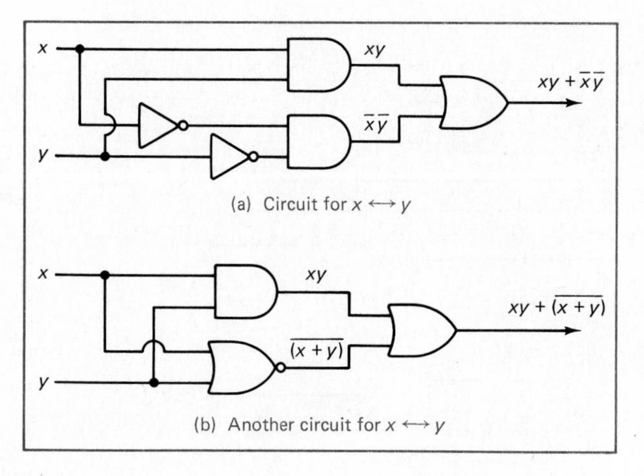

(a) Circuit for $x \longleftrightarrow y$

(b) Another circuit for $x \longleftrightarrow y$

**FIGURE 4.10**

## EXERCISES

**1.** Construct logic circuits producing each of the following.

(a) $x\bar{y} + \bar{x} + yz$

(b) $(\bar{x} + z)y + x\bar{y}$

(c) $(x + y)\overline{(x + y)}$

(d) $y(x + z\bar{w}) + \overline{(xz)}$

**2.** Two circuits for the biconditional were shown in Example 2. There are other ways to represent this operator. We have seen earlier that the biconditional is the negation of the X-OR: $x \leftrightarrow y \Leftrightarrow \overline{x \oplus y} \Leftrightarrow \overline{x\bar{y} + \bar{x}y}$. Draw a logic circuit for this final representation.

3. (a) Show that the biconditional, $x \leftrightarrow y$ can be additionally expressed as $(\bar{x} + y)(x + \bar{y})$.
   (b) Draw the logic circuit for this representation.

4. Because of associativity, we saw that *three* inputs can be fed simultaneously into AND and OR gates (see Figure 4.4). Is this also true regarding NAND and NOR gates? In other words, are the following diagrams ambiguous? You may use truth tables.

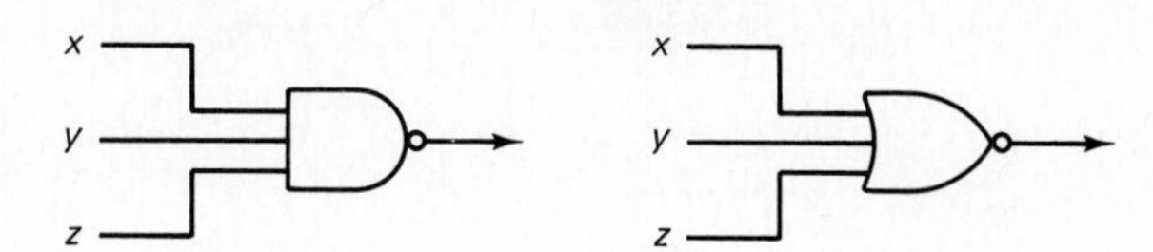


5. Write Boolean expressions corresponding to the following logic circuit diagrams.

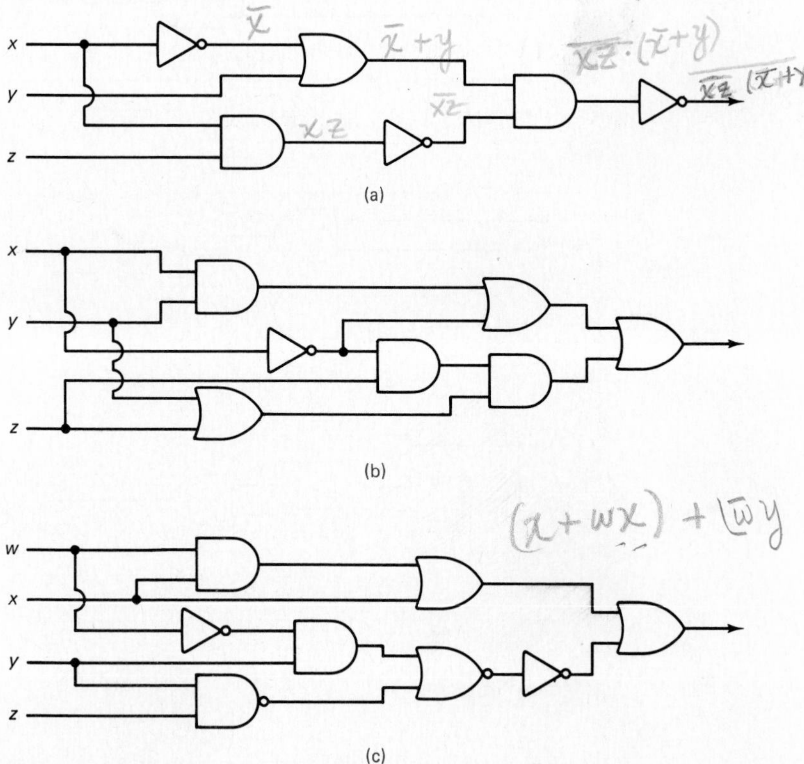


## 4.3 Some Logic Design Techniques

There are many ways to represent a Boolean expression and thus more than one way to design a circuit for a particular truth function. We saw several examples in the preceding section. In this section we will see some further examples of how gates may be altered for the purpose of designing circuits.

**NAND and NOR Conversions** As pointed out earlier (Section 1.9) all possible combinations of logical connectors can be written in terms of either the NAND operator or the NOR operator. As such, they have played a significant role in logic circuit design and have been given special designated symbols of Figures 4.6(b) and 4.7(b).

To show how some circuits can be expressed in terms of NAND and NOR gates solely, let us consider, for instance, the OR operator, $x + y$. In Example 1 of Section 1.9, we converted the logical $p \vee q$ to an expression in terms of the NAND operator alone. Let us mimic the conversion there in the following example.

**EXAMPLE 1** Convert the OR gate shown in Figure 4.11 to a circuit containing NAND gates only.

Now $x + y = \overline{\bar{x}\bar{y}}$ by De Morgan's law, which in turn equals $\overline{\overline{xx}\;\overline{yy}}$ by idempotence twice, which is the NAND of two NANDs.

The parts, $\overline{xx}$ and $\overline{yy}$, have the NAND gates shown in Figure 4.12(a). These can be combined through another NAND gate to obtain Figure 4.12(b).

**EXAMPLE 2** Convert the OR gate shown in Figure 4.13 to a circuit containing NOR gates only.

By following Example 4 of Section 1.9, we see that $x + y = \overline{\overline{x + y}}$ by involution, which in turn equals $\overline{\overline{(x + y)} + \overline{(x + y)}}$ by idempotence. Thus $x + y$ can be expressed as the NOR of two identical NORs, as shown in Figure 4.14.

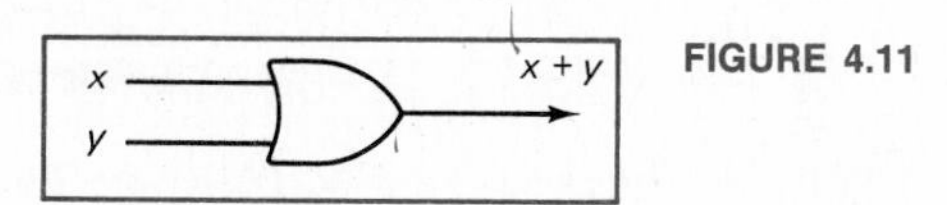


FIGURE 4.11

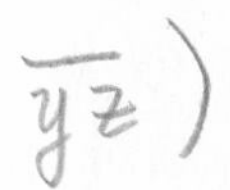

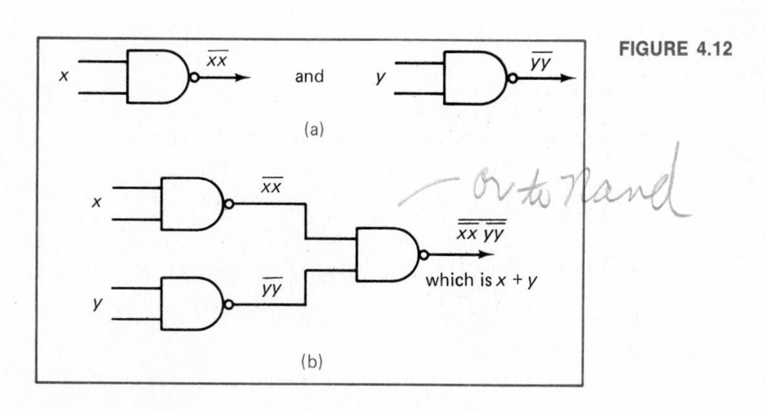


FIGURE 4.12

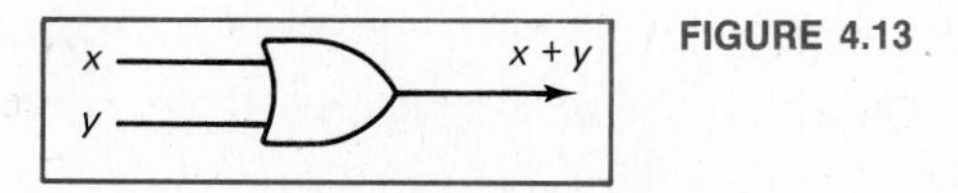


FIGURE 4.13

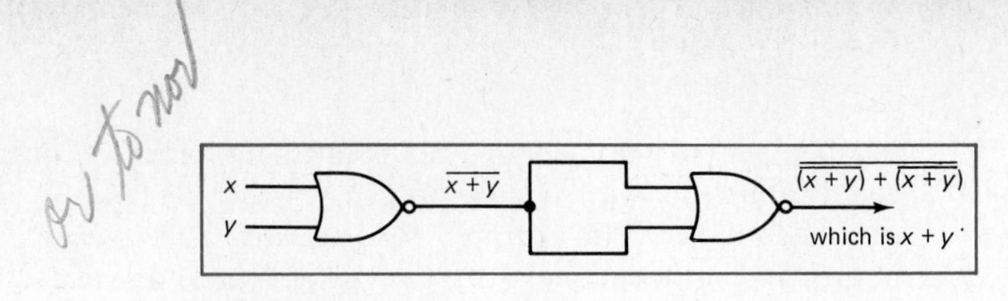


FIGURE 4.14

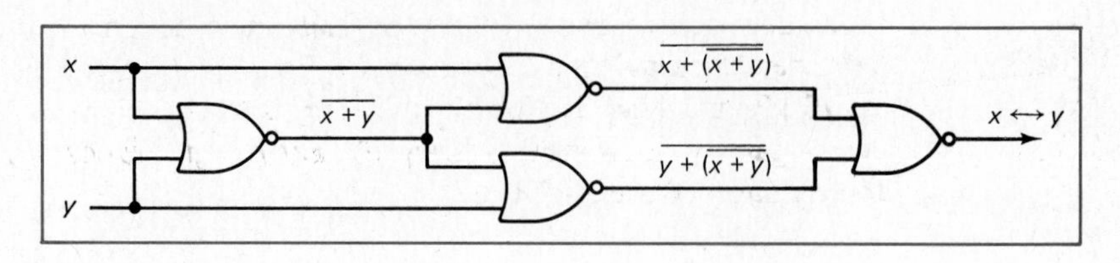


FIGURE 4.15

**EXAMPLE 3** In Example 2 of Section 4.2 we saw how to obtain a circuit for the biconditional, $x \leftrightarrow y$. Let us convert this circuit to one in terms of NOR gates only. In the example we presented two circuits and other ways were explored in Exercises 2 and 3 of Section 4.2. Any of these can be so converted but not all lead to a clean (i.e., without crossovers) circuit.

Let us take as our starting point the equivalence: $x \leftrightarrow y = xy + \bar{x}\bar{y}$.

$$
\begin{aligned}
x \leftrightarrow y = xy + \bar{x}\bar{y} &= xy + \overline{x+y} && \text{by De Morgan's law}\\
&= \overline{x+y} + xy && \text{by commutativity}\\
&= (\overline{x+y} + x)(\overline{x+y} + y) && \text{by distribution}\\
&= \overline{\overline{((\overline{x+y}) + x)} + \overline{((\overline{x+y}) + y)}} && \text{by De Morgan's law}
\end{aligned}
$$

and the expression is totally in terms of the NOR operator. Thus the circuit is as shown in Figure 4.15.

## EXERCISES

1. Express $xy$ in terms of NOR gates only. Observe that

$$xy = \overline{\bar{x} + \bar{y}} = \overline{(\overline{x+x}) + (\overline{y+y})}$$

Draw the circuit.

2. Express $xy$ in terms of NAND gates only and draw the circuit.

3. (a) By using idempotence once, obtain a NOR-gate representation for $\bar{x}$ and draw the circuit.

   (b) Express $\bar{x}$ in terms of a single NAND gate and draw the circuit.

4. Express $x \to y$ in terms of NAND gates alone; NOR gates alone. Recall that $x \to y$ is equivalent to $\bar{x} + y$. Draw the circuits.

5. By following the technique of Example 3, obtain an expression for $x \oplus y$ solely in terms of NAND gates and draw the circuit.

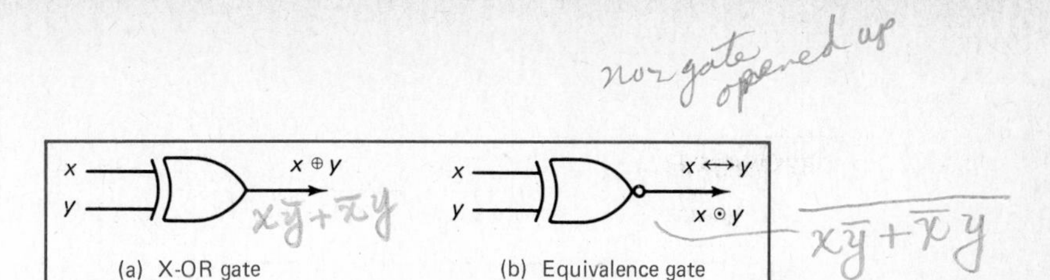


(a) X-OR gate (b) Equivalence gate

**FIGURE 4.16**

X-OR and Equivalence Gates Special single designations are also given to gates representing the biconditional [Figure 4.16(b)] and to its negation, the X-OR [Figure 4.16(a)]. The biconditional is commonly referred to as the *Equivalence* gate, with its symbol being $\odot$. For design purposes it is represented as the negation of the X-OR, $\oplus$.

Half-Adder and Full-Adder Let us see how to apply our study of logical gates to a practical design application. The basic arithmetic done by digital computers is performed on numbers in *base 2* (see Appendix A). The binary addition table is as shown in Table 4.1. Three of the four possible additions result in single-digit answers. The sum of $1 + 1$, however, has two digits, 10. The 1 in the sum of $1 + 1$ represents the *carry* bit to be added to the next higher digit, and the 0 is called the *sum* bit. The other three additions result in 0 carry bits.

The logic circuit that performs the addition of two bits is called a *half-adder*, while the circuit that performs this addition together with a previous carry is called a *full-adder*.

A half-adder needs two binary inputs (with values 0 or 1) and produces two binary outputs, one for the sum and one for the carry. The truth table that characterizes a half-adder, with $x$ and $y$ as input bits, is shown in Table 4.2.

**TABLE 4.1** Binary (Base 2) Addition Table

| + | 1 | 0 |
|---|---|---|
| 1 | 10 | 1 |
| 0 | 1 | 0 |

**TABLE 4.2** Truth Table for Half-Adder

| *Input bits* | | | |
|---|---|---|---|
| *x* | *y* | *Sum* | *Carry* |
| 1 | 1 | 0 | 1 |
| 1 | 0 | 1 | 0 |
| 0 | 1 | 1 | 0 |
| 0 | 0 | 0 | 0 |

Sum: X-OR
Carry: conjunction

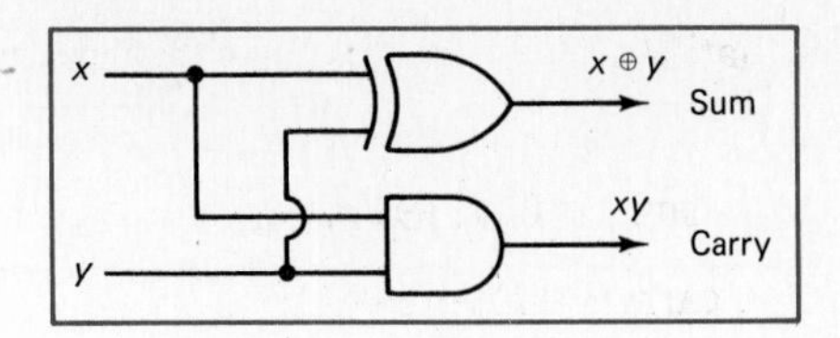


**FIGURE 4.17** Half-adder circuit.

It is obvious from this truth table that the *sum* column can be implemented by an X-OR gate; the *carry*, with an AND gate. Figure 4.17 shows this configuration.

If we implement the addition using a previous carry (the full-adder), we have a truth table with three input values: $x$ and $y$ as the input bits and the previous carry represented as $z$. This is shown in Table 4.3. From this table the Boolean expressions for the sum and carry can be identified:

$$\text{sum} = xyz + \bar{x}\bar{y}z + x\bar{y}\bar{z} + \bar{x}y\bar{z}$$

$$\text{carry} = xyz + x\bar{y}z + \bar{x}yz + xy\bar{z}$$

**TABLE 4.3** Truth Table for Full-Adder

| *Previous carry* $z$ | *Input bits* $x$ | $y$ | *Sum* | *Carry* |
|---|---|---|---|---|
| 1 | 1 | 1 | 1 | 1 |
| 1 | 1 | 0 | 0 | 1 |
| 1 | 0 | 1 | 0 | 1 |
| 1 | 0 | 0 | 1 | 0 |
| 0 | 1 | 1 | 0 | 1 |
| 0 | 1 | 0 | 1 | 0 |
| 0 | 0 | 1 | 1 | 0 |
| 0 | 0 | 0 | 0 | 0 |

**FIGURE 4.18** Full-adder circuit.

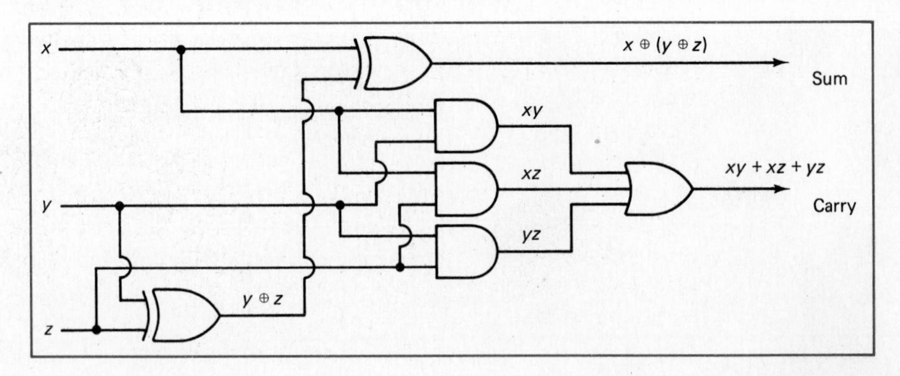

With a little bit of manipulation (see Exercises 7 and 8) these expressions can be reduced to

$$\text{sum} = x \oplus (y \oplus z)$$

$$\text{carry} = xz + yz + xy$$

The full-adder can be implemented by the circuit shown in Figure 4.18.

## EXERCISES

6. Show two other logic circuits using AND and OR gates for the half-adder of Figure 4.17.

7. (a) Show that the *sum* expression of the full-adder, $xyz + \bar{x}\bar{y}z + x\bar{y}\bar{z} + \bar{x}y\bar{z}$ is equivalent to $x \oplus (y \oplus z)$ (see Exercise 9 of Section 3.2).
   (b) Can this sum be written also as $(x \oplus y) \oplus z$?
   (c) From parts (a) and (b) can we say that X-OR is associative?

8. Show that the *carry* expression of the full-adder, $xyz + \bar{x}yz + x\bar{y}z + xy\bar{z}$, is equivalent to each of the following.
   (a) $xy + xz + yz$ (*Hint:* You may "join" the term $xyz$ as many times as you wish because of idempotence.)
   (b) $xy + (x\bar{y} + \bar{x}y)z = xy + (x \oplus y)z$
   (c) $xy + (x + y)z$

9. From Exercise 8(b) and (c) we see that $xy + (x \oplus y)z = xy + (x + y)z$. Does this mean that $(x \oplus y)z = (x + y)z$?

10. Design a circuit for the full-adder that uses fewer gates than Figure 4.18. Use the result from Exercise 8.

## 4.4 Simplification of Circuits

Consider the Boolean expression for a logic circuit, $y + xy\bar{z}$. Its diagram is shown at Figure 4.19. An equivalent expression can be obtained by the absorption law, namely just the input value $y$. (Show this reduction.) Obviously, this second expression will be universely accepted as "simpler" than the first one. Others are not so evident. For example, which is simpler: $x + \bar{y}$ or its equivalent $\overline{(\bar{x}y)}$? It is often desirable to find the simplest possible form for a logic circuit. This is not only an appealing mathematical exercise but is cost-effective in actual applications. In the design of such a circuit, the only requirement is that it have certain functional values as end results. Usually, this same value can be represented by a variety of circuits, some simpler than others. The cost involved in the design of circuits varies with the state of technology and it is not possible to determine completely which of equivalent expressions and circuits will actually

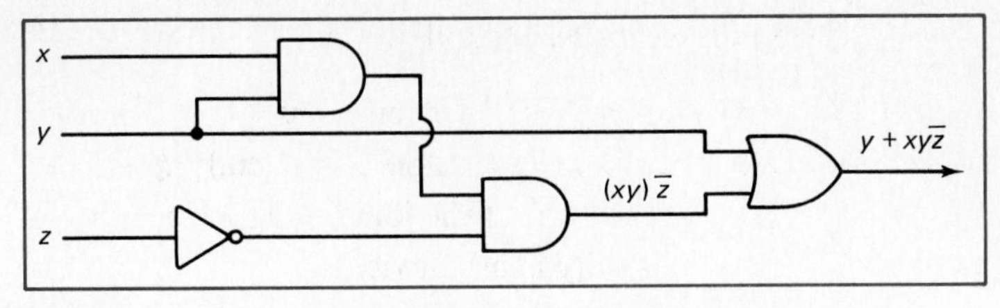


**FIGURE 4.19** Circuit for the expression $y + xy\bar{z}$.

be simplest and represent a minimum. What we seek in this section is a way to define "minimum" and then actually to realize it in a fast and practical manner.

To this end, then, let us define exactly what we mean by a Boolean expression.

**Definition** A Boolean expression, $E$, is any variable $x$, $y$, $z$, and so on, or any combination of variables using the operators $+$, $\cdot$, and $\bar{\ }$.

For example, the following are Boolean expressions:

$$x, \quad y, \quad \bar{x}, \quad \overline{(x + y)}, \quad \overline{(x(y + z))}, \quad (\overline{(\overline{xy\bar{z} + yz})xy}) + xy$$

To be precise with our definition, we may say that

1. a variable is an expression, and
2. if $E_1$ and $E_2$ are expressions, then so are $E_1 \cdot E_2$, $E_1 + E_2$, $\overline{E_1}$, and $\overline{E_2}$.

All possible Boolean expressions can be built up from these two statements. For example, statement 1 allows for variables $x$, $y$, and $z$ to be Boolean expressions. Then from statement 2, we see that, among others, $x + y$, $\bar{z}$, and $y \cdot z$ are Boolean expressions. By reapplying statement 2 we can obtain $((x + y) + \bar{z}) \cdot \overline{(y \cdot z)}$ as Boolean expression. This buildup can be extended indefinitely.

We further define a *literal* to be either a variable, say $x$, or its complement, $\bar{x}$. In a logic circuit the complement of a variable may be accomplished by an inverter. In practice, inverters of single variables will not be considered differently from input variables themselves. This convention stems from the fact that in designing circuits the complements of inputs chosen depend on arbitrary decisions made as to which one of variables and their complements will be represented by an input letter.

**Definition** A *fundamental product* is defined to be

(a) a literal, or
(b) if $P_1$ and $P_2$ are fundamental products, then so is $P_1 \cdot P_2$, provided that no literals in the product involve the same variable.

For example, these are fundamental products:

$$x, \quad xy, \quad xy\bar{z}, \quad \bar{x}\bar{y}\bar{z}w$$

These are not fundamental products:

1. $x + y$ (Here the join is used.)
2. $xyz\bar{x}$ (The variable $x$ is repeated.)
3. $xyz + wxy$ (The join is used.)
4. $\bar{x}yzy$ ($y$ is used twice.)

Notice in this second example that

$$xyz\bar{x} = (xyz)\bar{x} = \bar{x}(xyz) = (\bar{x}x)yz = 0 \cdot yz = 0$$

and in the fourth example

$$\bar{x}yzy = \bar{x}yyz = \bar{x}yz \qquad \text{(by idempotence)}$$

which is a fundamental product. It can be seen that any product will reduce to either 0 or to a fundamental product. In the literature fundamental products are sometimes referred to as *minterms.*

When fundamental products are joined through the + sign we have what is known as a *sum of products.* This will be fully defined below, but for now let us consider some examples of sums of products.

**EXAMPLE 1**

(a) $x + y + \bar{z}$. Each product joined is a literal and is in the simplest form possible.

(b) $x\bar{z} + xy\bar{z}$. By the absorption law this sum can be reduced: $x\bar{z} + xy\bar{z} = x\bar{z}$.

(c) $xy + z\bar{x} + xy\bar{z}w$. This sum can be rewritten as $xy + xy\bar{z}w + z\bar{x}$ and the first two products absorb to $xy$, leaving the result as $xy + z\bar{x}$.

This idea contained in Example 1 is our *first step* toward minimizing Boolean expressions. If any arbitrary expression can be put into a sum of products form (and it can, as seen below), it may be possible to further reduce it using the absorption law.

Now let us examine precisely what we mean by a *sum-of-products* (*S.O.P.*) form of a Boolean expression.

**Definition** A Boolean expression is said to be in S.O.P. form if it is

(a) a fundamental product, or

(b) if $E_1$ and $E_2$ are fundamental products, then $E_1 + E_2$ is in S.O.P. form provided that neither $E_1$ nor $E_2$ absorb into the other.

Thus from statement (a) we know that $x$, $yw$, and $y\bar{z}w$ are in S.O.P. form since they are fundamental products. From statement (b), then, we can state that $x + y\bar{z}w$ is in S.O.P. form, whereas $x + yw + y\bar{z}w$ is not. Why?

*Note:* The term "sum of products" arises from the nomenclature of Boolean algebra where the operations, join (+) and meet (·), are called by their arithmetic counterparts, sum and product, respectively. Another notation arises when the Boolean sum is called by its counterpart from logical propositions, the disjunction. When renamed in this fashion, our sum-of-products form is known as the "disjunctive normal form" (dnf). Both notations pervade the literature.

## EXERCISES

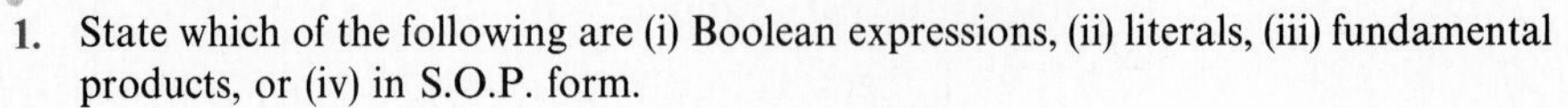

1. State which of the following are (i) Boolean expressions, (ii) literals, (iii) fundamental products, or (iv) in S.O.P. form.

| | | |
|---|---|---|
| (a) $y$ | (b) $y\bar{z} + w$ | (c) $\overline{(xyz)} + yz$ |
| (d) $x\bar{y}z + x\bar{y} + z$ | (e) $\bar{w}$ | (f) $(x + y)(y + z)(\bar{x} + w)$ |
| (g) $xy + yz + \bar{x}y$ | (h) $\bar{x} + \bar{x}yz + \bar{x}w\bar{z}$ | (i) $z + x + y$ |

What we have to do now is to examine how any Boolean expression can be put into a sum of products, which may then possibly be reduced by absorption.

To see how this may be done, let us consider the expression $\overline{(xy + \bar{z})}$ in Example 2.

**EXAMPLE 2**

$$\begin{aligned} \overline{(xy + \bar{z})} &= \overline{(xy)}\bar{\bar{z}} && \text{by De Morgan's law} \\ &= \overline{(xy)}z && \text{by involution} \\ &= (\bar{x} + \bar{y})z && \text{by De Morgan's law} \\ &= \bar{x}z + \bar{y}z && \text{by distribution} \end{aligned}$$

Thus we have a sum of products and the absorption law cannot be applied. Thus $\overline{(xy + \bar{z})}$ has as its S.O.P., $\bar{x}z + \bar{y}z$.

**EXAMPLE 3** Convert $z(\overline{(\bar{x} + z)} + \overline{(\bar{y} + \bar{z})})$ to its S.O.P.

$$\begin{aligned} z(\overline{(\bar{x} + z)} + \overline{(\bar{y} + \bar{z})}) &= z(\bar{\bar{x}}\bar{z} + \bar{\bar{y}}\bar{\bar{z}}) && \text{by De Morgan's law (twice)} \\ &= z(x\bar{z} + yz) && \text{by involution} \\ &= zx\bar{z} + zyz && \text{by distribution} \\ &= z\bar{z}x + zzy && \text{by commutativity} \\ &= 0 \cdot x + zy && \text{by complement and idempotence} \\ &= 0 + zy && \text{by boundness} \\ &= zy && \text{by identity} \end{aligned}$$

**EXAMPLE 4** Convert $x\bar{y}(\overline{\bar{y}\bar{w}}) + w\bar{y}$ to S.O.P. form.

$$\begin{aligned} x\bar{y}(\overline{\bar{y}\bar{w}}) + w\bar{y} &= x\bar{y}(y + w) + w\bar{y} && \text{by De Morgan's law and involution} \\ &= x\bar{y}y + x\bar{y}w + w\bar{y} && \text{by distribution} \\ &= 0 + x\bar{y}w + w\bar{y} && \text{by complementation} \\ &= x\bar{y}w + \bar{y}w && \text{by boundness} \end{aligned}$$

At this point we have a sum of products, but note that $x\bar{y}w$ absorbs into $\bar{y}w$, so we have the final form as $\bar{y}w$.

So we see that by using the axioms and theorems of Boolean algebra, we may always put an expression into a S.O.P. form and then further minimization may possibly be accomplished by absorption.

## EXERCISES

**2.** Convert each of the following Boolean expressions to its S.O.P. form.

(a) $(x + y)(x + z)$

(b) $\overline{(\bar{x} + z)}(y + x) + x\bar{y}$

(c) $\overline{(\overline{(x + y)} + y)}$

(d) $xyz + \overline{(\bar{x} + \bar{y})}x$

(e) $wxyz + \bar{w}\bar{x}\bar{y}\bar{z} + x + \bar{w}$

It may happen that some S.O.P. forms contain fundamental products that are superfluous, but this is not apparent by the absorption law. This can be illustrated by the following example.

**EXAMPLE 5** Consider the expression $\bar{x}yz + xy + \bar{x}y\bar{z}$ in S.O.P. We note that the absorption law is not applicable in any case. However, if we combine the first and third terms and use distribution, we get

$$\bar{x}y(z + \bar{z}) + xy$$

which reduces to

$$\bar{x}y + xy \qquad (\text{since } z + \bar{z} = 1)$$

which further reduces to

$$(\bar{x} + x)y = y$$

Another sequence of axioms and theorems can also be applied to reach the same result

$$\begin{aligned} \bar{x}yz + xy + \bar{x}y\bar{z} &= y(\bar{x}z + x + \bar{x}\bar{z}) \\ &= y(x + \bar{x}(z + \bar{z})) \\ &= y(x + \bar{x}) \\ &= y \end{aligned}$$

Thus there may be a number of ways by which the fundamental products can be combined in order to realize a possible reduction. Sometimes the appropriate laws to be used may be disguised and one may spend considerable time finding them.

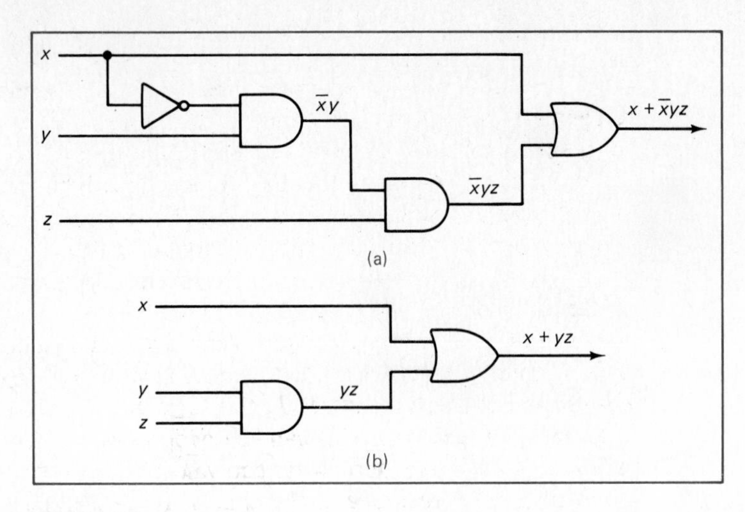


**FIGURE 4.20**

**EXAMPLE 6** As another example, consider the circuit in Figure 4.20(a). Is there a circuit simpler than the one depicted: $x + \bar{x}yz$? By applying the laws of distribution and complement, we get

$$x + \bar{x}yz = (x + \bar{x})(x + yz) = x + yz$$

and an equivalent circuit is shown in Figure 4.20(b).

We have reduced a circuit with three binary gates (and an inverter) to one with two binary gates. Still, however, all three input variables are used in the final result.

Pictorial Minimization To see the full significance of the minimization process we have been attempting, let us pull into full play the set theory model of Boolean algebra. Letting our variables be represented as *sets*, then, we observe that a sum of products is merely a *union* of *intersections* of sets. For example, the S.O.P. in two variables, $x + \bar{x}y$, can be interpreted as $x \cup (\bar{x} \cap y)$. Each fundamental product (which in this case now is a "fundamental" intersection) can be identified on the Venn diagram of Figure 4.21, in which $x$ is lightly shaded and $\bar{x} \cap y$ has the darker shading. The complete shading (union) is just $x + y$.

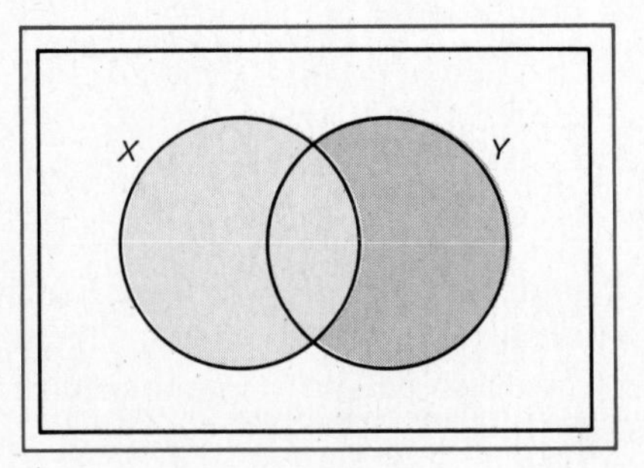


**FIGURE 4.21** $x \cup (\bar{x} \cap y)$ as a model for $x + \bar{x}y$.

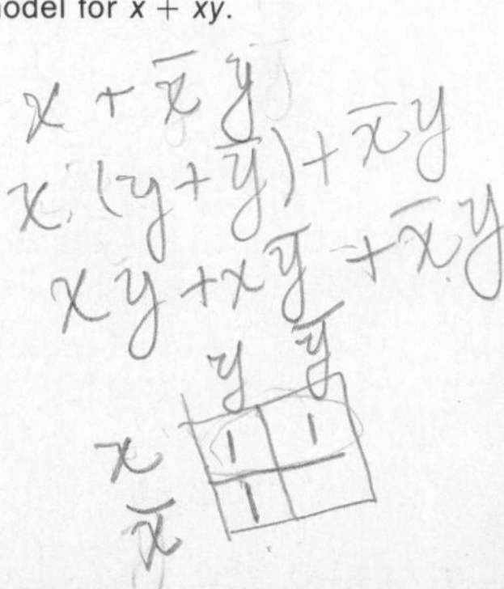

[*Note:* We see that $x + \bar{x}y$ reduces to $x + y$, because $x + \bar{x}y = (x + \bar{x})(x + y) = 1(x + y) = x + y$.]

To see how this works with three variables, consider the associated Venn diagram of three sets, $X$, $Y$, and $Z$, in Figure 4.22(a). The numbered regions correspond to intersections (fundamental products).

Adjacent regions in the diagram can be combined to form simpler terms. For instance, if we have both $x\bar{y}z$ and $\bar{x}\bar{y}z$ present (regions 3 and 7), they can be combined to form $\bar{y} \cap z$ (or $\bar{y}z$). Notice that the same result can be gotten from $x\bar{y}z + \bar{x}\bar{y}z = (x + \bar{x})\bar{y}z = \bar{y}z$. However, such reductions are not always so evident.

In order to prepare for this Venn diagram visualization, we expand our S.O.P.'s into what is known as *complete* sum-of-products form (known also as full disjunctive normal form). In this form *every* fundamental product has instances of all variables. This can be done as follows: For the case of three variables, $x$, $y$, and $z$, the term $xy$ is equivalent to $xy(z + \bar{z}) = xyz + xy\bar{z}$, and *each* term then represents *one* of the eight regions of the Venn diagram of Figure 4.22(a).

**EXAMPLE 7** To illustrate, consider the expression $xy + \bar{x}z + yz$. To put this expression into its complete S.O.P., we have

$$xy(z + \bar{z}) + \bar{x}z(y + \bar{y}) + yz(x + \bar{x})$$

which becomes

$$\underset{(a)}{xyz} + \underset{(b)}{xy\bar{z}} + \underset{(c)}{\bar{x}yz} + \underset{(d)}{\bar{x}\bar{y}z} + \underset{(e)}{xyz} + \underset{(f)}{\bar{x}yz}$$

When the terms are marked on a Venn diagram (by the indicated letters) we have Figure 4.23. Regions a (and e) and b combine to form $xy$ and regions c (and f) and d form $\bar{x}z$ (see the arrows in the figure). Thus the covering is minimally realized by $xy + \bar{x}z$ [or equivalently in set terms: $(x \cap y) \cup (\bar{x} \cap z)$].

Notice in Figure 4.22(a) that two regions are *adjacent* in the Venn diagram if they differ in exactly one literal which must be an uncomplemented variable

**FIGURE 4.22** Fundamental products as intersecting regions.

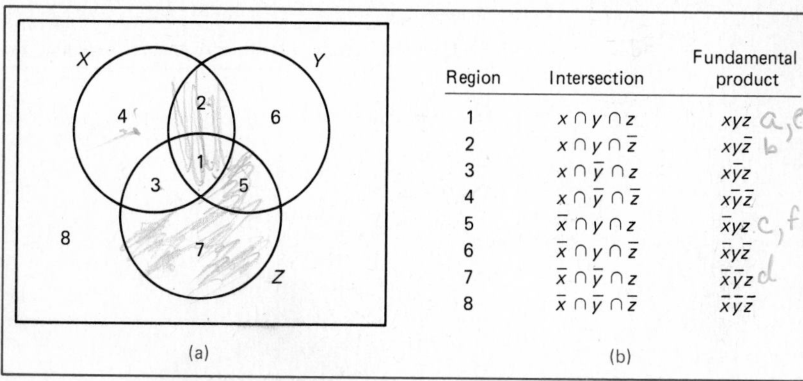


| Region | Intersection | Fundamental product |
|---|---|---|
| 1 | $x \cap y \cap z$ | $xyz$ |
| 2 | $x \cap y \cap \bar{z}$ | $xy\bar{z}$ |
| 3 | $x \cap \bar{y} \cap z$ | $x\bar{y}z$ |
| 4 | $x \cap \bar{y} \cap \bar{z}$ | $x\bar{y}\bar{z}$ |
| 5 | $\bar{x} \cap y \cap z$ | $\bar{x}yz$ |
| 6 | $\bar{x} \cap y \cap \bar{z}$ | $\bar{x}y\bar{z}$ |
| 7 | $\bar{x} \cap \bar{y} \cap z$ | $\bar{x}\bar{y}z$ |
| 8 | $\bar{x} \cap \bar{y} \cap \bar{z}$ | $\bar{x}\bar{y}\bar{z}$ |

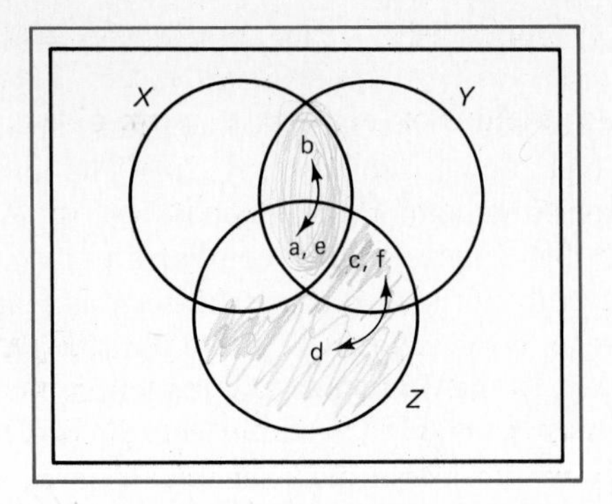


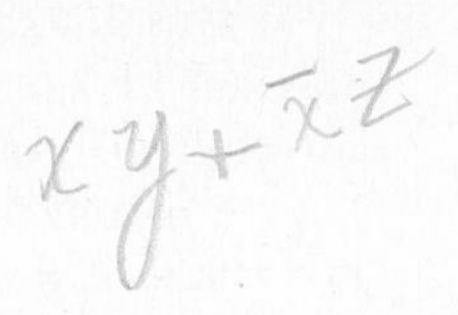


FIGURE 4.23 Minimal covering of a Boolean expression.

in one region (product) and its complement in the other region (product). For example, regions 1 and 2 differ only by $z$ and $\bar{z}$; hence the union of the two, $xyz$ and $xy\bar{z}$, becomes $xy$. Regions 4 and 8 differ in $x$ and $\bar{x}$, hence these products $x\bar{y}\bar{z}$ and $\bar{x}\bar{y}\bar{z}$ combine to form $\bar{y}\bar{z}$. Region 8 is adjacent not only to region 4 but also to regions 6 and 7. Regions 5 and 2 are not adjacent since $\bar{x}yz$ differs from $xy\bar{z}$ in two variables, $x$ and $z$.

Karnaugh Maps This covering of regions and combining of terms as illustrated above with Venn diagrams has been simplified graphically by a means known as Karnaugh maps. These are pictorial devices, as are Venn diagrams, for representing Boolean expressions written in complete sum-of-products form. In these maps all possible fundamental products are pictured together in a rectangle, rather than circles, composed of individual squares representing the products.

MAPS OF TWO VARIABLES In the case of two Boolean variables, $x$ and $y$, there are four possible fundamental products:

$$xy, \quad x\bar{y}, \quad \bar{x}y, \quad \bar{x}\bar{y}$$

The representation of this case is shown in Figure 4.24(a), where the top row

FIGURE 4.24 Karnaugh map for two variables and the associated Venn diagram.

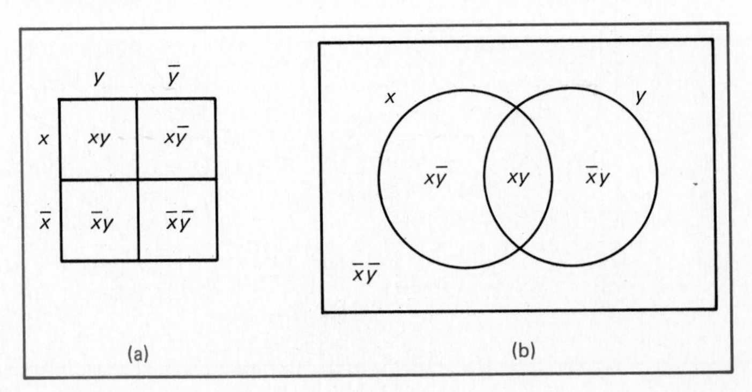

represents the two squares (regions) in $x$, the bottom row represents $\bar{x}$, the left column represents both squares in $y$, and the right column represents $\bar{y}$. These squares represent the four regions of the Venn diagram of two variables [Figure 4.24(b)].

Two squares are *adjacent* and thus can be reduced as in the Venn diagram iff they have a common side.

**EXAMPLE 8** For the expression $E_1 = xy + \bar{x}y$ we place 1's as shown in Figure 4.25(a). Since the fundamental products are located in adjacent squares, we circle them to indicate that these two may be reduced. It is seen that $E_1$ reduces to the single value, $y$. Figure 4.25(b) shows the 1's for the expression $E_2 = \bar{x}y + x\bar{y} + \bar{x}\bar{y}$. The circled adjacent squares indicate that $E_2$ reduces to $\bar{x} + \bar{y}$. The expression $E_3 = \bar{x}y + x\bar{y}$ is shown in Figure 4.25(c). Since the two squares checked are not adjacent, $E_3$ cannot be further reduced.

The information gained from Karnaugh maps for two variables may not be much more revealing than that gotten from an algebraic reduction. For instance, the expression $E_2$ of Example 8 may be reduced as shown in the following routine manipulation:

$$
\begin{aligned}
E_2 = \bar{x}y + x\bar{y} + \bar{x}\bar{y} &= \bar{x}y + \bar{x}\bar{y} + x\bar{y} && \text{by commutativity} \\
&= \bar{x}(y + \bar{y}) + x\bar{y} && \text{by distribution} \\
&= \bar{x} + x\bar{y} && \text{by complementation and identity} \\
&= \bar{x} + \bar{y} && \text{why?}
\end{aligned}
$$

MAPS OF THREE VARIABLES Boolean expressions in three variables have at most eight fundamental products (see Figure 4.22). These are

$$xyz, \quad xy\bar{z}, \quad x\bar{y}z, \quad \bar{x}yz, \quad \bar{x}y\bar{z}, \quad \bar{x}\bar{y}z, \quad x\bar{y}\bar{z}, \quad \bar{x}\bar{y}\bar{z}$$

In order to place them in a Karnaugh map so that adjacent squares differ by only one literal, complemented and noncomplemented, the rectangle of Figure 4.26(a) is used. The interpretation to be made is that the extreme left and right columns are to be thought of as adjacent as shown in Figure 4.26(b), where the map of part (a) has been "cut and pasted" to form a cylinder.

FIGURE 4.25

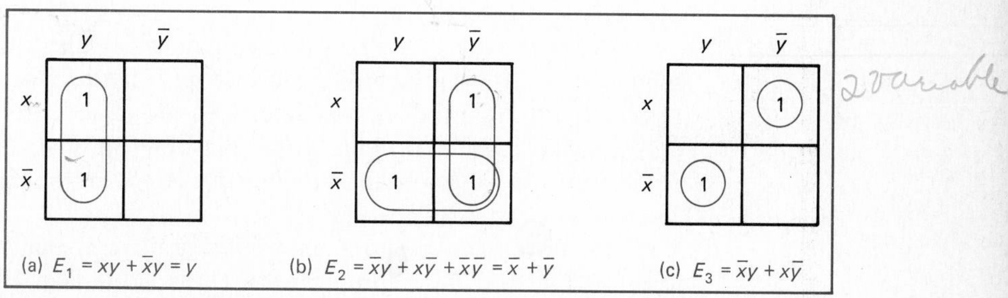


(a) $E_1 = xy + \bar{x}y = y$ (b) $E_2 = \bar{x}y + x\bar{y} + \bar{x}\bar{y} = \bar{x} + \bar{y}$ (c) $E_3 = \bar{x}y + x\bar{y}$

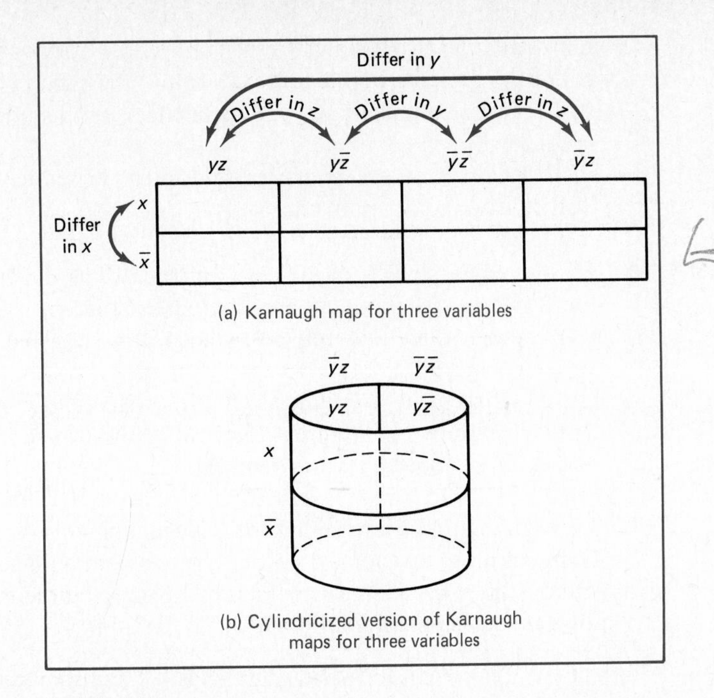


**FIGURE 4.26**

In the case of three variables the idea is to obtain a maximal number of adjacent squares. This grouping can be either by twos as in the case of a two-variable Karnaugh map or by fours. This is best explained by means of the illustration in Figure 4.27.

Figure 4.27(a) shows 1's in two squares. This represents the expression $E_4 = xyz + \bar{x}yz$. The covering indicates that $E_4$ reduces to $yz$.

$E_5 = xyz + xy\bar{z} + x\bar{y}\bar{z} + x\bar{y}z$ is shown at Figure 4.27(b). This is the entire $x$ row and thus this top row reduces expression $E_5$ to simply $x$.

$E_6$ in Figure 4.27(c) covers the four squares common to $\bar{z}$ and thus $E_6 = xy\bar{z} + x\bar{y}\bar{z} + \bar{x}y\bar{z} + \bar{x}\bar{y}\bar{z} = \bar{z}$.

$E_7$ covers the four squares common to $z$ [Figure 4.27(d)].

$E_8$ covers the four squares common to $\bar{y}$ [Figure 4.27(e)].

Figure 4.27(f) shows the covering for the expression $E_9 = xyz + xy\bar{z} + \bar{x}\bar{y}\bar{z} + \bar{x}\bar{y}z$. The pairs of two adjacent squares indicate the reduction: $E_9 = xy + \bar{x}\bar{y}$.

$E_{10} = xyz + xy\bar{z} + \bar{x}yz + \bar{x}\bar{y}\bar{z}$ (Figure 4.28) illustrates some interesting points about three-variable Karnaugh maps. There is one isolated square, corresponding to the product $\bar{x}\bar{y}\bar{z}$. Since it is not adjacent to any of the others, it must appear "as is" in the final S.O.P. The other three squares can be combined to form $xy$ and $yz$. Since the sum of these two covers the three squares, the final result is: $E_{10} = xy + yz + \bar{x}\bar{y}\bar{z}$.

To show that a minimum may not be unique, consider $E_{11} = xyz + xy\bar{z} + \bar{x}y\bar{z} + \bar{x}\bar{y}\bar{z} + x\bar{y}z$. One representation of circled adjacent squares is shown in

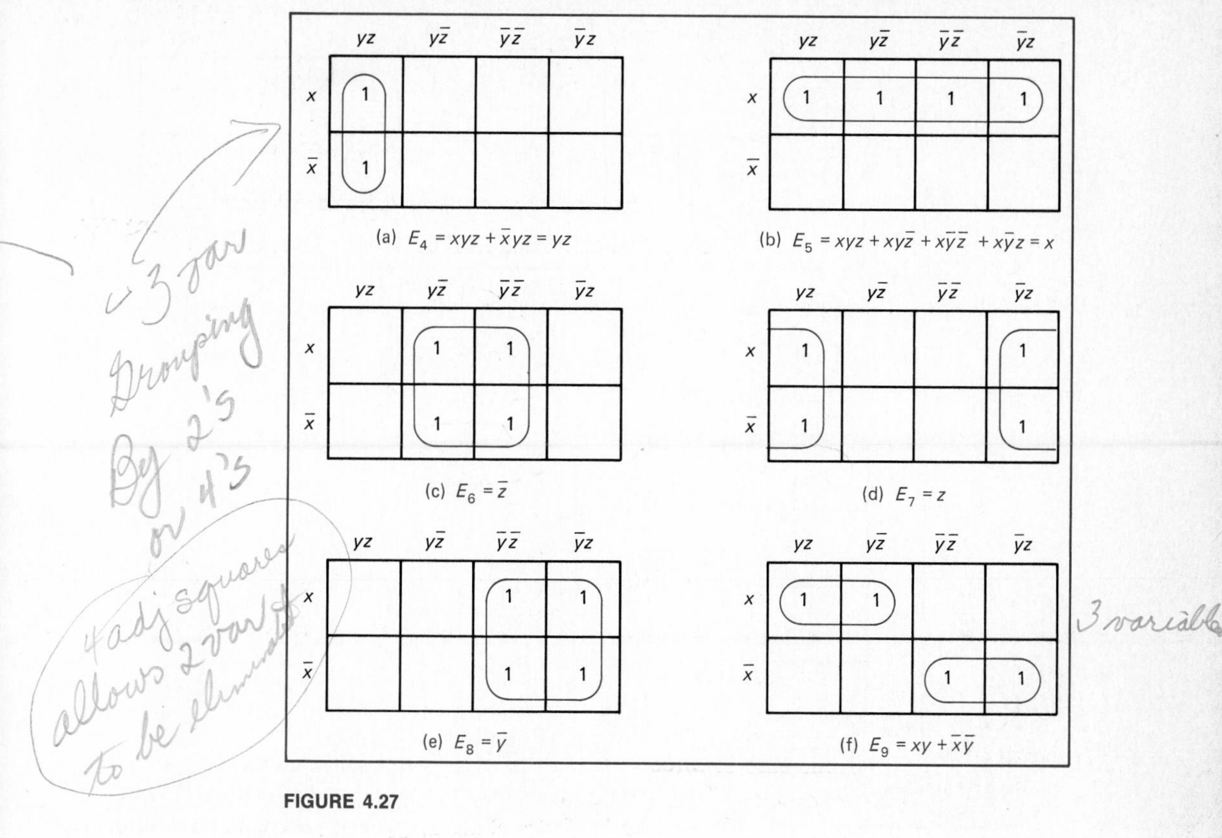


(a) $E_4 = xyz + \bar{x}yz = yz$

(b) $E_5 = xyz + xy\bar{z} + x\bar{y}\bar{z} + x\bar{y}z = x$

(c) $E_6 = \bar{z}$

(d) $E_7 = z$

(e) $E_8 = \bar{y}$

(f) $E_9 = xy + \bar{x}\bar{y}$

**FIGURE 4.27**

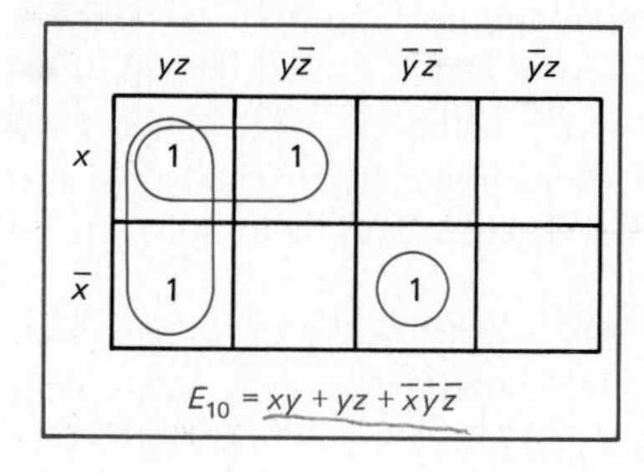


$E_{10} = xy + yz + \bar{x}\bar{y}\bar{z}$

**FIGURE 4.28**

Figure 4.29(a). This reduction is $E_{11} = xz + y\bar{z} + \bar{x}\bar{z}$. Figure 4.29(b) shows a different covering yielding $E_{11} = xz + xy + \bar{x}\bar{z}$. These two expressions are equivalent (check this with truth tables) and both represent a minimization of $E_{11}$. $E_{11}$ is said to have two minimal sums.

MAPS OF FOUR VARIABLES A Karnaugh map of four variables, *w*, *x*, *y*, and *z*, is illustrated in Figure 4.30. In this case there are 16 possible fundamental products. The diagram clearly indicates the possible coverings by adjacent cells.

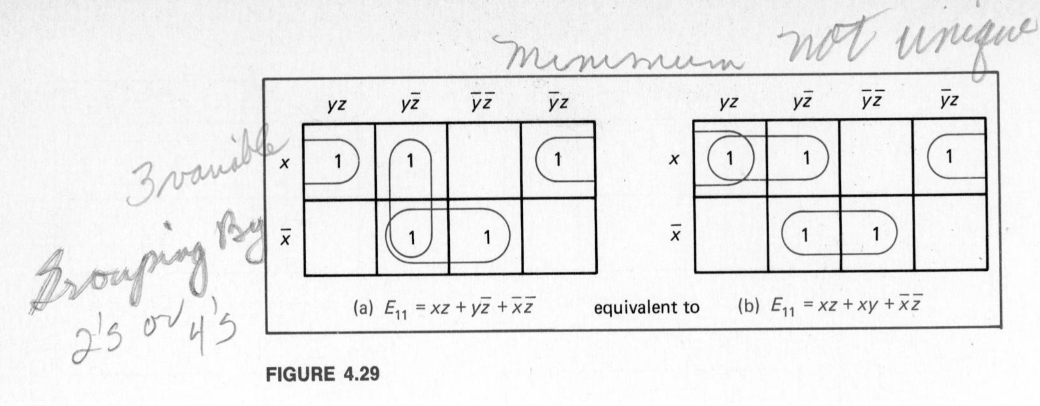


(a) $E_{11} = xz + y\bar{z} + \bar{x}\bar{z}$ equivalent to (b) $E_{11} = xz + xy + \bar{x}\bar{z}$

**FIGURE 4.29**

| | $yz$ | $y\bar{z}$ | $\bar{y}\bar{z}$ | $\bar{y}z$ |
|---|---|---|---|---|
| $wx$ | | | | |
| $w\bar{x}$ | | | | |
| $\bar{w}\bar{x}$ | | | | |
| $\bar{w}x$ | | | | |

**FIGURE 4.30** Karnaugh map for four variables.

If a four-variable map were checked as in Figure 4.31, the coverings indicated there show that the minimization yields $wx + w\bar{z} + \bar{w}\bar{x}z$.

Graphical representations for more than four variables are considerably more difficult to visualize. There are many other techniques that have been developed for the process of minimization of Boolean expressions. One will be explored in Section 4.5. Others you will encounter in further courses in switching theory. The purpose of these sections is to illustrate one of the power-

| | $yz$ | $y\bar{z}$ | $\bar{y}\bar{z}$ | $\bar{y}z$ |
|---|---|---|---|---|
| $wx$ | 1 | 1 | 1 | 1 |
| $w\bar{x}$ | | 1 | 1 | |
| $\bar{w}\bar{x}$ | 1 | | | 1 |
| $\bar{w}x$ | | | | |

**FIGURE 4.31** $wx + w\bar{z} + \bar{w}\bar{x}z$.

ful applications of Boolean algebra to the field of electronic and computing devices.

## EXERCISES

3. Form Karnaugh maps of each of the following Boolean expressions and from them find minimal sums.

   (a) $xy + x\bar{y}$

   (b) $\bar{x}\bar{y} + xy$

   (c) $xy + xz + \bar{y}z$

   (d) $xyz + x\bar{y}z + x\bar{y}\bar{z} + \bar{x}\bar{y}\bar{z} + \bar{x}y\bar{z}$

   (e) $y + x\bar{y}z$ (see Example 7)

   (f) $xyz + \overline{(\bar{x} + \bar{y})}x$

   (g) $(x + y)(x + z)$

   (h) $\overline{(xyz)}$

   (i) $xyz + y\bar{w} + x\bar{z}w + \bar{x}\bar{y}\bar{z}$ (there are three minimal sums)

   (j) $wxy\bar{z} + w\bar{x}y\bar{z} + w\bar{x}yz + \bar{w}x\bar{y}\bar{z} + \bar{w}xy\bar{z} + \bar{w}xyz + \bar{w}x\bar{y}z$

4. Minimize the expression

$$((x + \bar{y}) \rightarrow z) + \bar{x}\bar{y}\bar{z}$$

by first expanding it in S.O.P. form and then constructing its Karnaugh map.

5. An alternative method of converting a Boolean expression, $E$, to its complete sum-of-products form with no repeated terms is the following. First construct the truth table for $E$; then "pick off" the rows where $E$ has the value 1 (or T). This is the method explored in Section 1.3.

   For example, the final column of the truth table for $x(y + \overline{xz})$ is as shown in Table 4.4. There are three 1's in it and they correspond to the fundamental products: $xyz$, $xy\bar{z}$, and $x\bar{y}\bar{z}$. This sum of products: $xyz + xy\bar{z} + x\bar{y}\bar{z}$ eventually reduces to $xy + x\bar{z}$.

   (a) Show this by a Karnaugh map.

   (b) Find the six fundamental products for the expression in Exercise 4 by this method.

**TABLE 4.4**

| $x$ | $y$ | $z$ | $x \cdot (y + \overline{xz})$ |
|---|---|---|---|
| 1 | 1 | 1 | 1 |
| 1 | 1 | 0 | 1 |
| 1 | 0 | 1 | 0 |
| 1 | 0 | 0 | 1 |
| 0 | 1 | 1 | 0 |
| 0 | 1 | 0 | 0 |
| 0 | 0 | 1 | 0 |
| 0 | 0 | 0 | 0 |

## 4.5 The Quine–McClusky Method

When the number of variables in a Boolean expression does not exceed four, the map method of the previous section is a convenient means of minimization. This method becomes cumbersome, however, for more variables. A useful technique for simplifying expressions with a large number of variables is the *tabulation* method, also known as the *Quine–McClusky* method (Q–M for short) after its developers.

The method is not as visual as the Karnaugh map, is quite mechanical (thus subject to human error), and is suitable for machine computation. Algorithms can be and have been developed to carry it out.

The minimization is presented in two parts. The first step finds all the terms that are candidates for inclusion in the final minimal form. These terms are known as *prime implicants.* The second step chooses the terms to include in the final form.

As with the map method, the expression to be minimized should be written in its complete sum-of-products form so that each fundamental product contains instances of each and all variables, either complemented or uncomplemented. We will explain the process by means of the following examples.

**EXAMPLE 1** For the sake of simplicity let us begin with an expression in three variables. To this end consider the expression

$$E = xyz + xy\bar{z} + \bar{x}yz + \bar{x}y\bar{z} + \bar{x}\bar{y}\bar{z},$$

and its representation by column E of 1's and 0's in the truth table of Table 4.5.

The essence of the Q–M method is to group together the terms that can possibly be combined to reduce to simpler terms. As with the Karnaugh map method we can reduce two terms iff they differ from each other by only *one* variable (complemented in one term and uncomplemented in the other). This is accomplished very nicely if the terms are displayed in a tabular form in decreasing numbers of 1's in their bit representation. (See Table 4.6.)

**TABLE 4.5**

| $x$ | $y$ | $z$ | $E$ | *Fundamental Products* | *Bit Representation* |
|---|---|---|---|---|---|
| 1 | 1 | 1 | 1 | $xyz$ | 111 |
| 1 | 1 | 0 | 1 | $xy\bar{z}$ | 110 |
| 1 | 0 | 1 | 0 | | |
| 1 | 0 | 0 | 0 | | |
| 0 | 1 | 1 | 1 | $\bar{x}yz$ | 011 |
| 0 | 1 | 0 | 1 | $\bar{x}y\bar{z}$ | 010 |
| 0 | 0 | 1 | 0 | | |
| 0 | 0 | 0 | 1 | $\bar{x}\bar{y}\bar{z}$ | 000 |

**TABLE 4.6**

| | | |
|---|---|---|
| 1. $xyz$ | 111 | Three 1's |
| 2. $xy\bar{z}$ | 110 | Two 1's |
| 3. $\bar{x}yz$ | 011 | |
| 4. $\bar{x}y\bar{z}$ | 010 | One 1 |
| 5. $\bar{x}\bar{y}\bar{z}$ | 000 | Zero 1's |

Thus *step one* is to make just such a table with terms grouped by the numbers of 1's in the terms. See Table 4.6.

*Note:* Some textbooks show the tabulation in increasing numbers of 1's. This is the same as our table in reverse order. All one has to do is read from bottom to top.

The combinings (*joinings*) of terms can only be done from one group (level) to the next lowest, since these would be the only possibilities for a single variable difference. It can be seen in Table 4.7 that the term 111 can be joined first with the term 110. (This, of course, represents the sum $xyz + xy\bar{z}$.) As these two terms are joined, the $z$ variable "drops out" ($xyz + xy\bar{z} = xy$) and the reduced bit expression is written as 11–, with the dash to indicate the place of the removed variable. The numbers of the terms used are also indicated by the reduced bit expression. One continues this process of combining and places a check mark by each term that is used in combination with another. Terms may be combined more than once. See column A of Table 4.7.

The process is repeated on this list of reduced terms, but note that only those terms can be combined that have dashes in the same place. See column B of Table 4.7.

We have proceeded as far as we can, and see that the unchecked terms are 0 – 0 in column A and – 1 – (twice) in column B. Thus the prime implicants are $\bar{x}\bar{z}$ and $y$.

*Step two* is to determine if we need to include both of these prime implicants in the final minimal form. This selection process is done by yet another

**TABLE 4.7**

| | | A | | B | |
|---|---|---|---|---|---|
| | | *Terms Used* | *Reduced Expression* | *Terms Used* | *Reduced Expression* |
| 1. $xyz$ | 111 ✓ | (1, 2) | 1 1 – ✓ | [(1, 2), (3, 4)] | –1 – |
| | | (1, 3) | – 1 1 ✓ | [(1, 3), (2, 4)] | – 1 – |
| 2. $xy\bar{z}$ | 110 ✓ | (2, 4) | – 1 0 ✓ | | |
| 3. $\bar{x}yz$ | 011 ✓ | (3, 4) | 0 1 – ✓ | | |
| | | (4, 5) | 0 – 0 | | |
| 4. $\bar{x}y\bar{z}$ | 010 ✓ | | | | |
| 5. $\bar{x}\bar{y}\bar{z}$ | 000 ✓ | | | | |

**TABLE 4.8**

| | | Original Terms | | | |
|---|---|---|---|---|---|
| | 1 | 2 | 3 | 4 | 5 |
| | $xyz$ | $xy\bar{z}$ | $\bar{x}yz$ | $\bar{x}y\bar{z}$ | $\bar{x}\bar{y}\bar{z}$ |
| | 111 | 110 | 011 | 010 | 000 |
| Prime Implicants $y = -1-$ | (✓) | (✓) | (✓) | ✓ | |
| $\bar{x}\bar{z} = 0-0$ | | | | ✓ | (✓) |

table. In Table 4.8 each column corresponds to one of the terms in the original expression, and the rows correspond to the prime implicants found in the previous step.

For each prime implicant, a check is placed in the columns of the original terms accounted for by the prime implicant. For example, both prime implicants are accounted for by the fourth term (column 4). What we seek is a *covering* of the original expression by some combination of the prime implicants. Columns that contain only *one* check dictate that the corresponding prime implicant must be included in the final sum. See the circled checks; these are termed *essential prime implicants.* It is seen that in this example, both prime implicants are needed in the final minimization. Thus the minimal sum is the join of the two prime implicants: $\bar{x}\bar{z} + y$.

Let us check this example with the Karnaugh map of Figure 4.32. We observe from the map that the minimal sum is indeed $\bar{x}\bar{z} + y$, as we obtained with the Q–M method.

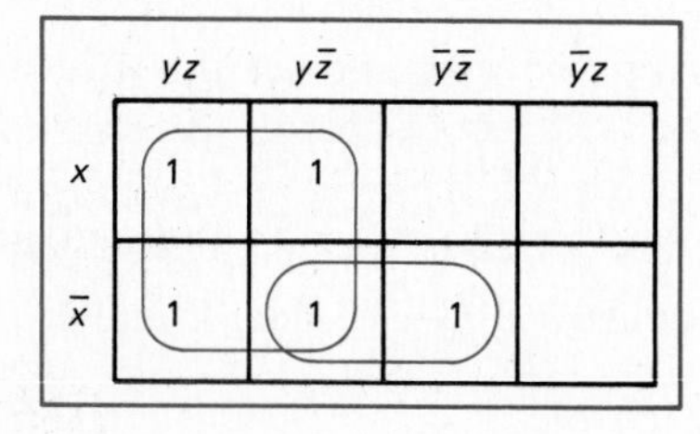


**FIGURE 4.32**

**EXAMPLE 2** The expression $E_{11}$ in Figure 4.29:

$$xyz + xy\bar{z} + \bar{x}y\bar{z} + x\bar{y}z + \bar{x}\bar{y}\bar{z}$$

was seen to have two minimal sums. Let us see how this is handled by the Q–M method.

The tabulation of *step one* is shown in Table 4.9, where the original terms are displayed in decreasing numbers of 1's and the combinations and reductions are carried out in column A. As this reduction can be carried no further, we have as the prime implicants: $xy$, $xz$, $y\bar{z}$, and $\bar{x}\bar{z}$.

TABLE 4.9

| | | | A |
|---|---|---|---|
| 1. $xyz$ | 111 ✓ | (1, 2) | 1 1 – |
| 2. $xy\bar{z}$ | 110 ✓ | (1, 3) | 1 – 1 |
| 3. $x\bar{y}z$ | 101 ✓ | (2, 4) | – 1 0 |
| 4. $\bar{x}y\bar{z}$ | 010 ✓ | (4, 5) | 0 – 0 |
| 5. $\bar{x}\bar{y}\bar{z}$ | 000 ✓ | | |

TABLE 4.10

| | | Original Terms | | | | |
|---|---|---|---|---|---|---|
| | | 1 | 2 | 3 | 4 | 5 |
| | | $xyz$ | $xy\bar{z}$ | $x\bar{y}z$ | $\bar{x}y\bar{z}$ | $\bar{x}\bar{y}\bar{z}$ |
| | | 111 | 110 | 101 | 010 | 000 |
| Prime Implicants | $xy = 11-$ | ✓ | ✓ | | | |
| | $xz = 1-1$ | (✓) broken circle | | (✓) | | |
| | $y\bar{z} = -10$ | | ✓ | | ✓ | |
| | $\bar{x}\bar{z} = 0-0$ | | | | (✓) broken circle | (✓) |

Now, proceeding to *step two*, we form Table 4.10 with columns as original terms and rows as prime implicants. In this table we find the essential prime implicants to be those corresponding to the circled checks, namely $xz$ and $\bar{x}\bar{z}$. These essential prime implicants also cover the first and fourth terms (see the broken circles). Consequently, we need only to cover the second term. This covering can be done by including *either* the prime implicant $xy$ or the prime implicant $y\bar{z}$. Thus we get the final minimal covering to be the join of the *essential* prime implicants with either of the other two. Thus we have *two* minimal sums:

$$\underbrace{xz + \bar{x}\bar{z}}_{\text{essential}} + xy \quad \text{or} \quad \underbrace{xz + \bar{x}\bar{z}}_{\text{essential}} + y\bar{z}$$

**EXAMPLE 3** Minimize the following seven-term Boolean expression $E$ in four variables by using the Q–M method.

$$E = wxy\bar{z} + w\bar{x}yz + \bar{w}xyz + w\bar{x}y\bar{z} + \bar{w}xy\bar{z} + \bar{w}x\bar{y}z + \bar{w}x\bar{y}\bar{z}$$

*Step one:* Table 4.11 shows the tabular groupings of the terms in bit representation and the two columns resulting from combining terms. The prime implicants (resulting from the unchecked terms) are $wy\bar{z}$, $xy\bar{z}$, and $w\bar{x}y$ from column A; and $\bar{w}x$ from column B.

**TABLE 4.11**

| | | A: Terms Used | A: Reduced Expression | B: Terms Used | B: Reduced Expression |
|---|---|---|---|---|---|
| 1. $wxy\bar{z}$ | 1110 ✓ | (1, 4) | 1–10 | [(3, 5), (6, 7)] | 01–– |
| 2. $w\bar{x}yz$ | 1011 ✓ | (1, 5) | –110 | [(3, 6), (5, 7)] | 01–– |
| 3. $\bar{w}xyz$ | 0111 ✓ | (2, 4) | 101– | | |
| 4. $w\bar{x}y\bar{z}$ | 1010 ✓ | (3, 5) | 011– ✓ | | |
| 5. $\bar{w}xy\bar{z}$ | 0110 ✓ | (3, 6) | 01–1 ✓ | | |
| 6. $\bar{w}x\bar{y}z$ | 0101 ✓ | (5, 7) | 01–0 ✓ | | |
| 7. $\bar{w}x\bar{y}\bar{z}$ | 0100 ✓ | (6, 7) | 010– ✓ | | |

**TABLE 4.12**

| Prime Implicants | 1 $wxy\bar{z}$ | 2 $w\bar{x}yz$ | 3 $\bar{w}xyz$ | 4 $w\bar{x}y\bar{z}$ | 5 $\bar{w}xy\bar{z}$ | 6 $\bar{w}x\bar{y}z$ | 7 $\bar{w}x\bar{y}\bar{z}$ |
|---|---|---|---|---|---|---|---|
| $wy\bar{z}$ | ✓ | | | ✓ | | | |
| $xy\bar{z}$ | ✓ | | | | ✓ | | |
| $w\bar{x}y$ | | (✓) | | (✓) | | | |
| $\bar{w}x$ | | | (✓) | | (✓) | (✓) | (✓) |

(Columns: Original Terms)

*Step two:* We form Table 4.12 with original terms as columns and prime implicants as rows. From the circled checks in the table we see that $w\bar{x}y$ and $\bar{w}x$ are the essential prime implicants. These two cover terms 2, 3, 4, 5, 6, and 7. We can cover the first term by using either prime implicant, $wy\bar{z}$ or $xy\bar{z}$. Thus we have two minimal sums: either $w\bar{x}y + \bar{w}x + wy\bar{z}$ or $w\bar{x}y + \bar{w}x + xy\bar{z}$. If you worked Exercise 3(j) of Section 4.4, you should have found the same two minimal sums.

## EXERCISES

Minimize the following Boolean expressions by using the Quine–McClusky method.

1. $xyz + xy\bar{z} + x\bar{y}z + \bar{x}y\bar{z} + \bar{x}\bar{y}\bar{z}$
2. $xy\bar{z} + x\bar{y}\bar{z} + x\bar{y}z + \bar{x}y\bar{z} + \bar{x}\bar{y}z + \bar{x}\bar{y}\bar{z}$
3. $xyz + x\bar{y}\bar{z} + \bar{x}y\bar{z} + \bar{x}\bar{y}z$
4. $xyz + xy\bar{z} + x\bar{y}z + \bar{x}yz + \bar{x}y\bar{z} + \bar{x}\bar{y}\bar{z}$
5. $wxyz + wxy\bar{z} + wx\bar{y}z + wx\bar{y}\bar{z} + w\bar{x}y\bar{z} + \bar{w}\bar{x}yz + w\bar{x}\bar{y}\bar{z} + \bar{w}\bar{x}\bar{y}\bar{z}$ (see Figure 4.31)

6. $wxyz + wxy\bar{z} + w\bar{x}yz + \bar{w}xyz + \bar{w}x\bar{y}z + w\bar{x}y\bar{z}$

7. $wxyz + wx\bar{y}z + \bar{w}xyz + w\bar{x}y\bar{z} + \bar{w}x\bar{y}z + \bar{w}\bar{x}y\bar{z} + w\bar{x}\bar{y}\bar{z} + \bar{w}\bar{x}\bar{y}\bar{z}$

8. $w\bar{x}yz + \bar{w}xyz + wx\bar{y}\bar{z} + w\bar{x}\bar{y}z + \bar{w}x\bar{y}z + \bar{w}\bar{x}\bar{y}z$

## Summary and Selected References

The application of the laws of logic and Boolean algebra is the cornerstone of the study of digital logic design. More than cursory treatments from a mathematical standpoint can be found in Birkhoff and Bartee (1970), Gill (1976), and Fisher (1977). From a svstem design standpoint, the reader is referred to Dietmeyer (1978), Tocci (1980), Kohavi (1978), Lewin (1983), Booth (1978), and Mano (1979).

# Functions, Recursion, and Induction

# 5

## 5.1 Introduction

All of you at some time or other in previous mathematics courses have encountered the term *function*. It is one of the most central concepts in mathematics and is certainly not limited to our treatment of discrete mathematical structures. Many times it is referred to in a particular context and different notations and definitions are prevalent in the literature.

The word itself appears about as often in everyday life and in connection with nonmathematical subjects as it does in mathematics. One encounters it in such a variety of settings as: "interest rates are a function of the status of the economy," "the price of oranges is a function of the weather in Florida," "the existence of a particular animal characteristic is a function of evolutionary change," and "the output of a computer program is a function of the input."

Each of these instances of the word "function" expresses an implied relationship between (at least) two changing events (variables), one of the variables depending upon the other in some significant way.

In this chapter we introduce the basic notions and terminology of functions that are deemed important for further study in computer science. First, we examine the direct connection with sets, and, in particular, note that a function is a special type of relation. We then introduce the modern terminology associated with functions. Next, we study two function concepts that are so important in computer science: parenthesis-free (Polish) notation as functions of two variables, and recursive functions. Finally, we explore mathematical induction as a method of proof of discrete systems and show its direct connection with recursion.

## 5.2 Functions

Mathematically, functions are treated as particular kinds of *relations* (see Section 2.8). Recall that a relation from some set $A$ to another set $B$ is a subset of the Cartesian product, $A \times B$, of the two sets. Functions are commonly identified by letters such as $f$, $g$, $h$, and so on; however, almost any symbol is acceptable. Thus a function, $f$, becomes a set of ordered pairs, each of the form $\langle a, b \rangle$ for $a \in A$ and $b \in B$. We say that $\langle a, b \rangle \in f$ to indicate that this particular pair belongs to the function (relation), $f$.

Not all relations are functions. There are two severe restrictions placed on a relation for it to constitute a function:

1. For a relation from set $A$ to set $B$ to be a function, *each* $a \in A$ has to be used as a *first* element of an ordered pair. We say that the function is *defined* on set $A$ if every $a \in A$ is so used. The set $A$ is referred to as the *domain* of the function and the set $B$ is the *codomain*.
2. Once an element $a \in A$ is used in an ordered pair, it cannot be used again. That is, there is "assigned" to every $a \in A$ a *unique* $b \in B$. The set of assigned values in $B$ is called the *image* or *range* of the function. The range need not be the same set as the codomain, as will be shown in some of the examples below.

The discussion and restrictions above lead to the formal definition of a function:

**Definition** A function, $f$, from set $A$ to set $B$ is a relation from $A$ to $B$ such that for *every* $a \in A$ there is *one and only one* $b \in B$ such that the ordered pair $\langle a, b \rangle \in f$.

The symbolism ordinarily used to denote a function is

$$f: A \rightarrow B$$

which is to be read as: "$f$ is a function from set $A$ to set $B$" or "the function $f$ *maps* set $A$ to set $B$." If the ordered pair $\langle a, b \rangle$ is in $f$, we say that $b$ is the *value* or *image* of $a$, and write $f(a) = b$ (read as "$f$ of $a$ is $b$"). The value $b$ is the unique assignment made by $f$ to the element $a \in A$.

The element $a$ is referred to as the *argument* of the function. Also, $a$ is the *preimage* of $b$. Additionally, the elements $a$ and $b$ are referred to as the independent and dependent variables, respectively. We see, then, that a function can be viewed as a *rule* of assignment that specifies which particular $b \in B$ is associated with each $a \in A$.

The discussion and associated terminology above can be shown pictorially as in Figure 5.1.

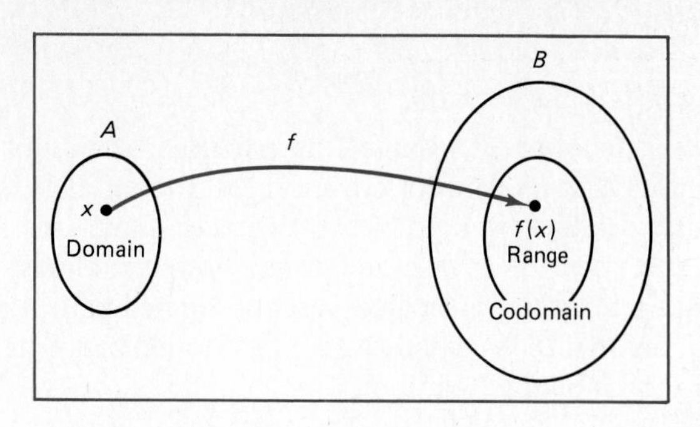


**FIGURE 5.1** Pictorial description of $f: A \to B$.

**EXAMPLE 1** We are all familiar with the function that maps (or sends) each real number $x$ to its square, $x^2$. Here the domain is the set of real numbers, designated by $R$, as is the codomain. The function, then, maps $R$ to $R$. If we let $f$ designate the function, we have

$$f: R \to R$$

defined by

$$f(x) = x^2 \qquad \text{for } x \in R$$

Some ordered pairs of $f$ are

$$\langle 0,0 \rangle, \quad \langle 3,9 \rangle, \quad \langle 5.21, 27.1441 \rangle, \quad \langle -3,9 \rangle, \quad \langle \tfrac{1}{2}, \tfrac{1}{4} \rangle$$

We know that the graph of $f$ can be pictured on coordinate axes as shown in Figure 5.2. We note that $f$ is never negative; hence the *range* of $f$ is the set of nonnegative real numbers and is *not* the same as the codomain, which is all real numbers.

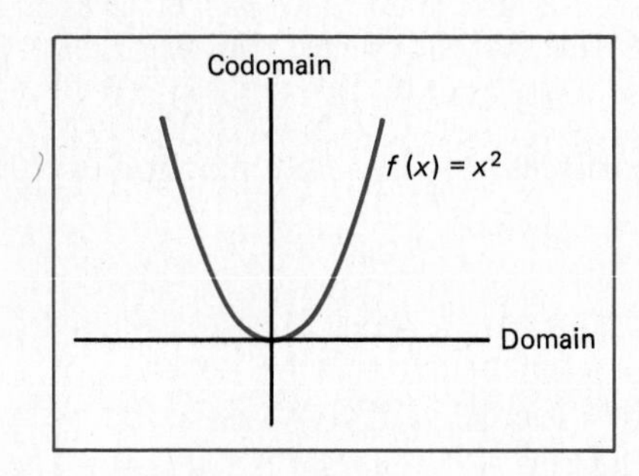


**FIGURE 5.2**

**EXAMPLE 2** The function $g(x) = 2x + 3$ has as its domain, codomain, and range the set of all real numbers, $R$. Thus we can write

$$g: R \to R$$

defined by

$$g(x) = 2(x) + 3 \qquad \text{for all } x \in R$$

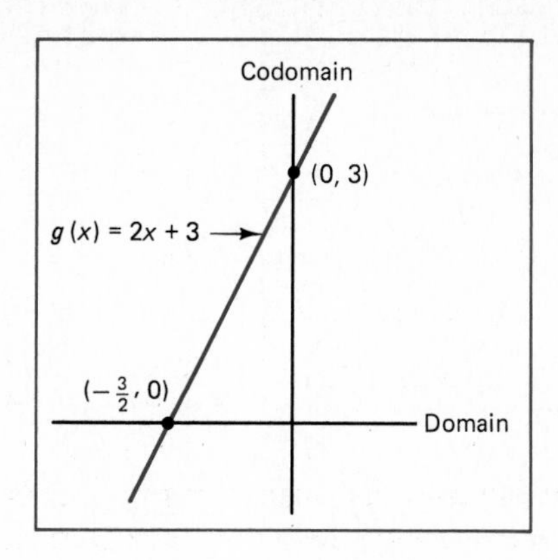

FIGURE 5.3

Some ordered pairs of $g$ are

$$\langle 0, 3\rangle, \quad \langle 1, 5\rangle, \quad \langle -\tfrac{3}{2}, 0\rangle, \quad \langle -4, -5\rangle$$

The graph of this linear function is shown in Figure 5.3.

**EXAMPLE 3** Consider the function $h: M \to N$ that maps the set $M = \{a, b, c\}$ to the set $N = \{d, e, f\}$, defined by

$$h(a) = d \qquad h(b) = e \qquad h(c) = d$$

The ordered pairs of $h$ are

$$\langle a, d\rangle, \quad \langle b, e\rangle, \quad \langle c, d\rangle$$

This mapping can be pictorially represented by Figure 5.4. The fact that the element $d \in N$ is used twice does not violate the restriction of uniqueness. Each element of $M$ has a unique image in $N$. Notice also that $f \in N$ is not the image of any element of $M$. Thus the range of $h$ does not equal the codomain.

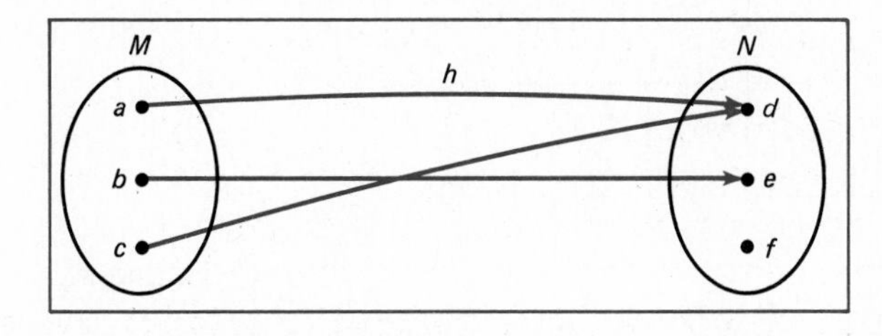

FIGURE 5.4

**EXAMPLE 4** The relation from $P$ to $Q$ pictured in Figure 5.5 is *not* a function. There are *two* violations:

1. Element $b \in P$ has no image in $Q$.
2. Element $a \in P$ has two images, $d$ and $e$.

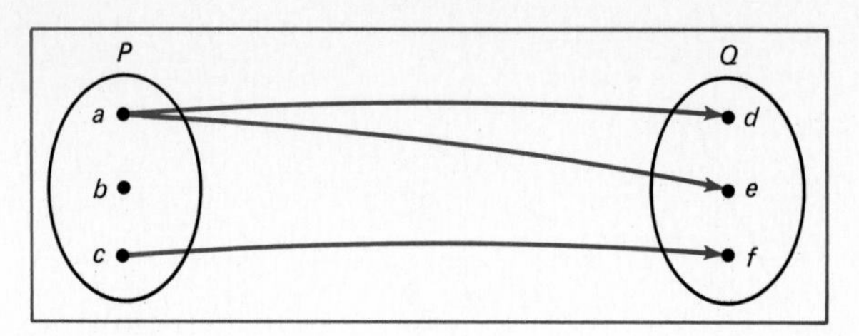


FIGURE 5.5

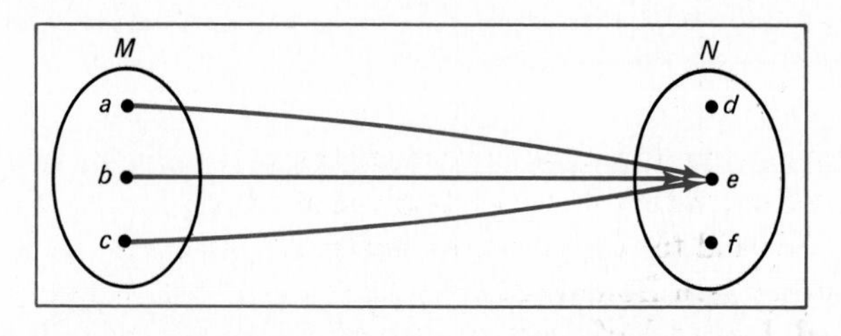


FIGURE 5.6

**EXAMPLE 5** The function from $M$ to $N$ pictured in Figure 5.6 is an example of what is called a *constant* function: the image element is the same (or is constant) for all the elements in the domain.

Another constant function with which you are probably familiar is one similar to

$$f(x) = 3 \qquad \text{for } x \in R$$

Its graph on a coordinate axis is shown in Figure 5.7.

**EXAMPLE 6** We looked at the *relation* "is a multiple of" on a set $T = \{2, 3, 4, 5, 6\}$ in Table 2.14. Another pictorial description is that of Figure 5.8. It should be clear that this relation is *not* a function. Why is it not?

FIGURE 5.7

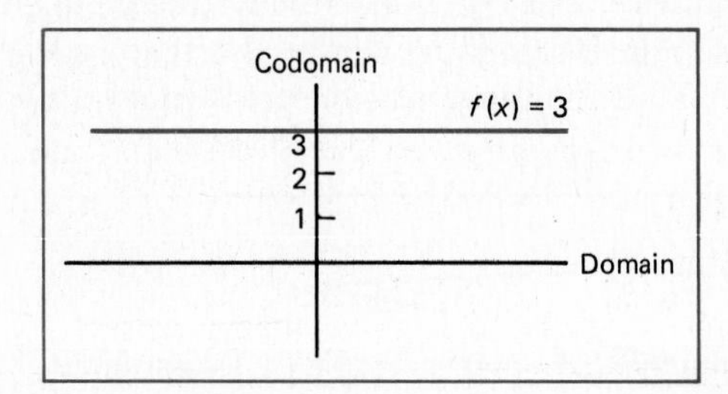


FIGURE 5.8

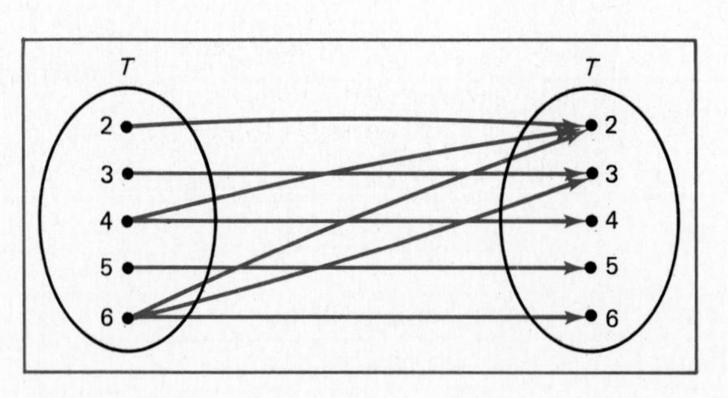

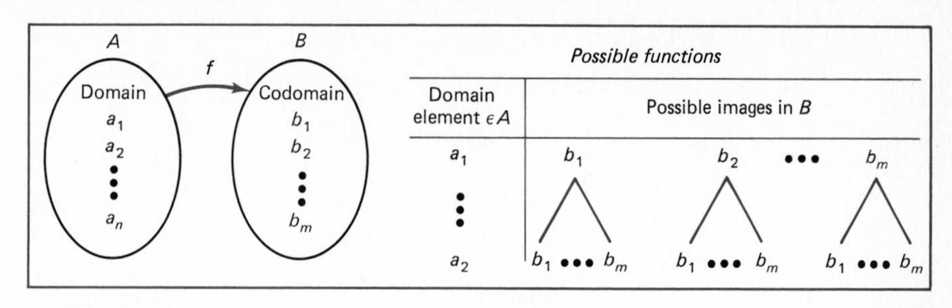


**FIGURE 5.9**

**EXAMPLE 7** An interesting counting question is to determine *how many* functions there are from one set to another. Let us suppose that the domain set $A$ has $n$ elements: $a_1, a_2, \ldots, a_n$; and the codomain set has $m$ elements: $b_1, b_2, \ldots, b_m$.

It can be seen from Figure 5.9 that $a_1 \in A$ can have $m$ possible distinct images in $B$. For *each* of these choices for $a_1$, there will be $m$ possible images for $a_2$, yielding $m \cdot m = m^2$ possible functions. For a third element, $a_3$, the $m$ possible images yields $m \cdot m \cdot m = m^3$ functions. This can be extended to the $n$ elements of $A$, obtaining

$$\underbrace{m \cdot m \cdots m}_{n \text{ times}} = m^n$$

possible functions. An exact proof of this result will be given in Section 5.7.

**EXAMPLE 8** Suppose that a student has been assigned three programming problems and knows four programming languages appropriate for the problems. How many ways can the student program his three problems?

In this example we see that we wish to *assign* a programming problem to a programming language. By using the notion from Example 7, we thus have a function from the set of problems to the set of languages. The number of elements in the domain would be *three;* and the codomain has four elements. Thus we get $4^3 = 64$ possible ways for the student to program his assigned problems.

This example will be explored further in Exercise 6.

Special Functions We consider here some special classes of functions that have broad application to many areas of mathematics and computer science.

1. If the function $f: A \to B$ has the property that its range is the *same* set as the codomain, we say that that $f$ is *surjective.* We have that for each $b \in B$ there is an $a \in A$ such that $f(a) = b$. Stated informally: $B$ is completely "used up." The function is "on" all of $B$. Such functions are also referred to as *onto* functions and can be specified by

$$f: A \xrightarrow{\text{onto}} B$$

or

$$f: A \xrightarrow{\text{sur}} B$$

The function

$$g: \{a, b, c, d\} \rightarrow \{e, f\}$$

defined by

$$g(a) = e \qquad g(b) = f \qquad g(c) = f \qquad g(d) = e$$

is surjective.

Also the function

$$h: R \rightarrow R$$

defined by

$$h(x) = -x \qquad \text{for } x \in R$$

is surjective. When a function is not "onto," we usually say that it is "into."

2. If a function $f: A \rightarrow B$ maps distinct elements of $A$ to *distinct* elements of $B$ we say that $f$ is *one-to-one* (1-1) or *injective*. An injection never maps two different domain elements to the same element of the codomain. Thus the functions of Examples 3 and 5 are *not* injections, whereas $g(x) = 2x + 3$ of Example 2 is.

A nonmathematical injective function is a teacher's seating chart which assigns each student in a class to a particular desk in the classroom. The set of students constitutes the domain and the desks the codomain. Every student is assigned a desk, but no student is assigned to more than one desk. Why is this function not necessarily a surjection? What is the range?

Another injective function is the rule of marriage in a monogomous society in which each married person is "assigned" to one and only one other married person.

Injective (1-1) functions can be specified by

$$f: A \xrightarrow{\text{1-1}} B$$

A formal description of injective functions is the following: $f: A \rightarrow B$ is 1-1 when for every $a, a' \in A$, if $f(a) = f(a')$, then $a = a'$.

3. If function $f: A \rightarrow B$ is *both* injective and surjective (1-1 and onto), we say that it is a *bijection*. For a function to be bijective, the domain and codomain have to have the same number of elements. Why?

Such functions can be specified by

$$f: A \xrightarrow[\text{onto}]{\text{1-1}} B$$

A bijective relationship between two sets is commonly known as a "one-to-one correspondence."

Bijective functions are important in that there is automatically defined a *reverse* mapping from the codomain to the domain that is *also* a bijection. This is possible since each codomain element has a unique domain element assigned to it. These domain elements then become unique assignments for the codomain elements and thus the reverse mapping is a function. This reverse function of

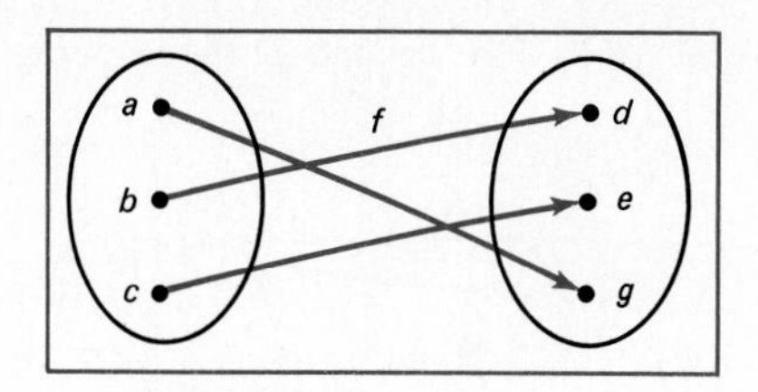


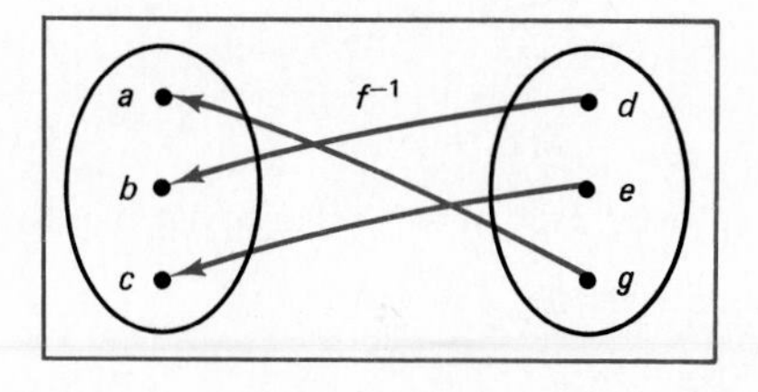


**FIGURE 5.10** A bijection and its inverse.

a bijection $f$ is known as the *inverse* of $f$, and is designated by $f^{-1}$. A function that has an inverse is said to be invertible. The function $f: \{a, b, c\} \to \{d, e, g\}$ defined by the set of ordered pairs $\{\langle a, g\rangle, \langle b, d\rangle, \langle c, e\rangle\}$ is a bijection. Its inverse $f^{-1}$ consists of the pairs $\{\langle g, a\rangle, \langle d, b\rangle, \langle e, c\rangle\}$. These two functions are shown in Figure 5.10. It is seen that

$$f^{-1}(d) = b$$

$$f^{-1}(e) = c$$

$$f^{-1}(g) = a$$

## EXERCISES

1. Determine which of the following relations are functions. The domain under consideration is the set $\{a, b, c\}$ and the codomain is the set $\{d, e, f\}$.

   (a) $\{\langle a, d\rangle, \langle b, e\rangle, \langle c, f\rangle\}$

   (b) $\{\langle a, e\rangle, \langle b, e\rangle, \langle c, e\rangle\}$

   (c) $\{\langle a, d\rangle, \langle a, e\rangle, \langle b, d\rangle, \langle c, e\rangle\}$

   (d) $\{\langle b, e\rangle, \langle c, e\rangle\}$

   (e)

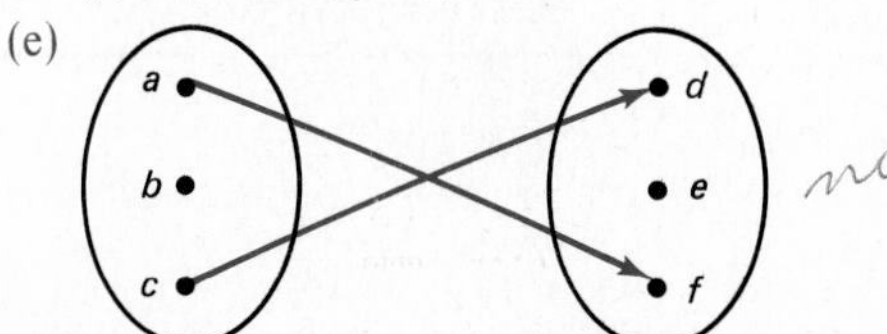


   (f)

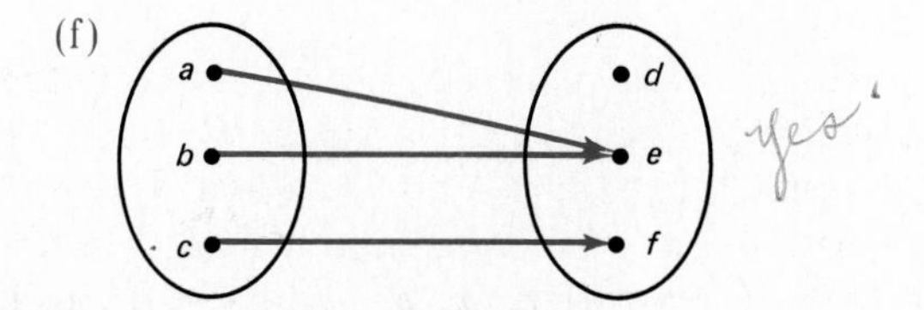

2. Let $T = \{2, 3, 4, 5, 6\}$ and let $f: T \to T$ be defined by the mapping pictured.

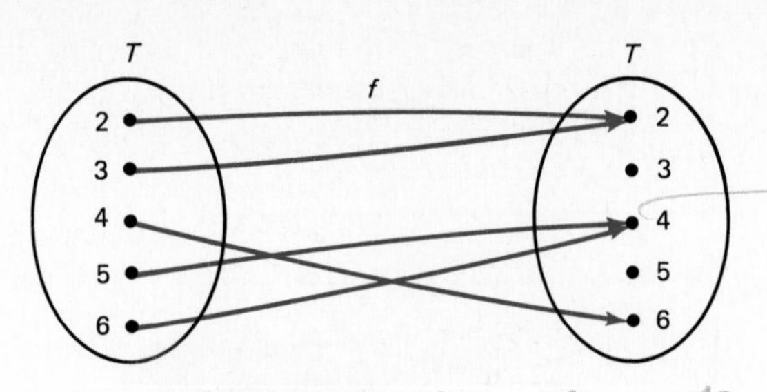


Write $f$ as a set of ordered pairs. Is $f$ surjective, injective, bijective?

3. Show examples of functions that are
   (a) Surjections and not injections.
   (b) Injections and not surjections.
   (c) Neither surjections nor injections.
   (d) Bijections and not injections.
   (e) "Into" and not 1-1.

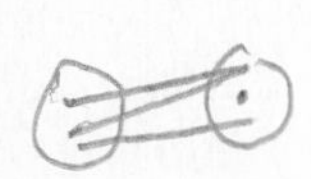

4. Let $P$ be the set of computer programs run yesterday at your local computing center. Determine which of the following assignments defines a function on $P$.
   (a) To each program assign a programming language.
   (b) To each program assign an execution time.
   (c) To each program assign members of a programming team.

5. Is the function $f(x) = 2x$ an injection? A surjection? A bijection?

6. In Example 8 we wished to find how many possible ways (functions) the student could program his assignments. Suppose we make the restriction that each programming language would be used only once. That is, if one of the problems were programmed in language one ($L_1$), then $L_1$ could not be used again. How many ways is this possible? The assignments can be shown by the following tree diagram (partial), in which some of the elements are left out.
   (a) Finish the tree diagram and determine how many functions are possible.

Possible mappings

Domain element $\epsilon P$ | Possible images $\epsilon L$

$P_1$: $L_1$ $L_2$ $L_3$ $L_4$

$P_2$: $L_2$ $L_3$ $L_4$ | $L_1$ $L_3$ $L_4$ | $L_1$ $L_2$ $L_4$ | $L_1$ $L_2$ $L_3$

$P_3$: $L_3$ $L_4$ $L_2$ $L_4$ $L_2$ $L_3$

(b) Of which of the special types of functions mentioned in this section is this an example: injection, surjection, or bijection?

(c) Another way to illustrate this counting example is to note that $P_1$ can be assigned to four images, or languages. For each of these four, $P_2$ has *three* possible assignments; and then $P_3$ has *two*. How many is this? Show that this is the same as obtained from the tree.

7. With $T$ defined as in Exercise 2, how many possible functions are there from $T$ to $T$? How many injections? How many surjections?

8. Suppose that you had five cans of different-color paint and wished to paint four pieces of furniture.

   (a) How many possible ways are there to paint the individual pieces of furniture?

   (b) How many ways can you paint the furniture so that no two pieces of furniture had the same color?

apostol p37

## 5.3 Composition of Functions

If we have two functions

$$f: B \to C \qquad \text{and} \qquad g: A \to B$$

?? not always

as shown in Figure 5.11, it is possible to define a new function that maps set $A$ directly to set $C$. This new function is called the *composition* of $f$ and $g$ and is symbolized by $f \circ g$ ($f$ circle $g$).

For an element $x \in A$, its image under $g$ is $g(x)$ in $B$. The image, then, of this element under function $f$ is $f(g(x))$, which is in set $C$. Thus, for $x \in A$, we have

$$f \circ g(x) = f(g(x))$$

Notice that the *domain* of $f$ is the *range* of $g$.

do ①

FIGURE 5.11

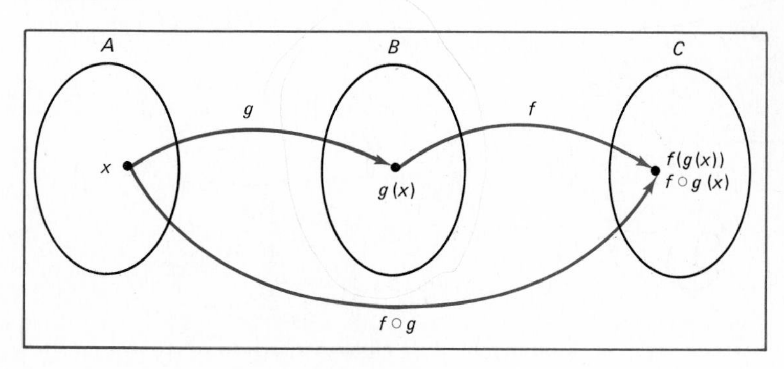

**EXAMPLE 1** Let $A = \{a, b, c\}$, $B = \{d, e\}$, and $C = \{l, m\}$. Define $g: A \to B$ by

$$g(a) = d \qquad g(b) = d \qquad g(c) = e$$

and define $f: B \to C$ by

$$f(d) = m \qquad f(e) = l$$

These functions are pictured in Figure 5.12.

The function $f \circ g: A \to C$ can then be seen to be defined as

$$f \circ g(a) = f(g(a)) = f(d) = m$$
$$f \circ g(b) = f(g(b)) = f(d) = m$$
$$f \circ g(c) = f(g(c)) = f(e) = l$$

If we consider $g \circ f$, that is, $g(f(x))$, for $x \in C$, it should readily be seen that this is not a function. (Why is it not?) So we can conclude that, in general,

$$f \circ g \neq g \circ f$$

**EXAMPLE 2** Consider two functions, $f$ and $g$, on the real numbers $R$:

$$f(x) = x^2 + 3 \qquad \text{and} \qquad g(x) = \sqrt{x}$$

Then

$$f \circ g(x) = f(g(x)) = f(\sqrt{x}) = (\sqrt{x})^2 + 3 = x + 3$$

and

$$g \circ f(x) = g(f(x)) = g(x^2 + 3) = \sqrt{x^2 + 3}$$

So we see again that, in general, $f \circ g \neq g \circ f$.

FIGURE 5.12

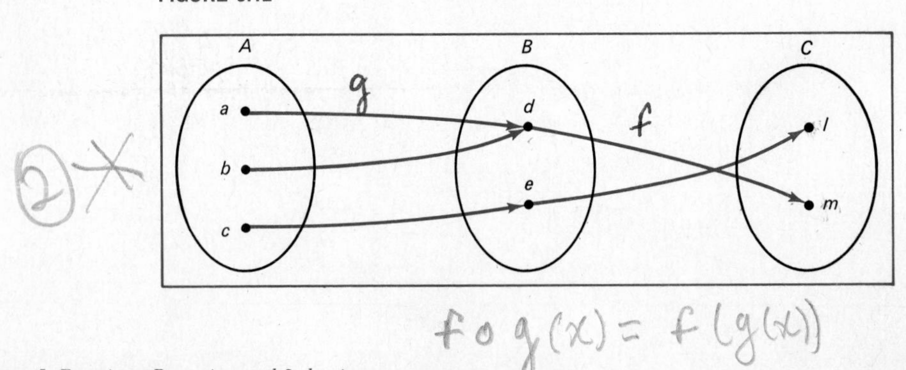

## EXERCISES

1. Let $f(x) = 1/\sqrt{x+3}$ and $g(x) = x^2 + 3x + 4$. What is $f \circ g(x)$? $g \circ f(x)$?

2. For sets $A = \{1, 2, 3\}$, $B = \{4, 5, 6\}$, and $C = \{7, 8, 9, 10\}$, define the function $f$: $A \to B$ by the ordered pairs $\{\langle 1, 5\rangle, \langle 2, 4\rangle, \langle 3, 4\rangle\}$; and define the function $g: B \to C$ by the ordered pairs $\{\langle 4, 9\rangle, \langle 5, 7\rangle$, and $\langle 6, 8\rangle\}$.

   (a) Draw a pictorial description of these two functions and the composite function, $g \circ f: A \to C$.

   (b) List the ordered pairs of $g \circ f$.

3. For the functions in Exercise 2, find $f^{-1}$ and $g^{-1}$ if they exist.

## 5.4 Functions with More Than One Argument

It is possible for a function, $f$, to have more than one argument. When there are two arguments, we say that $f$ is *binary*. Functions with more than two arguments are called *n-ary* (3-ary, 4-ary, and so on, for three, four, or more arguments). Those of you who have studied three-dimensional analytic geometry are familiar with such functions. Functions with two arguments or variables are also known as *bivariate functions*.

A binary function with arguments $a$ and $b$ can be designated by: $f(a, b) = c$, meaning that the value (image), $c$, in the range of $f$, is assigned by $f$ to the pair $\langle a, b\rangle$. The arguments themselves originate from the Cartesian product of two sets, and the mapping is to a third set. The domain is a set of ordered pairs. To designate this binary mapping, we use the notation

$$f: A \times B \to C$$

defined by

$$f(a, b) = c \qquad \text{for } \langle a, b\rangle \in A \times B \text{ and } c \in C$$

*Note:* The resulting triple $\langle a, b, c\rangle$ is in $A \times B \times C$.

An $n$-ary function with three arguments can be written as

$$f: A \times B \times C \to D$$

defined by

$$f(a, b, c) = d \qquad \text{for } \langle a, b, c\rangle \in A \times B \times C \text{ and } d \in D$$

A binary function with both arguments from the same set, say $A$, and with codomain, $B$, would be designated as

$$f: A \times A \to B$$

or alternatively as

$$f: A^2 \to B$$

The codomain can be the same set as the domain, as in

$$f: A^2 \rightarrow A$$

**EXAMPLE 1** Let us consider a function $f$ of the form

$$f: A^2 \rightarrow A$$

To this end, let $A = \{0, 1\}$ and the function $f: A^2 \rightarrow A$ be defined as follows:

$$f(0, 0) = 0$$

$$f(0, 1) = 0$$

$$f(1, 0) = 0$$

$$f(1, 1) = 1$$

with its accompanying table:

| | | $y$ | |
|---|---|---|---|
| | $f(x, y)$ | 0 | 1 |
| $x$ | 0 | 0 | 0 |
| | 1 | 0 | 1 |

This function can be conveniently represented in tabular form (Table 5.1). As can be seen, this function gives rise to the familiar truth table for the Boolean "meet." It is also the "carry" function discussed in Chapter 4.

Examples of binary functions abound. For instance, the operation of addition of two real numbers, $x$ and $y$, yields another real number. As such, addition can be thought of as a function of two arguments (variables), thus binary.

If we designate the symbol $+$ to represent this addition function, and $R$ as the set of real numbers, then we have the function to indicate addition:

$$+: R \times R \rightarrow R$$

**TABLE 5.1**

| Domain element | | |
|---|---|---|
| $x$ | $y$ | $f(x, y)$ |
| 1 | 1 | 1 |
| 1 | 0 | 0 |
| 0 | 1 | 0 |
| 0 | 0 | 0 |

where $+(x, y) = x + y$ for $x$, $y$, and $x + y \in R$. Thus we have, for example:

$$+(7, 13) = 20$$

$$+(5, -11) = -6$$

$$+(a, b) = a + b, \quad \text{etc.}$$

It should be obvious that subtraction, multiplication, division, and exponentiation are also binary functions on the real numbers. Intersection and union are binary functions on sets, and conjunction and disjunction are binary functions on logical propositions.

Parenthesis-Free (Polish) Notation By using the notation that we have developed above for expressions involving binary functions, an interesting fact emerges. If we consider the expression: $(a + b) - c$, we may first write the indicated addition, $a + b$, as the addition function $+(a, b)$ or just $+ab$, expressing the fact that the symbol $+$ is a function of the two arguments $a$ and $b$. Thus we have $(+ab) - c$. Now the subtraction function has two arguments, $+ab$ and $c$, and the entire expression can be written as $- + abc$.

The need for parentheses is *gone*, for the evaluation is unambiguous! To evaluate, one finds the first operator that has two variables (arguments) immediately following it:

$$- \quad \underset{\uparrow}{+} \quad \underset{\uparrow}{\underbrace{a \quad b}} \quad c$$

operator two arguments

This operation can be performed: $-(a + b)c$ and then the next operator, $-$, has two arguments, $(a + b)$ and $c$. Performing the subtraction yields $(a + b) - c$.

We, of course, are accustomed to writing operator symbols between the arguments in an expression, such as $a + b$, $p \wedge q$, $A \cup B$, and so on. This positional arrangement is known as *infix* notation. What we did in the preceding two paragraphs was to rewrite our familiar infix expression with the operator symbol preceding the arguments. This formulation is known as *prefix* notation. A Polish mathematician and logician, Jan Lukasiewicz (1878–1956), was the discoverer of this notation and in his honor this formulation is known as Polish (prefix) notation. His research actually dealt with logical operators and the formulation in this context is identical with the example above.

One can rewrite the logical expression

$$(p \wedge q) \rightarrow (r \vee (p \leftrightarrow q))$$

in Polish prefix notation as follows:

| | |
|---|---|
| $(p \wedge q) \rightarrow (r \vee (p \leftrightarrow q))$ | |
| $\wedge pq \rightarrow (r \vee \leftrightarrow pq)$ | rewrite the $\wedge$ and $\leftrightarrow$ in prefix |
| $\wedge pq \rightarrow (\vee r \leftrightarrow pq)$ | rewrite the $\vee$ in prefix |
| $\rightarrow \wedge pq \vee r \leftrightarrow pq$ | rewrite the $\rightarrow$ in prefix |

Notice that the last operation (function) to be performed becomes the first one written in the final form.

An analogous situation occurs when one places the operator *after* the arguments: as in $ab+$ to represent $a + b$. This formulation is referred to as *postfix notation* or as *reverse Polish notation* (RPN). This is widely used in computations on some calculators.

The whole topic of prefix and postfix notation is used in the design of compilers to recognize arithmetic expressions and in their evaluation. You will encounter such applications in courses on data structures and compiler construction. We will encounter this notation again in Chapter 7.

**EXAMPLE 2** Convert the prefix expression

$$\wedge r \leftrightarrow \rightarrow pqr$$

to infix notation.

1. $\wedge r \leftrightarrow \rightarrow pqr$
2. $\wedge r \leftrightarrow (p \rightarrow q)r$     $\rightarrow$ has two arguments
3. $\wedge r[(p \rightarrow q) \leftrightarrow r]$     $\leftrightarrow$ has two arguments
4. $r \wedge [(p \rightarrow q) \leftrightarrow r]$     $\wedge$ has two arguments

**EXAMPLE 3** Convert

$$a + (b * ((c/d) \uparrow a))$$

to postfix notation. (We will use $*$ for multiplication and $\uparrow$ for exponentiation.)

As we work from "inside out," we note that the innermost expression, $(c/d)$, can be rewritten in postfix notation as $cd/$. Thus the first step in the conversion is

1. $a + (b * (cd/\uparrow a))$

We next note that $cd/\uparrow a$ can be rewritten as $cd/a\uparrow$. Thus the next step in the conversion is

2. $a + (b * cd/a\uparrow)$

The expression $b * cd/a\uparrow$ can be rewritten as $bcd/a\uparrow*$, yielding

3. $a + (bcd/a\uparrow*)$

Finally we obtain

4. $abcd/a\uparrow*+$

**EXAMPLE 4** Convert the postfix expression

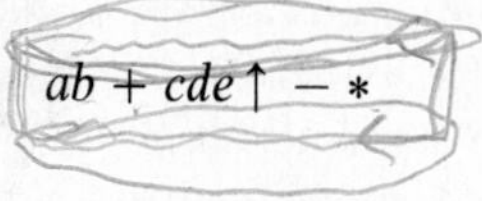

$$ab + cde\uparrow - *$$

to infix notation.

1. $ab + cde\uparrow - *$
2. $(a + b)c(d\uparrow e) - *$     $+$ and $\uparrow$ each have two arguments
3. $(a + b)[c - (d\uparrow e)] *$     $-$ has two arguments
4. $(a + b) * [c - (d\uparrow e)]$     $*$ has two arguments

## EXERCISES

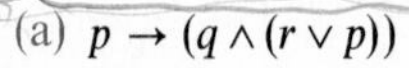

1. Convert each expression to prefix and postfix notation.
   (a) $p \rightarrow (q \wedge (r \vee p))$
   (b) $((p \leftrightarrow q) \rightarrow (r \wedge (p \vee q)))$
   (c) $A * ((B + (C/D)) + E)$
   (d) $(A + (B/C)) - (D\uparrow A)$   ($\uparrow$ is used for exponentiation)
   (e) $p \wedge (q \vee (r \rightarrow (s \leftrightarrow t)))$
   (f) $(((p \wedge q) \vee r) \rightarrow s) \leftrightarrow t$
2. Convert each prefix expression to equivalent infix notation.
   (a) $\rightarrow \wedge p \vee qrp$
   (b) $\uparrow + AB * CD$
   (c) $\vee \wedge p \leftrightarrow r \rightarrow qpr$

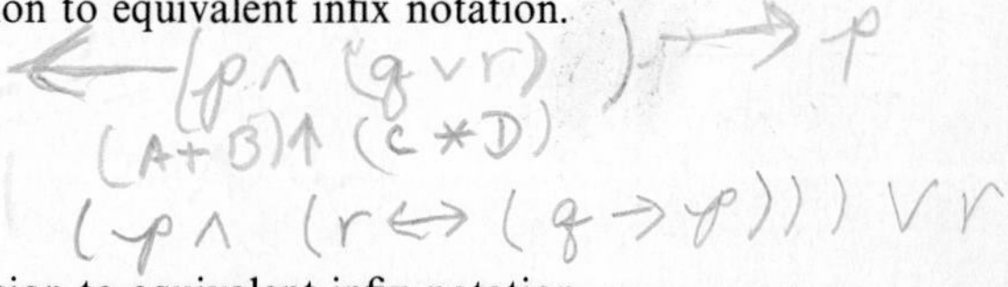

3. Convert each postfix expression to equivalent infix notation.
   (a) $AB + CD * \uparrow$
   (b) $pqr \vee \wedge p \rightarrow$
   (c) $ABCD - \uparrow * A + D/$

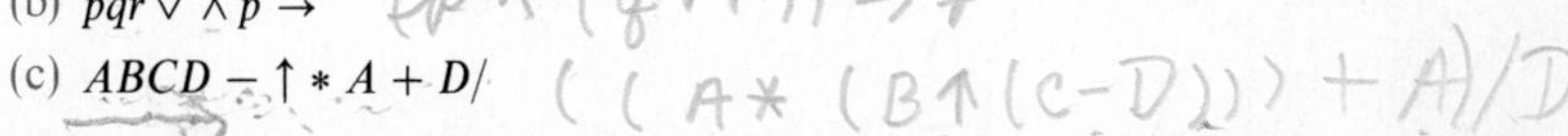

4. Show informally that the variables in any expression stay in the same order whether expressed in infix, prefix, or postfix notation.

## 5.5 Recursive Functions

Some important functions that have wide application in computer science and arise often are ones that are defined *recursively* or by *recursion.* Such definitions rely on self-reference in that these functions seem to be defined in terms of themselves. One may tend to think that the definitions are circular, but this is really not the case. A circular definition would be of the following form. You want to know what a "srangle" is; someone defines it using the word "warge"; and then the word "srangle" turns up in the definition of "warge." Or perhaps the word "srangle" is used to define "srangle."

On the other hand, recursion is a technique wherein the definition employs a *simplified version* of the object or function being defined. A recursive definition of a function is one that uses the function in question but is not really defined in terms of "itself." As an example from outside mathematics, cellular growth may be defined recursively. Descriptions of cells can be given in terms of previous cells of an almost identical nature. New cells are generated from older "simpler" cells but having almost identical structure.

There are two steps in the process of defining a function recursively. *First*, there is a *generating rule* that describes the growth or computation of the function directly in terms of previously known values of the function. *Second*, a *termination* value where the process ends is always defined.

Let us give a recursive definition of a function that you already know how to evaluate. Consider the function

$$f_1(n) = 2^n \qquad \text{for } n \in \{0, 1, 2, 3, \ldots\}$$

For the first four elements in the domain we have

$$f_1(0) = 2^0 = 1$$
$$f_1(1) = 2^1 = 2$$
$$f_1(2) = 2^2 = 4$$
$$f_1(3) = 2^3 = 8$$

by direct *explicit* computation by exponentiation. For instance, we also have $f_1(10) = 2^{10} = 1024$ and nothing has to be known about the values of the function for arguments less than 10.

On the other hand, consider the two-part function $f_2$ defined by

$$(1) \quad f_2(0) = 1$$
$$(2) \quad f_2(n) = 2 \cdot f_2(n-1) \qquad \text{for } n > 0$$

It becomes easy to compute $f_2(0)$. Its value is 1 by definition. But what about $f_2(3)$ or $f_2(6)$? There is no explicit definition for these. Also, when we look at part (2) it seems as though $f_2$ is defined "in terms of" $f_2$ circularly! But this is not the case. $f_2(n)$ is defined in terms of $f_2(n-1)$. This is the generating rule that describes the evaluation in terms of a previous value.

The key to the definition is that there must be a way to *know* the previous value. This is the purpose of part (1) of the definition. This is the termination value mentioned above. As we keep passing down from $n$ to $n-1$ to $n-2$, and so on, we arrive at the *sim*plest case of $n$, namely 0, and this is where the process stops.

To illustrate, let us compute $f_2(5)$. From part (2) we obtain $f_2(5) = 2 \cdot f_2(4)$. Now we are one level lower than before. Again from part (2), $f_2(4) = 2 \cdot f_2(3)$.

Plunging even deeper, we obtain

$$f_2(3) = 2 \cdot f_2(2)$$
$$f_2(2) = 2 \cdot f_2(1)$$
$$f_2(1) = 2 \cdot f_2(0)$$

We have finally arrived at the *simplest* case. From part (1) we know that $f_2(0) = 1$, so that $f_2(1) = 2 \cdot 1$. As we work our way back up we finally arrive at $f_2(5) = 2 \cdot 2 \cdot 2 \cdot 2 \cdot 2 \cdot 1 = 32$.

The evaluation above can be viewed as

$$f_2(5) = 2 \cdot \underbrace{f_2(4)}_{2 \cdot \underbrace{f_2(3)}_{2 \cdot \underbrace{f_2(2)}_{2 \cdot \underbrace{f_2(1)}_{2 \cdot \underbrace{f_2(0)}_{1}}}}} = 32$$

The function $f_2$ gives identical results as $f_1$. However, $f_2$ gives an evaluating mechanism to produce new values from known old ones. This is the heart of a recursive definition.

The process can be summarized:

1. $f_2(0) = 1$: This is called the *terminating* condition in which the function has a known specified value assigned to the "simplest" argument. (It is also known as the *basis* or even *initial* condition.)
2. $f_2(n) = 2 \cdot f_2(n - 1)$: This is the *generating rule* wherein the value of the function at one argument is defined in terms of the value of the same function at a smaller (simpler) argument. [The rule could have been stated: $f_2(n + 1) = 2 \cdot f_2(n)$. It should be readily determined that this is essentially the same statement as originally given.]

The subject of recursion is treated delightfully by Douglas Hofstadter in his Pulitzer Prize-winning book *Gödel, Escher, Bach.* He gives many examples both within and outside mathematics. Of the many accounts of recursion in life with which he deals, perhaps the one on page 127 is one that most readers will find familiar:

> One of the most common ways in which recursion appears in daily life is when you postpone completing a task in favor of a simpler task, often of the same type. Here is a good example. An executive has a fancy telephone and receives many calls on it. He is talking to A when B calls. To A he says, "Would you mind holding for a moment?" Of course he doesn't really care if A minds; he just pushes a button, and switches to B. Now C calls. The same deferment happens

to B. This could go on indefinitely, but let us not get too bogged down in our enthusiasm. So let's say the call with C terminates. Then our executive "pops" back up to B, and continues. Meanwhile, A is sitting at the other end of the line, drumming his fingernails against some table, and listening to some horrible Muzak piped through the phone lines to placate him. . . . Now the easiest case is if the call with B simply terminates, and the executive returns to A finally. But it *could* happen that after the conversation with B is resumed, a new caller—D—calls. B is once again pushed onto the stack of waiting callers, and D is taken care of. After D is done, back to B, then back to A. This executive is hopelessly mechanical, to be sure—but we are illustrating recursion in its most precise form.

Other examples of recursively defined mathematical functions are:

**EXAMPLE 1** (*The Factorial Function*) You are probably familiar with the *factorial* of a positive integer $n$, symbolized by $n!$ This is evaluated as $n! = n(n-1)(n-2)\cdots 3 \cdot 2 \cdot 1$. Thus $5! = 5 \cdot 4 \cdot 3 \cdot 2 \cdot 1 = 120$. This notation occurs in a variety of settings, most notably, in counting (combinatorics) and probability. For application purposes, $0!$ is defined to be 1, so that the function is defined on all nonnegative integers $\{0, 1, 2, 3, \ldots\}$.

We give here a recursive definition of the factorial function as $F(n)$ defined as follows:

$$(1) \quad F(0) = 1$$

$$(2) \quad F(n) = n \cdot F(n-1) \qquad \text{for } n > 0$$

In this case the "simplest" argument is 0 and we terminate the recursion when this argument is reached by the generating rule (2).

Let us compute $F(5)$:

$$F(5) = 5 \cdot F(4)$$
$$F(4) = 4 \cdot F(3)$$
$$F(3) = 3 \cdot F(2)$$
$$F(2) = 2 \cdot F(1)$$
$$F(1) = 1 \cdot F(0)$$
$$F(0) = 1$$

so

$$F(5) = 5 \cdot 4 \cdot 3 \cdot 2 \cdot 1 \cdot 1 = 120$$

The factorial function and some of its uses will be discussed in Chapter 8.

**EXAMPLE 2** The function $h(n)$ defined by

$$(1) \quad h(0) = 0$$

$$h(1) = 1$$

$$(2) \quad h(n) = h(n-1) + h(n-2) \qquad \text{for } n > 1$$

has two initial (terminating) conditions.

Compute $h(6)$:

$$h(6) = h(5) + h(4)$$
$$h(5) = h(4) + h(3)$$
$$h(4) = h(3) + h(2)$$
$$h(3) = h(2) + h(1)$$
$$h(2) = h(1) + h(0)$$

We can now compute $h(2)$ since we know $h(1)$ and $h(0)$ explicitly from the initial conditions

$$h(2) = 1 + 0 = 1$$
$$h(3) = 1 + 1 = 2$$
$$h(4) = 2 + 1 = 3$$
$$h(5) = 3 + 2 = 5$$
$$h(6) = 5 + 3 = 8$$

giving the sequence 0, 1, 1, 2, 3, 5, 8.

It should readily be apparent that the next number in the sequence is 13 (the sum of the two previous numbers in the sequence, $5 + 8$). The next number is $8 + 13 = 21$.

This is the Fibonacci sequence, with which some of you will be familiar.

**EXAMPLE 3** Suppose we had a dumb machine (or person or other calculating device) that only knew how to add two positive integers, $x$ and $y$. Give it two numbers as input and it would add them together right away and give the result. How could we utilize this "power" to create a multiplication function on the machine that could take two integers, $x$ and $y$, and produce the product $x \cdot y$?

We can define such a function recursively. Let us *define* multiplication by 0 (zero) directly to be 0; or as a function $M$, we will have $M(x, 0) = 0$, by definition.

As the generating rule for multiplication we can define the product $M(x, y)$ to be $M(x, y) = M(x, y - 1) + x$. Our recursive generating rule defines multiplication by $y$ in terms of the next lower $y$: $y - 1$, and uses the fact that it knows how to add.

See how it works: $M(4, 3)$ should be 12. So

$$\begin{aligned} M(4, 3) &= M(4, 2) + 4 \\ &= (M(4, 1) + 4) + 4 \\ &= ((M(4, 0) + 4) + 4) + 4 \\ &= ((0 + 4) + 4) + 4 \\ &= (4 + 4) + 4 \\ &= 8 + 4 \\ &= 12 \end{aligned}$$

as it should be.

**EXAMPLE 4** Determining the value of a dollar investment after a number of years of compounding at an interest rate of $i$ percent is a recursive procedure. This is called the future value (FV) of the investment, and after $n$ years will be symbolized as FV($n$). The value of the investment after $n$ years, FV($n$), depends strictly on its value in the *preceding* year, FV($n - 1$), and the interest rate, $i\%$, by the following generating rule:

$$\text{FV}(n) = \text{FV}(n-1) + (i\%)\text{FV}(n-1) = (1 + i\%)\text{FV}(n-1)$$

The terminating (initial) condition is the original amount of the investment at year zero: FV(0).

Thus if we invested \$1000 at 10%, we would have as the recursive function

$$\text{FV}(0) = 1000$$
$$\text{FV}(n) = \text{FV}(n-1) + (0.1)\text{FV}(n-1) = (1.1)\text{FV}(n-1)$$

Let us compute FV(3).

$$\text{FV}(3) = (1.1)\text{FV}(2)$$
$$\text{FV}(2) = (1.1)\text{FV}(1)$$
$$\text{FV}(1) = (1.1)\text{FV}(0) \quad \text{or} \quad (1.1)(1000) = 1100$$

Thus

$$\text{FV}(2) = (1.1)(1100) = 1210$$

and

$$\text{FV}(3) = (1.1)(1210) = 1331$$

## EXERCISES

1. For the recursive factorial function $F$ of Example 1, compute $F(7)$, $F(1)$, and $F(0)$.
2. For the recursive function $h$ of Example 2, compute $h(8)$.
3. Given the recursive function

$$g(0) = 0$$
$$g(n) = n + g(n-1) \qquad \text{for } n > 0$$

Compute $g(5)$.

4. Given the recursive function

$$h(0) = 10$$
$$h(n+1) = h(n) + 1 \qquad \text{for } n > 0$$

Compute $h(7)$.

5. Given the recursive function POWER designed to raise $m$ to the $n$th power

$$\text{POWER}\,(m, 0) = 1$$
$$\text{POWER}\,(m, n) = m * \text{POWER}\,(m, n - 1)$$

Use POWER to raise 2 to the fourth power. $m$ will be 2 and $n$ will be 4.

6. For the future-value function, FV, of Example 4
   (a) Compute FV(5) for FV(0) = \$1000 and $i\% = 0.10$.
   (b) Compute FV(4) for FV(0) = \$2000 and $i\% = 0.12$.

7. *Indirect recursion* is accomplished when one procedure $P$ contains a reference to another procedure $Q$, which itself contains a reference to $P$. Compute $P(4)$ and $Q(4)$ for the following indirectly recursive function.

$$P(0) = 0 \qquad\qquad Q(0) = 2$$
$$P(n) = n + Q(n - 1) \qquad Q(n) = 3 + P(n - 1)$$

8. In the first example of this section we could have expressed the generating rule as follows: Let $a_n$ be the $n$th value of the function, $a_{n-1}$ be the $(n - 1)$st value, and so on. Thus we would have the rule: $a_n = 2 \cdot a_{n-1}$. We still have the initial value $a_0 = 1$. When written in this fashion we have what is commonly known as a *recurrence relation.* Formally, a recurrence relation is a formula that expresses the $n$th term, $a_n$, of a sequence in terms of one or more of the previous values, $a_0, a_1, \ldots, a_{n-1}$. Rewrite each of the following recursive definitions as a recurrence relation.
   (a) The factorial function of Example 1.
   (b) The Fibonacci sequence of Example 2.
   (c) The future-value formula of Example 4.

## 5.6 Recursion in Programming

Recursion is a characteristic of many programming languages. It is quite a powerful technique and undoubtedly you will see examples of it later in your career. Procedures such as subroutines as well as functions can be defined recursively in that they *call themselves* either directly, or indirectly, by means of another procedure. One has to be careful in implementing such a procedure in that it must have some terminating criteria to avoid its calling itself infinitely. This could cause considerable expenditure of time and space.

Standard FORTRAN and BASIC do not support recursion, but it is permitted in ALGOL, PL/1, Pascal, APL, LISP, SNOBOL, and other languages.

To understand how it might work in a programming environment, let's consider the recursive factorial function illustrated in Example 1 of Section 5.5. The function was defined as

$$F(0) = 1$$
$$F(n) = n \cdot F(n - 1) \qquad \text{for } n > 0$$

```
procedure F(n)
  if n = 0 then return (1);
  else return (n * F(n − 1))
  end procedure;
```

**FIGURE 5.13** A recursive function for factorial.

We will implement this in Figure 5.13 (in a language similar to Pascal) for $N \geq 0$. In this function we see that if the value input to $F$ is the simplest value, namely 0, the defined value, 1, is returned to the calling program. If the value input is greater than 0, then the call to $F$ is made repeatedly (recursively) until this simplest value is reached. At this time the "unwinding" process continues until the function is evaluated.

An intuitive way to consider recursive procedures is to visualize laid out in the memory of the computer as many copies of the procedure as are needed. This is illustrated in Figure 5.14 with an input value of 4. The *actual* implementation is not as described in Figure 5.14 usually, but in future courses you will learn how it is done.

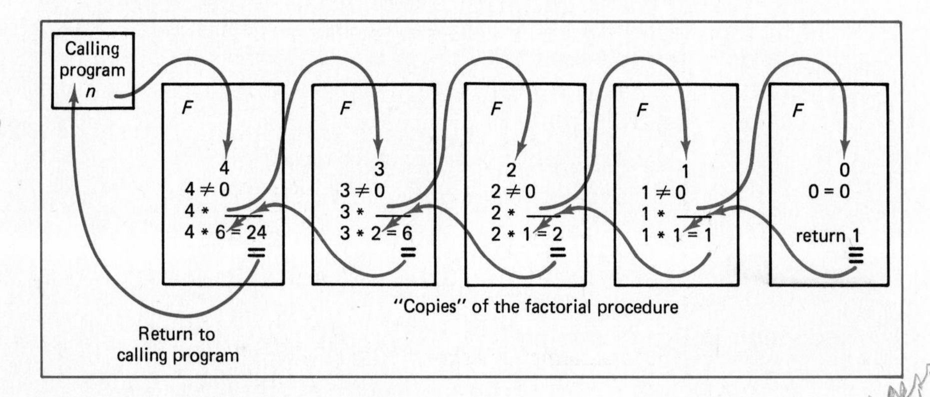


**FIGURE 5.14** "Copies" of the factorial procedure.

## 5.7 Mathematical Induction

A technique of proof that is used considerably in computer science is that of *mathematical induction.* This is a method that closely parallels recursion, and as in recursion, the process is limited to discrete values from the set $N = \{0, 1, 2, 3, \ldots\}$, the nonnegative integers. In the proof, one is to demonstrate that a given theorem or property (or proposition) of $N$ is true for all $n \in N$. This property can be designated by $P$, and to state that it is in terms of values from $N$, we state it as $P(n)$ *for all* $n \in N$.

As is recursion, mathematical induction is a two-stage process. Recursion provides, first, the definition of a function, $f$, for the *simplest* value of the domain, usually $n = 0$ or $n = 1$. Thus we have a (constant) value for $f(0)$ or $f(1)$.

Second, we are given a means of computing $f$ at $n$ [$f(n)$] when we *know* the value at $n - 1$ [$f(n - 1)$].

The method of mathematical induction provides the proof of a theorem or property whose domain is $N$ when we can do the following. First, we prove the theorem or property to be true at the *simplest value* in the domain (usually 0 or 1). That is, we prove that $P(0)$ or $P(1)$ is true, and in a sense, have a definition for this smallest value as in recursion. Second, we prove that the theorem or property is true at some arbitrary value of $n$, say $k$, or $P(k)$ *if* it is true for $P(k - 1)$, the previous value of $n$, namely $k - 1$. In this step we *assume* that $P(k - 1)$ is true and then $P(k)$ is to be proven or generated explicitly in terms of $P(k - 1)$. [Recursion also gives $f(k)$ in terms of $f(k - 1)$.]

The process suggests that after we have shown that $P$ is true for the smallest intended value, say $n = 1$, then it is shown to be true for the next larger: $n = 1 + 1 = 2$, and then for the next one: $n = 2 + 1 = 3$, and so on for all $n \in N$. It should be observed that this proof technique is inherently discrete and thus is applicable to many areas of computer science.

Formally, we state the discussion above as the

□ Principle of Mathematical Induction  Let $P(n)$ be a proposition or property defined for each $n \in N = \{0, 1, 2, 3, \ldots\}$. Then $P(n)$ is true for every $n \in N$ if

1. $P(1)$ [or $P(0)$] is true, and
2. $P(k)$ is true whenever $P(k - 1)$ is true.

$$[P(k - 1) \Rightarrow P(k).]$$

*Note:* Step 2 is sometimes stated as

$$P(k) \Rightarrow P(k + 1)$$

All either statement declares is that when $P$ is true for some $n \in N$, then it is also true for the *next* $n$.

Further discussion of the two steps above is warranted:

1. "$P(1)$ is true" is known as the *basis* of the induction. This is, of course, analogous to the *termination* value given by recursive definitions.
2. "$P(k - 1) \Rightarrow P(k)$" is known as the *inductive step*. Using the language of propositional calculus and logical implication, it means that "*if* $P(k - 1)$ is true, *then* $P(k)$ is true," or $P(k - 1) \Rightarrow P(k)$. The assumption that $P(k - 1)$ is true is known as the *inductive hypothesis* (IH).

   In using the alternative terminology, "$P(k) \Rightarrow P(k + 1)$," the assumption of the truth of $P(k)$ is the IH; and then this is used to prove $P(k + 1)$.

The analogy that best motivates the induction process is to imagine a row of dominoes placed side by side so that if one falls, it knocks over its neighbor.

The *basis step* is that the first domino in the row is knocked over. The *inductive step* is that if any one of the dominoes is knocked over, the *next one* is knocked over also. One can conclude that the property of "falling" will be true about all the dominoes when

5

1. the first one falls, and
2. if the $k$th domino falls, then the $(k + 1)$st domino will fall also, for all the dominoes.

**EXAMPLE 1** Let us illustrate this principle of induction to prove a statement made earlier (Theorem 2.5) regarding the power set of some given set. There we stated that if a set $A$ has $n$ elements, then there are $2^n$ possible subsets of $A$—the power set, $P(A)$. This theorem can be established very effectively by means of induction.

First let us make a few observations.

If set $A$ has zero elements (thus $A = \varnothing$), then there is only one element in $P(A)$, namely $\varnothing$.

If set $A$ has one element, then there are two possible subsets, or elements of $P(A)$: $A$ itself, and $\varnothing$.

We can carry this process further: If $\#(A) = 2$, then we can establish that $\#P(A) = 4$; if $\#(A) = 3$, then $\#P(A) = 8$; and so on.

We can tabulate our observations as shown in Table 5.2 and see what might be an emerging pattern. The column under $\#P(A)$ *seems to be* powers of 2, and furthermore the power used is $\#(A)$ and we are getting $2^{\#(A)}$.

Can we say that this will always be true? That is, for example, if $\#(A) = 10$, can we assume that $\#P(A) = 2^{10}$, which equals 1024? Even if it is true (by very tedious enumeration), can we generalize and state that the next term in the column will be $2^{11}$?

We will show how to use mathematical induction to prove that the pattern observed is what we think it is.

**TABLE 5.2**

| $\#(A)$ | $\#P(A)$ |
|---|---|
| 0 | 1 |
| 1 | 2 |
| 2 | 4 |
| 3 | 8 |
| 4 | 16 |

**Proof** The two stages of the proof are as follows:

**Basis Step.** Show that the property is true for $k = 0$. Here we have a set $A$ with no elements, the null set. We have seen that $\#P(A) = 2^0 = 1$. Thus the *basis* is established.

**Induction Hypothesis.**

If $\#(A) = k$, then $\#P(A) = 2^k$.

**Inductive Step.** Show that:

If $\#(A) = k + 1$, then $\#P(A) = 2^{k+1}$.

We do this as follows. We know from the IH that there are $2^k$ subsets from a $k$-element set, $A$. Let us "add" one more distinct element to $A$, obtaining a $(k + 1)$-element set. It should be evident that this new element can be excluded from the $2^k$ subsets (of the $k$-element set), giving us the *same* $2^k$ subsets, or it can be added to these $2^k$ subsets, giving us $2^k$ *more* subsets in the power set. Thus we have $2^k + 2^k = 2 \cdot 2^k = 2^{k+1}$ subsets, which is what we were to show.

Here we have shown that *if* the number of subsets of a $k$-element set is $2^k$, *then* the number of subsets of a $k + 1$-element set is $2^{k+1}$. That is, *if* the property is true for one given integral value of $n$, *then* it is true for the *next greater* one, and thus is true for all integral values, $n$.

**EXAMPLE 2** In Chapter 3 (and Chapters 1 and 2) we established the idempotence law for the Boolean "join," namely $x + x = x$. Let us use mathematical induction to show that this law generalizes to

6

$$\underbrace{x + x + \cdots + x}_{n \text{ terms}} = x.$$

**Proof** **Basis Step.** The simplest case here is $k = 2$, and we know that

$$\underbrace{x + x}_{2 \text{ terms}} = x$$

Thus the basis is established.

**Induction Hypothesis.**

$$\underbrace{x + x + \cdots + x}_{k \text{ terms}} = x$$

We are to show that

$$\underbrace{x + x + \cdots + x}_{k+1 \text{ terms}} = x.$$

**Inductive Step.** We may merely "join" one more $x$ to both sides of the IH:

$$\underbrace{x + x + \cdots + x}_{k \text{ terms}} + x = x + x$$

The left side reduces to $x + x$ by the IH and the right side is $x + x = x$. Thus the law is proved.

**EXAMPLE 3** In Example 7 of Section 5.2, we counted the number of functions from one set to another. There we stated that if a domain set $A$ had $n$ elements and the codomain set $B$ had $m$ elements, the number of possible functions was $m^n$. Let us prove this by induction on $n$, the number of domain elements.

**Proof** **Basis Step.** Let $n = 1$. Then there is only one element, $a_1 \in A$, with $m$ possible images in $B$. Thus the number of possible functions is $m = m^1$. Thus the basis is established (see Figure 5.15).

**Induction Hypothesis.** Let $n = k$. We assume that there are $m^k$ functions from $A$ to $B$ (see Figure 5.15).

**Inductive Step.** Let us add one more element $a_{k+1}$, to the domain. We show there are $m^{k+1}$ functions from $A$ to $B$. Now, for each of the $m^k$ functions (from the IH) the new domain element, $a_{k+1}$, can be mapped to $m$ images, yielding $m \cdot m^k = m^{k+1}$ functions, thus proving the theorem.

**EXAMPLE 4** It can be observed that if we add together a series of consecutive odd numbers, starting with 1, as follows:

$$1 = 1$$

$$1 + 3 = 4$$

$$1 + 3 + 5 = 9$$

$$1 + 3 + 5 + 7 = 16$$

$$1 + 3 + 5 + 7 + 9 = 25$$

an interesting pattern begins to emerge. The sums obtained *seem to be* consecutive *perfect squares* ($n^2$). Furthermore, if the number of addends on the left of each equation above is $M$, then the sum is $M^2$. Can we say that this will always be true? That is, for example, does the sum of the first 15 consecutive odd numbers equal $15^2$, or 225? We easily check this arithmetic to find out. But what about the sum of the first 25,000 consecutive odd numbers? Is the sum $(25{,}000)^2$? Even if it is, can we say anything in general, such as "the sum of the first 25,001 odd numbers is $(25{,}001)^2$" without verifying it?

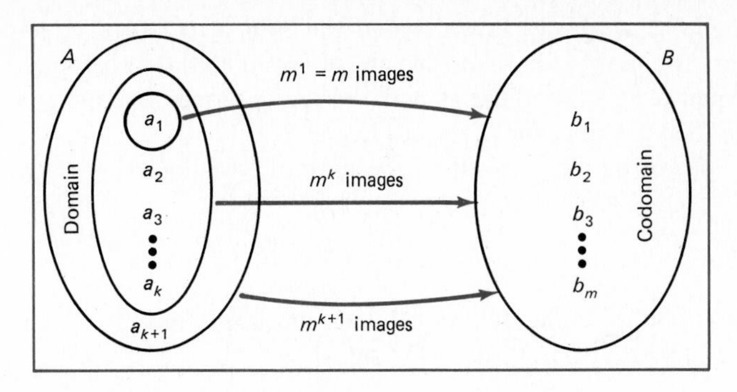


**FIGURE 5.15**

**Proof** We use the technique of mathematical induction to prove such a statement. We know that odd numbers can be written as $2k - 1$ for $k \in \{1, 2, 3, \ldots\}$, so that the odd numbers are:

| *Integers, k* | 1 | 2 | 3 | 4 | 5 | ⋯ | $k$ |
|---|---|---|---|---|---|---|---|
| *Odd Numbers,* $2k - 1$ | 1 | 3 | 5 | 7 | 9 | ⋯ | $2k - 1$ |

What we have observed above and would like to show by mathematical induction is that "the sum of the first $k$ odd numbers is equal to $k^2$." This is the proposition $P$ on $N$ that we wish to prove.

The two stages of the proof are as follows.

**Basis Step.** Show that $P(1)$ is true.

Let $k = 1$. Then $2k - 1 = 1$, which $= 1^2$. Therefore, $P(1)$ is true and the *basis* is established.

**Induction Hypothesis.**

$$1 + 3 + 5 + \cdots + (2k - 1) = k^2$$

**Inductive Step.** Show that

$$1 + 3 + 5 + \cdots + (2k - 1) + \underbrace{(2k + 1)}_{(k+1)\text{st odd}} = (k + 1)^2$$

To do this we assume the IH to be true and then add the next odd integer $(2k + 1)$ to both sides:

$$\begin{array}{rl} \text{IH:} & 1 + 3 + 5 + \cdots + (2k - 1) = k^2 \\ \text{Add } 2k + 1\text{:} & \qquad\qquad + (2k + 1) \qquad + (2k + 1) \\ \hline & 1 + 3 + 5 + \cdots + (2k - 1) + (2k + 1) = k^2 + (2k + 1) \end{array}$$

The right side reduces: $k^2 + 2k + 1 = (k + 1)^2$

Thus: $1 + 3 + 5 + \cdots + (2k + 1) = (k + 1)^2$

which is what we were to show.

Here we have shown that *if* the sum of the first $k$ consecutive odd integers is equal to $k^2$, then the sum of the first $(k + 1)$ consecutive odd integers is equal to $(k + 1)^2$. That is, *if* the property is true for one given integral value of $k$, then it is true for the *next greater* one.

By using the $\sum$ notation, we have proved that

$$\sum_{k=1}^{n} (2k - 1) = n^2$$

The symbol $\sum$ is merely a shorthand notation to indicate a *sum*. It is the capital Greek letter sigma ($\sum$, or S, for sum).

**EXAMPLE 5** Prove by mathematical induction that $n^2 > 2n + 1$ for all integers greater than 2.

**Proof**

**Basis Step.** Since the smallest integral value of concern here is 3, the basis will be established for $k = 3$. The left side is $3^2 = 9$, and this is indeed greater than the right side, $2 \cdot 3 + 1 = 7$, thus establishing the basis.

**Induction Hypothesis.** $k^2 > 2k + 1$ for $k > 2$.

**Inductive Step.** Show that $(k + 1)^2 > 2(k + 1) + 1 = 2k + 3$. Now $(k + 1)^2 = k^2 + 2k + 1$, which is greater than $(2k + 1) + (2k + 1)$ by the IH. This, in turn, equals $2k + 2k + 2$, which is greater than $2k + 3$, which is the desired result. This sequence of equalities and inequalities is as follows:

$$(k + 1)^2 = \underbrace{k^2 + 2k + 1 > 2k + 1}_{\text{IH}} + 2k + 1 = 2k + \underbrace{2k + 2 > 2k + 3}_{2k + 2 > 3}$$

It should be stressed that *both* parts of the argument must be established by mathematical induction, or we have proved nothing.

To illustrate, it is possible to assume a false proposition to be true for $n = k$ and then show it also to be true for $n = k + 1$, but to find *no* case where it is true. Consider the *false* proposition on real numbers (which is not true for any $n$)

$$n = n + 1$$

Now use this as the inductive hypothesis; so IH: $n = n + 1$. We are to prove that it is true for the next $n$, $n + 1$. To this end, substitute $n + 1$ for $n$ in the IH, getting

$$n + 1 = (n + 1) + 1$$

Now the IH is that $n = n + 1$, so add 1 to both sides: $n + 1 = n + 2$, which is what we are to prove.

It is also possible to have an expression that is true for $n = 1$, but to have no way to prove the induction step. As an example, the expression $n^2 - n + 41$ is equal to 41, a prime number, when $n = 1$.

When $n = 2$ we have $2^2 - 2 + 41 = 43$, a prime.

When $n = 3$ we have $3^2 - 3 + 41 = 47$, a prime.

We may continue this process for quite a few numbers and will get primes. So we may make the generalization that $n^2 - n + 41$ will always yield a prime number. Primes will be generated for $n = 1, 2, \ldots, 40$. But for $n = 41$ we have $41^2 - 41 + 41 = 41^2$, which is obviously not a prime.

## EXERCISES

1. By mathematical induction prove the following.

   (a) The generalized idempotence law for the Boolean "meet":

   $$\underbrace{x \cdot x \cdot x \cdots x}_{n \text{ terms}} = x$$

   (b) The generalized De Morgan's laws

   (i) $\overline{x_1 x_2 x_3 \cdots x_n} = \overline{x_1} + \overline{x_2} + \overline{x_3} + \cdots + \overline{x_n}$

   (ii) $\overline{x_1 + x_2 + x_3 + \cdots + x_n} = \overline{x_1}\,\overline{x_2}\,\overline{x_3} \cdots \overline{x_n}$

2. Prove the algebraic rule for exponentiation:

   $$(ab)^n = a^n b^n$$

   by mathematical induction. The basis step is $(ab)^0 = a^0 b^0$, which is true since $1 = 1 \cdot 1$.

3. (a) Suppose that a fruit fly colony doubles every day. If we start with an initial population of 100 flies (days = 0), show by mathematical induction that after $n$ days there will be $100 \cdot 2^n$ flies in the colony.

   (b) Suppose that the colony triples instead of doubles every day. Find how many flies an original colony of 100 will become after $n$ days.

4. Prove the following by mathematical induction:

   (a) That $n^2 + n$ is always even for all positive integers $n$. The IH is that $k^2 + k$ is even and you are to show that $(k + 1)^2 + (k + 1)$ is even.

   (b) That $n^3 + 5n$ is divisible by 6 for all positive integers $n$. Use the result of part (a).

   (c) That $2^n > n$ for all positive integers. The IH is that $2^k > k$ and the inductive step $2^{k+1} > k + 1$ follows from the IH.

   (d) That $2^n > n^2$ for all integers greater than 4. Here the basis is $k = 5$. Use the result of Example 5 to show that $2^{k+1} > (k + 1)^2$, which $= k^2 + 2k + 1$.

5. Prove by mathematical induction:

   (a) $1 + 2 + 3 + \cdots + n = \frac{1}{2}(n)(n + 1)$, or

   $$\sum_{k=1}^{n} k = \tfrac{1}{2}n(n + 1)$$

   This equality states that the sum of the first $n$ positive integers is equal to $\frac{1}{2}(n)(n + 1)$.

(b) $2 + 4 + 6 + \cdots + 2n = n(n + 1)$, or

$$\sum_{k=1}^{n} 2k = n(n + 1)$$

(c) $4 + 8 + 12 + \cdots + 4n = 2n(n + 1)$, or

$$\sum_{k=1}^{n} 4k = 2n(n + 1)$$

(d) $2^0 + 2^1 + 2^2 + \cdots + 2^n = 2^{n+1} - 1$

## Summary and Selected References

The subject of functions underlies all of mathematics and readers are referred to practically any mathematics textbook. In this chapter we stressed the modern notation of functions and the concept that prefix and postfix notation, which are so prevalent in computer science, can be viewed as functions of two variables. Especially good treatments of discrete functions as they relate to computer science can be found in Stanat and McAllister (1977), Stone (1973), and Sahni (1981). An elementary treatment is given in Kapps and Bergman (1975).

Recursive functions and recursion in general tend to be one of the most powerful tools of computer scientists, and the intricacies and delicacies are often poorly understood by beginners in the field. The subject is definitively portrayed in the remarkable book by Hofstadter (1979).

Mathematical induction is covered in a variety of introductory mathematics courses, often with poor results. It is the most important method of proof in discrete systems and thus should be understood thoroughly. The relationships between induction and recursion and their importance to computer science are covered in Stanat and McAllister (1977), Wand (1980), and Beckman (1980).

# Relations and Their Graphs

# 6

## 6.1 Introduction

As we saw in Chapter 2, an appealing way to visualize abstract relations is pictorially. There we examined a particular relation, "is adjacent to," on the set of states $S = \{\text{Missouri, Illinois, Arkansas, Indiana, Kentucky}\}$. The diagram of the relation as shown on a portion of the U.S. map in Figure 2.19(b) is repeated here as Figure 6.1.

The information obtained from the map about the relation "is adjacent to" can also be obtained from a diagram, where the states are represented by points or small circles which are joined to others by line segments or curves if they are adjacent. The points in Figure 6.2 are placed so that they are oriented geographically. The same information about adjacencies can be obtained from different orientations as shown in Figure 6.3. States adjacent to one another in Figure 6.2 are still adjacent in Figure 6.3.

Thus the *relation represented* is what is important, not the physical or geometric location of the constituent parts of the structure. We state that the structures are equivalent (or isomorphic). We will have more to say about equivalence after a few terms have been introduced.

There are many occasions to represent a structure pictorially as we have done here. We are often confronted with diagrams showing connecting points, city maps showing connections by streets, state maps showing connections between cities and towns, and various other networks and charts. All these diagrams in one way or another show relations and connections among constituent elements of various discrete structures.

| | Missouri | Illinois | Arkansas | Indiana | Kentucky |
|---|---|---|---|---|---|
| Missouri | | × | × | | × |
| Illinois | × | | | × | × |
| Arkansas | × | | | | |
| Indiana | | × | | | × |
| Kentucky | × | × | | × | |

**FIGURE 6.1** Relation "is adjacent to" on the map of Figure 2.19(b).

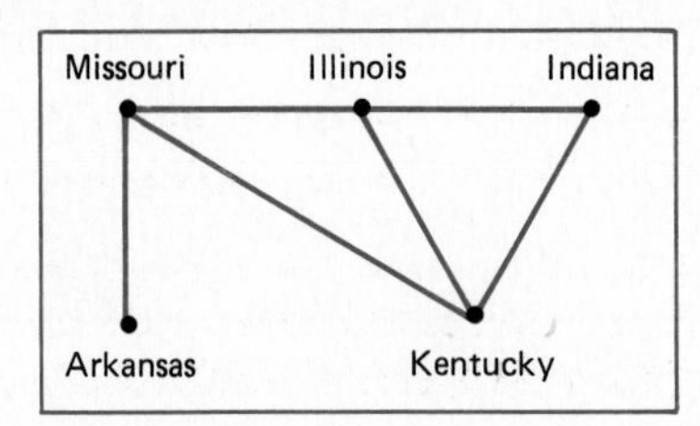


**FIGURE 6.2**

When we utilize such structures in an attempt to clarify or simplify matters, it gives rise to the mathematical structure known as a *graph*. This is not a graph in the same sense as that used in algebra or analytic geometry, in which you plotted lines and curves on a coordinate system. Rather, the graphs we will study here are simple geometric figures consisting of points and connecting lines or curves.

Theoretically, graphs are composed of two sets. One set has as its members the distinct elements of the structures, such as the states in the figures above. Each element of this set is referred to as a *vertex* (plural, *vertices*), a *point*, or a *node* of the graph. The variety of terminology is due to the fact that the theory eminates from several sources, each with its own nomenclature. The number of vertices of a graph will be called the *order* of the graph. We will use $V = \{v_1, v_2, \ldots, v_n\}$ to represent a set of vertices of order $n$. In the text we consider finite graphs only.

**FIGURE 6.3** Different orientations of Figure 6.2.

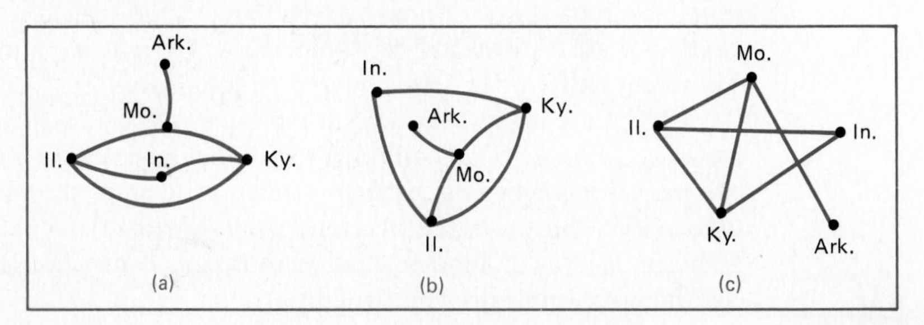

The other set indicates the relation present among the vertices and is shown by sets of unordered pairs of vertices. Thus the neighboring states can be identified by, for example, (Missouri, Arkansas), which represents the same pair as (Arkansas, Missouri). These connections are referred to variously as *edges, lines,* and *arcs*. The number of edges present in a graph will be called the *size* of the graph. Thus $E = \{e_1, e_2, \ldots, e_k\}$ is an indication of the edges of a graph whose size is $k$. Each $e_i \in E$ is a pair $(v_l, v_m)$ indicating that the vertex $v_l$ is related to $v_m$ in a symmetrical fashion; that is, this is the same pair as $(v_m, v_l)$. (Recall that a symmetric relation $R$ is such that, for elements $a$ and $b$, if $a R b$, then $b R a$.)

For our map of states above, we have a graph of order 5 and size 6 and

$V = \{$Missouri, Illinois, Arkansas, Indiana, Kentucky$\}$

$E = \{$(Missouri, Illinois), (Missouri, Arkansas), (Missouri, Kentucky) (Illinois, Indiana), Indiana, Kentucky), (Illinois, Kentucky)$\}$

It should be noted that relations represented by graphs are, in general, not transitive: for Kentucky "is adjacent to" Missouri and Missouri "is adjacent to" Arkansas, but Kentucky "is *not* adjacent to" Arkansas; nor are they either antisymmetic or reflexive. We will see later that reflexivity may be possible; however, when this is the case, the structure is not technically a graph.

Graphs are important mathematical structures of a *discrete* nature. Each vertex represents a distinct unit. The edges tell the story of the underlying structure. As such, graphs are used as mathematical models for a variety of situations. The theory of graphs is important not only as a mathematical topic but also as an integral part of computer science.

As you will learn in other courses, graphs are utilized by computer scientists as an aid in storing and searching data and in aspects of machine theory. Additionally, the theory is relevant in that many problems to be solved by computers can be phrased in terms of graph theory. We will examine both of these aspects later in this chapter and in Chapter 7.

For now, let us examine closely the nature of graphs through definitions of important concepts and establish some theorems for later use.

## 6.2 Graph Definitions

A graph with its sets of vertices and edges, $V$ and $E$, respectively, will be designated by $G(V, E)$, or just by $G$ if there is no ambiguity.

Two vertices, $v_1$ and $v_2 \in V$, are said to be *adjacent* if there is an edge $(v_1, v_2) \in E$.

If a vertex, $v$, is the end point of an edge, $e$, we say that $e$ is *incident* with $v$ (also, $v$ is incident with $e$).

The number of edges incident with a vertex, $v$, is known as the *degree* of $v$, denoted by $\deg(v)$. A vertex is said to be even or odd depending on whether its degree is an even or an odd number.

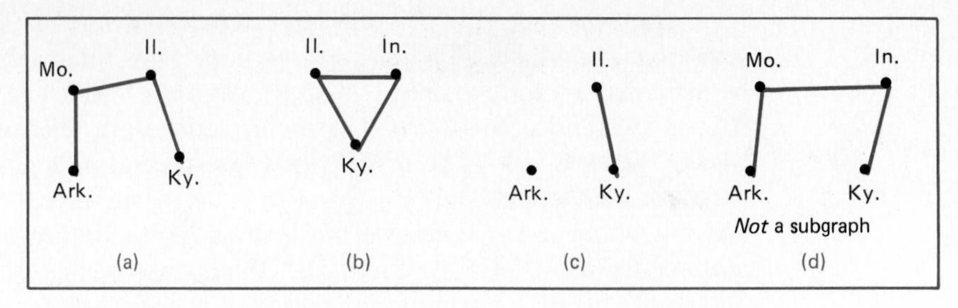


**FIGURE 6.4** Subgraphs of Figure 6.2.

In Figure 6.2 above (or Figure 6.3) we see that

$$\deg(\text{Missouri}) = \deg(\text{Illinois}) = \deg(\text{Kentucky}) = 3$$

$$\deg(\text{Indiana}) = 2$$

$$\deg(\text{Arkansas}) = 1$$

If a graph $G_1(V_1, E_1)$ is such that there is another graph $G_2(V_2, E_2)$ and $V_1 \subseteq V_2$ and $E_1 \subseteq E_2$ where every edge in $E_1$ is incident with vertices in $V_1$, we say that $G_1$ is a *subgraph* of $G_2$ generated by the vertices of $V_1$. Three subgraphs of our graph of adjacent states are shown in Figure 6.4(a) to (c). Figure 6.4(d) is not a subgraph because the edge (Missouri, Indiana) is not an edge of the original graph.

If we allow for multiple edges and reflexivity as shown in the graph of Figure 6.5, we have what is known as a *multigraph.* Here we see that edges $e_1$ and $e_2$ both join vertices $v_1$ and $v_2$. We also have an edge, $e_5$, that joins vertex $v_4$ to itself (reflexive). Edge $e_5$ is referred to as a *loop.*

The vertex $v_7$ has no edges incident with it. This is perfectly permissible in a graph. (We would still have a graph of a relation with our adjacency structure of states above even if California were added to the set of states.) A vertex incident with no edges is known as an *isolated point.* It has degree zero.

Further, in Figure 6.5 we observe that there is no way to "get to" $v_5$ from, say, $v_3$, or from $v_7$ by following the edges of the graph. We say that this structure, then, is not *connected.* We will define this term precisely below.

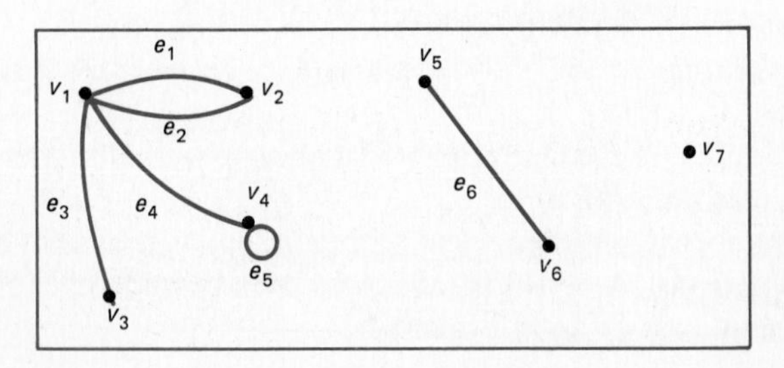


**FIGURE 6.5** Disconnected multigraph.

The multiple edges and loop at $v_4$ in Figure 6.5 contribute to the degrees of the vertices. In the multigraph

$$\deg(v_1) = 4$$
$$\deg(v_4) = 3$$
$$\deg(v_2) = 2$$
$$\deg(v_3) = \deg(v_5) = \deg(v_6) = 1$$
$$\deg(v_7) = 0$$

With this very rudimentary beginning and with the few concepts established so far, we are already able to prove two theorems. This tends to illustrate the depth of the theory. Not many concepts are needed to establish some rather far-reaching theories about graphs in general.

□ Theorem 6.1 The sum of the degrees of all the vertices of a graph $G(V, E)$ equals twice the number of edges.

**Proof** Let $V = \{v_1, v_2, \ldots, v_n\}$. When computing the sum of the degrees $\sum_{i=1}^{n} \deg(v_i)$, each edge in $E$ is counted twice, once for each of the two vertices incident to the edge. Thus this sum of degrees is twice the number of edges.

In the graph of Figure 6.5 we note that the sum of all the degrees of the vertices is $4 + 3 + 2 + 1 + 1 + 1 + 0 = 12$ and there are six edges. Therefore, we see that Theorem 6.1 applies to multigraphs as well as graphs.

□ Theorem 6.2 If a graph $G(V, E)$ is finite (i.e., has a finite number of vertices), the number of odd vertices is even.

**Proof** The sum of all the degrees of $V$ is even from Theorem 6.1. (This sum is 2 times the number of edges, therefore is even.) Let this sum be $2k$, where $k$ is the number of edges in $E$.

Let $V_0$ be the set of even vertices, and $V_1$ be the set of odd vertices. We are to show that the number of vertices in $V_1$ is even.

Now the sum of the degrees of $V_0$ is even, since the sum of any number of even numbers is even. Let this sum be $2m$. Thus the sum, $p$, of the degrees of the vertices in $V_1$ must be

$$p = 2k - 2m = 2(k - m)$$

which is even.

Adjacency Matrices One descriptive way to display a graph $G(V, E)$ is by its *adjacency matrix* $M$ (see Appendix B). This will be an $n \times n$ matrix where $n$ is the order of $G$. The rows and columns of $M$ are labeled by the vertices of $V$ and

$$M_{ij} = M_{ji} = \begin{cases} 1, & \text{iff } v_i \text{ is adjacent to } v_j \\ 0, & \text{otherwise} \end{cases}$$

| | Missouri | Illinois | Arkansas | Indiana | Kentucky |
|---|---|---|---|---|---|
| Missouri | 0 | 1 | 1 | 0 | 1 |
| Illinois | 1 | 0 | 0 | 1 | 1 |
| Arkansas | 1 | 0 | 0 | 0 | 0 |
| Indiana | 0 | 1 | 0 | 0 | 1 |
| Kentucky | 1 | 1 | 0 | 1 | 0 |

**FIGURE 6.6** Adjacency matrix for graph of Figure 6.2.

The adjacency matrix for the graph at Figures 6.2 and 6.3 is shown in Figure 6.6. The fact that it is symmetric should be obvious.

The adjacency matrix $M$ for a multigraph has as its $M_{ij}$ entries the number of edges that join the vertices $i$ and $j$. The adjacency matrix for a *graph* is seen to be a special case of this, for no more than one edge joins any two vertices; hence the entries are 0 or 1. The matrix $M$ for the multigraph at Figure 6.5 is shown in Figure 6.7.

Notice that if you are given an adjacency matrix of a graph $G$, it is easy to determine the degree of vertex, $v_i$, by simply adding the elements in either the $i$th row or column since the matrix is symmetric.

The (4, 4) element of Figure 6.7 is worthy of note. Every edge will be counted twice, once when it "leaves" a vertex and once when it "enters" one. In this case, edge $e_5$ enters and leaves the same vertex; hence it must be counted twice.

Isomorphism Earlier we mentioned the equivalence of graph structures. Let us examine this concept briefly. We say that two graphs are equivalent if there is a bijective (1–1 and onto) function that maps the vertices of one of the graphs onto the vertices of the other one such that corresponding edges can be mapped bijectively also. That is, equivalence implies that a pair of vertices are adjacent in one of the graphs iff the corresponding pair of vertices is adjacent in the other graph. This bijective correspondence is known as an *isomorphism* and the two graphs are said to be *isomorphic*.

To be isomorphic, two graphs must, of course, be of the same order and size. But this is not sufficient. Additionally, to name a few other properties

**FIGURE 6.7** Adjacency matrix for the multigraph of Figure 6.5.

| | $v_1$ | $v_2$ | $v_3$ | $v_4$ | $v_5$ | $v_6$ | $v_7$ |
|---|---|---|---|---|---|---|---|
| $v_1$ | 0 | 2 | 1 | 1 | 0 | 0 | 0 |
| $v_2$ | 2 | 0 | 0 | 0 | 0 | 0 | 0 |
| $v_3$ | 1 | 0 | 0 | 0 | 0 | 0 | 0 |
| $v_4$ | 1 | 0 | 0 | 2 | 0 | 0 | 0 |
| $v_5$ | 0 | 0 | 0 | 0 | 0 | 1 | 0 |
| $v_6$ | 0 | 0 | 0 | 0 | 1 | 0 | 0 |
| $v_7$ | 0 | 0 | 0 | 0 | 0 | 0 | 0 |

of isomorphism, corresponding vertices have the same degree, corresponding subgraphs must be isomorphic, and the adjacency matrices of two isomorphic graphs are identical (after a possible rearrangement of the rows and columns).

We will not delve any deeper here into the problem of isomorphism except to state that determining whether two graphs are isomorphic can be a tedious and time-consuming project, especially if the orders and sizes are relatively large. The problem is explored further in Exercises 10 and 11.

## EXERCISES

1. For graphs $H$ and $J$:

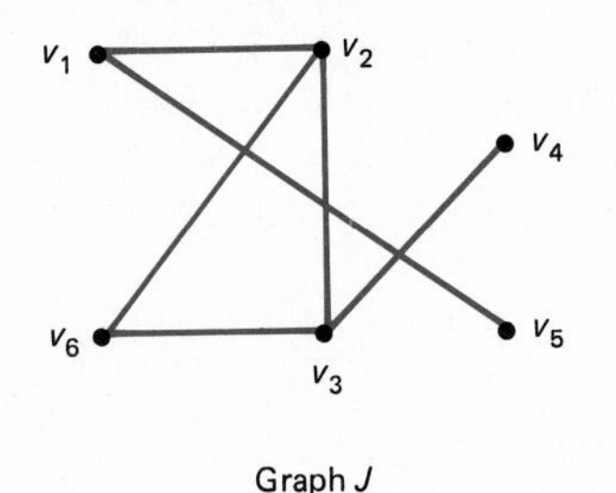


Graph $J$

Graph $H$

   (a) Draw three other orientations of each graph.
   (b) Give the degree of each vertex of the graphs.
   (c) State the orders and sizes of the graphs.
   (d) Show that Theorems 6.1 and 6.2 are true about the graphs.
   (e) Construct adjacency matrices for the graphs.
   (f) Display four subgraphs of each of the graphs.

2. Repeat Exercise 1 for the multigraph K.

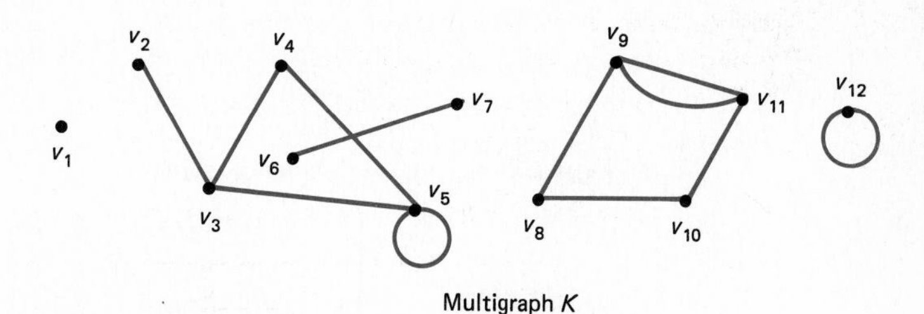


Multigraph $K$

3. Give (draw) examples, if possible, of graphs having
   (a) All odd vertices.
   (b) All even vertices.
   (c) Exactly two even vertices.
   (d) Exactly two odd vertices.

4. Can the adjacency matrix $M$ of a graph have a 1 in any $(i, i)$ position, that is, on the main diagonal? What about a multigraph?

5. Another matrix representation of a graph is known as its *incidence matrix*. Let a graph have order $n$ and size $m$. Then the incidence matrix, $I$, is an $n \times m$ matrix. The rows of $I$ will represent the vertices and the columns will represent edges. The $(i, j)$ entries will be such that

$$I_{ij} = \begin{cases} 1, & \text{if vertex } i \text{ is incident with edge } j \\ 0, & \text{otherwise} \end{cases}$$

The incidence matrix of the graph $G$:

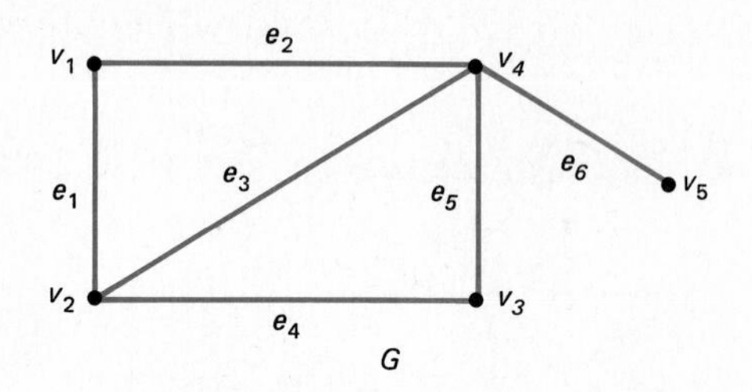


will be

$$I = \begin{array}{c} \\ v_1 \\ v_2 \\ v_3 \\ v_4 \\ v_5 \end{array} \begin{array}{|cccccc|} e_1 & e_2 & e_3 & e_4 & e_5 & e_6 \\ \hline 1 & 1 & 0 & 0 & 0 & 0 \\ 1 & 0 & 1 & 1 & 0 & 0 \\ 0 & 0 & 0 & 1 & 1 & 0 \\ 0 & 1 & 1 & 0 & 1 & 1 \\ 0 & 0 & 0 & 0 & 0 & 1 \\ \hline \end{array}$$

Construct incidence matrices of the graphs $H$, $J$, and $K$ of Exercises 1 and 2. Are incidence matrices symmetric?

6. Draw the graph (multigraphs) represented by the adjacency matrices

(a) $\begin{bmatrix} 0 & 1 & 1 & 0 & 1 & 0 \\ 1 & 0 & 0 & 0 & 1 & 1 \\ 1 & 0 & 0 & 1 & 0 & 1 \\ 0 & 0 & 1 & 0 & 0 & 0 \\ 1 & 1 & 0 & 0 & 0 & 1 \\ 0 & 1 & 1 & 0 & 1 & 0 \end{bmatrix}$  (b) $\begin{bmatrix} 0 & 1 & 0 & 0 & 1 \\ 1 & 0 & 0 & 0 & 1 \\ 0 & 0 & 0 & 1 & 0 \\ 0 & 0 & 1 & 0 & 0 \\ 1 & 1 & 0 & 0 & 2 \end{bmatrix}$

7. Draw the graph having the incidence matrix

$$\begin{bmatrix} 1 & 1 & 0 & 1 & 0 & 0 & 0 & 0 \\ 0 & 1 & 1 & 0 & 1 & 0 & 0 & 0 \\ 0 & 0 & 0 & 0 & 0 & 1 & 0 & 1 \\ 0 & 0 & 0 & 0 & 1 & 1 & 1 & 0 \\ 0 & 0 & 0 & 1 & 0 & 0 & 1 & 1 \\ 1 & 0 & 1 & 0 & 0 & 0 & 0 & 0 \end{bmatrix}$$

8. The *complement* $\bar{G}$ of a graph $G = (V, E)$ is a graph having the same vertices as $G$, but with the property that two distinct vertices of $\bar{G}$ are adjacent iff the same two vertices of $G$ are *not* adjacent. We will designate $\bar{G}$ as the pair $(V, \bar{E})$.

(a) Draw the complement of the graph of Figure 6.2.

(b) Draw the complement of graph $H$ of Exercise 1.

(c) For a graph $G = (V, E)$ whose order is $n$, show that the number of edges in $E \cup \bar{E}$ is $n(n-1)/2$.

9. The *union* of two graphs, $G_1 = (V_1, E_1)$ and $G_2 = (V_2, E_2)$, is another graph, $G_3 = (V_3, E_3)$, where the new vertex set $V_3 = V_1 \cup V_2$, and the new edge set $E_3 = E_1 \cup E_2$. Similarly, the *intersection* of the graphs $G_1$ and $G_2$ is a graph $G_4 = (V_4, E_4)$ consisting of only those vertices and edges that are common to $G_1$ and $G_2$; that is, $V_4 = V_1 \cap V_2$ and $E_4 = E_1 \cap E_2$. Find the union and intersection of the graphs $G_1$ and $G_2$.

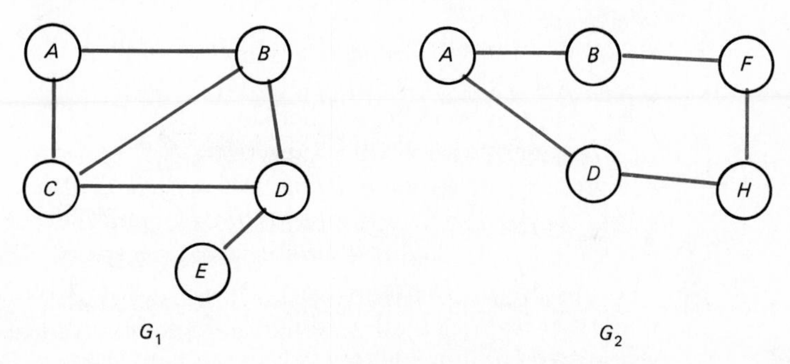


10. Graphs (a) and (b) below each have the same order (6) and the same size (9), and each vertex is of degree 3. Yet they are *not* isomorphic. Explain why.

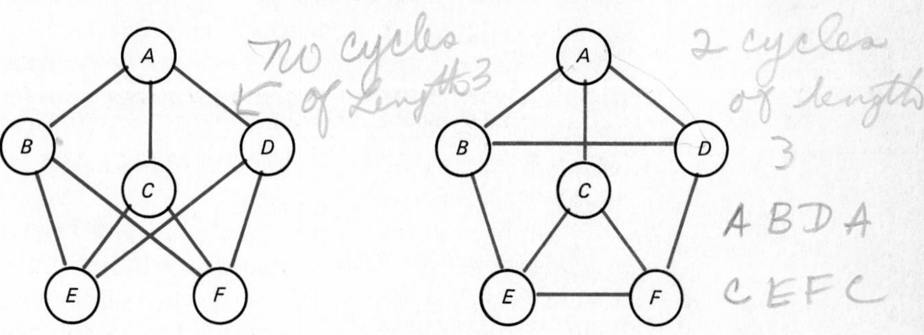


11. One approach that can be used to determine isomorphism between two graphs is by examining their complements (see Exercise 8). Two graphs $G_1$ and $G_2$ will be isomorphic iff their complements, $\bar{G}_1$ and $\bar{G}_2$, are isomorphic. Construct the complements of the two graphs of Exercise 10 and explain why they are not isomorphic.

## 6.3 Walks and Connectivity

Any alternating sequence of vertices and incident edges in a graph (multigraph) is known as a *walk*. A walk begins and ends with vertices. The listing can be done by alternating vertices and edges or just by listing the sequence of adjacent

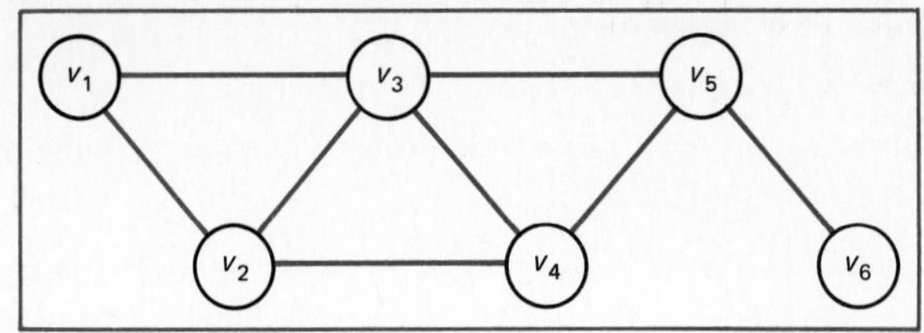


**FIGURE 6.8**

vertices. The *length* of the walk is the number of edges. A walk is said to be *closed* if the beginning and ending vertices are the same. A walk of length $n$ is said to be an *n-walk.*

For example, $v_2v_3v_4v_5v_4v_2v_1$ is a walk in the graph of Figure 6.8. Its length is 6. Observe that the edge $(v_4, v_5)$ appears twice in the walk. The walk $v_5v_4v_2v_3v_4v_5$ is a closed walk of length 5.

A *path* is a walk with *distinct* vertices. For example, $v_1v_2v_3v_4v_5$ is a path in Figure 6.8. The theory permits *closed paths* also, which allows for the beginning and ending vertices to be the only nondistinct vertices. Thus $v_2v_3v_5v_4v_2$ is a closed path. A closed walk with distinct edges is also known as a *cycle.*

A walk with distinct edges is termed a *trail.* Even though the edges are the important items here, the listing can be done by vertices. Thus $v_1v_2v_3v_4v_5v_3v_1$ is a (closed) trail in Figure 6.8 even though vertex $v_3$ appears twice.

In the multigraph of Figure 6.5 we noted that there was no way to "get to" $v_5$ from $v_3$. Let us state this fact now more formally. Indeed, there is *no path* in the graph from $v_5$ to $v_3$. This idea leads to a very important graph concept, that of connectedness. A graph (multigraph) is said to be *connected* iff there is a path (or walk) between every pair of vertices. A vertex is said to be *reachable* from another vertex iff there is a path joining the two vertices. Thus we see formally that the multigraph of Figure 6.5 is *disconnected,* whereas the graph at Figure 6.8 is connected.

On occasion it is instructive to "remove" from a graph $G$ a vertex $v$ together with all its incident edges. The resulting graph, $G'$, is noted by $G - \{v\}$. Figure 6.9 shows several of these deletions from Figure 6.8. These vertex deletions drastically alter the structure of the graph. If the structure is altered in deleting a vertex $v$ so that an originally connected graph becomes disconnected, the vertex $v$ is known as a *cut point.* From Figure 6.9(c), we see that $v_5$ is a cut point of the graph of Figure 6.8. In the graph of states shown in Figure 6.2, the state (vertex) Missouri is the only cut point.

An analogous situation occurs when deleting edges. When an edge, $e$, is deleted from a graph $G$, we have $G - \{e\}$. Notice, however (Figure 6.10), that we do not delete any incident vertices. Any edge which upon deletion disconnects a connected graph is known as a *bridge.* The edge $(v_5, v_6)$ is a bridge.

A *component* (sometimes connected component) of a graph $G(V, E)$ is a connected subgraph $G'(V', E')$ of $G$ such that no vertex in $V'$ is connected in $G$ to any vertex outside $V'$. Another way to state this is: A connected subgraph $G'$ of a graph $G$ is a *component* of $G$ if it is not contained in any larger (order greater than that of $G'$) connected subgraph of $G$. For example, the graph of Figure

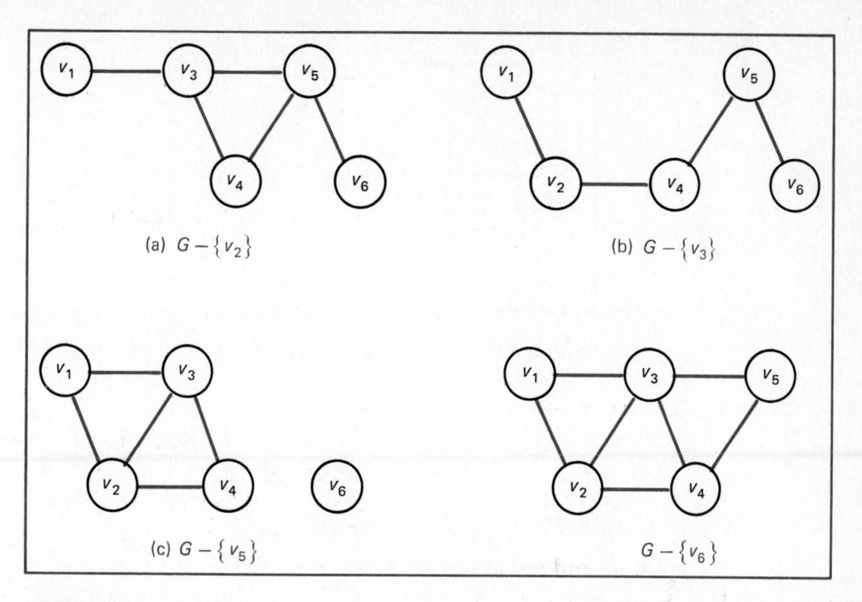


**FIGURE 6.9**

6.11 has three components. Figures 6.9(c) and 6.10(b) each have two components, whereas Figure 6.2 has one component. This concept also holds for multigraphs. It should be clear that any graph (multigraph) is connected if and only if it consists of only one component.

Walk and Connection Matrices Adjacency matrices are very useful for answering questions concerning walks and connectivity. Consider the graph and its adjacency matrix, $M$, in Figure 6.12. As stated previously, each 1 in the

**FIGURE 6.10**

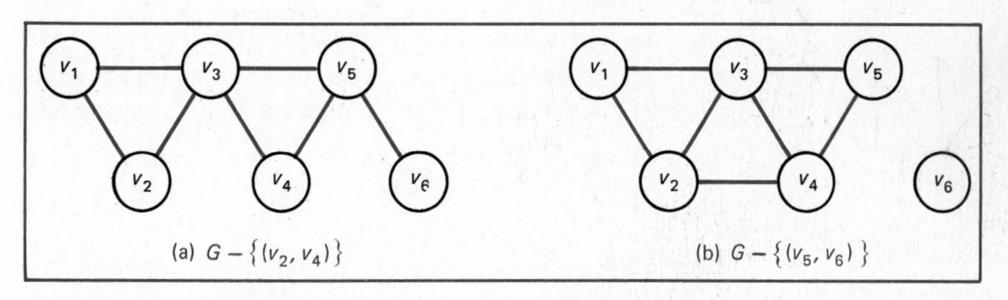


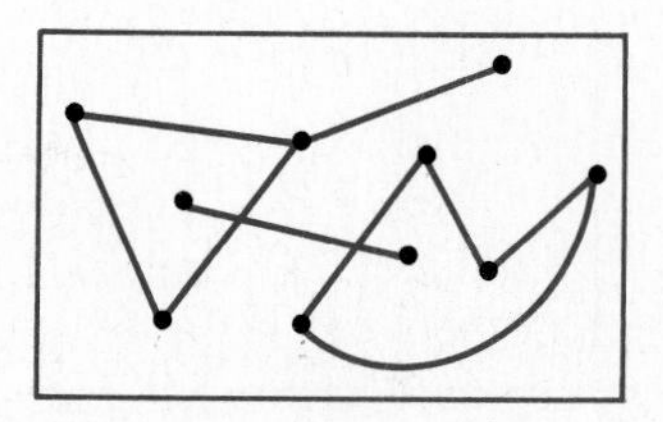

**FIGURE 6.11** Graph with three components.

(a)

| | A | B | C | D | E |
|---|---|---|---|---|---|
| A | 0 | 1 | 1 | 0 | 0 |
| B | 1 | 0 | 1 | 1 | 0 |
| M = C | 1 | 1 | 0 | 1 | 1 |
| D | 0 | 1 | 1 | 0 | 0 |
| E | 0 | 0 | 1 | 0 | 0 |

(b)

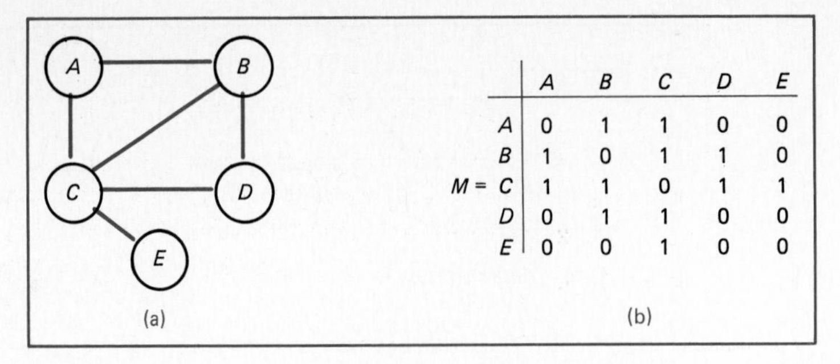

**FIGURE 6.12** Graph and its adjacency matrix.

matrix indicates that the corresponding vertices are adjacent. Another way to state this is: There is a 1-walk between the vertices. If the matrix is squared (by usual matrix multiplication) as in Figure 6.13, the elements obtained will be the number of 2-walks between the vertices.

For instance, the (1, 4) position of $M^2$ of Figure 6.13 has the value 2. This means that there are two 2-walks between vertices $A$ and $D$. They are seen from the graph: $A$–$B$–$D$ and $A$–$C$–$D$. There are no 2-walks between vertices $E$ and $C$.

The entries along the main diagonal of $M^2$

$$\begin{bmatrix} 2 & & & & \\ & 3 & & & \\ & & 4 & & \\ & & & 2 & \\ & & & & 1 \end{bmatrix}$$

indicate two pieces of information: (1) they represent the degrees of the vertices and (2) they represent the number of 2-walks of a vertex with itself. The (2, 2) position has the value 3. These are the three 2-walks from $B$ to itself, namely, $B$–$A$–$B$, $B$–$C$–$B$, and $B$–$D$–$B$.

Carrying our matrix multiplication further, we find $M^3$ and $M^4$ to be as in Figure 6.14. These matrices give the number of 3-walks and 4-walks, respectively, among the vertices. From $M^3$ we see that there are six 3-walks between vertices $C$ and $D$. They are $C$–$A$–$B$–$D$, $C$–$A$–$C$–$D$, $C$–$B$–$C$–$D$, $C$–$E$–$C$–$D$, $C$–$D$–$C$–$D$, and $C$–$D$–$B$–$D$.

**FIGURE 6.13**

$$M^2 = M \times M = \begin{bmatrix} 0 & 1 & 1 & 0 & 0 \\ 1 & 0 & 1 & 1 & 0 \\ 1 & 1 & 0 & 1 & 1 \\ 0 & 1 & 1 & 0 & 0 \\ 0 & 0 & 1 & 0 & 0 \end{bmatrix} \cdot \begin{bmatrix} 0 & 1 & 1 & 0 & 0 \\ 1 & 0 & 1 & 1 & 0 \\ 1 & 1 & 0 & 1 & 1 \\ 0 & 1 & 1 & 0 & 0 \\ 0 & 0 & 1 & 0 & 0 \end{bmatrix} = \begin{bmatrix} 2 & 1 & 1 & 2 & 1 \\ 1 & 3 & 2 & 1 & 1 \\ 1 & 2 & 4 & 1 & 0 \\ 2 & 1 & 1 & 2 & 1 \\ 1 & 1 & 0 & 1 & 1 \end{bmatrix}$$

$$M^3 = \begin{bmatrix} 2 & 5 & 6 & 2 & 1 \\ 5 & 4 & 6 & 5 & 2 \\ 6 & 6 & 4 & 6 & 4 \\ 2 & 5 & 6 & 2 & 1 \\ 1 & 2 & 4 & 1 & 0 \end{bmatrix}$$

(a)

$$M^4 = \begin{bmatrix} 11 & 10 & 10 & 11 & 6 \\ 10 & 16 & 16 & 10 & 6 \\ 10 & 16 & 22 & 10 & 4 \\ 11 & 10 & 10 & 11 & 6 \\ 6 & 6 & 4 & 6 & 4 \end{bmatrix}$$

(b)

**FIGURE 6.14** Third and fourth powers of matrix of Figure 6.12.

It should be easy to see that if any of the powers of matrix $M$ contain all nonzero entries, then the matrix is connected. How far do we need to find powers to establish connectivity? Since the number of vertices is finite, if the graph is connected, then walks among all the vertices will exist. If there are $n$ vertices, then each vertex should be reachable from any other in no more than $n-1$ walks. Hence for our graph above of order 5, $M^4$ is as far as we need to compute.

The discussion above leads to the following:

□ Theorem 6.3 Let $M$ be the adjacency matrix of graph $G$ of order $n > 1$. Then the $(i, j)$ entry of $M^p$ gives the number of $p$-walks between vertices $v_i$ and $v_j$.

If the Boolean operations of *and* and *or* are used to compute the powers of the adjacency matrix of order $n$, we can obtain what is known as the *connection* matrix. We use the *and* (conjunction) operation on the 0/1 entries instead of multiplication and the *or* (disjunction) operation instead of addition. We then compute $M \vee M^2 \vee \cdots \vee M^{n-1}$ to find if 1-walks or 2-walks or $\cdots (n-1)$-walks exist.

Thus to determine if the graph of order 4 in Figure 6.15(a) is connected [matrices larger than Figure 6.15(b) are not always so easy to "see"], we find

**FIGURE 6.15** Graph with its connection matrices.

(a)

$$M = \begin{array}{c|cccc} & A & B & C & D \\ \hline A & 0 & 1 & 0 & 0 \\ B & 1 & 0 & 0 & 1 \\ C & 0 & 0 & 0 & 0 \\ D & 0 & 1 & 0 & 0 \end{array}$$

(b)

$$\text{Boolean } M^2 = \begin{bmatrix} 1 & 0 & 0 & 1 \\ 0 & 1 & 0 & 0 \\ 0 & 0 & 0 & 0 \\ 1 & 0 & 0 & 1 \end{bmatrix}$$

(c)

$$\text{Boolean } M^3 = \begin{bmatrix} 0 & 1 & 0 & 0 \\ 1 & 0 & 0 & 1 \\ 0 & 0 & 0 & 0 \\ 0 & 1 & 0 & 0 \end{bmatrix}$$

(d)

$$M \vee M^2 \vee M^3 = \begin{bmatrix} 1 & 1 & 0 & 1 \\ 1 & 1 & 0 & 1 \\ 0 & 0 & 0 & 0 \\ 1 & 1 & 0 & 1 \end{bmatrix}$$

(e)

the Boolean $M^2$ and $M^3$ and then compute $M \vee M^2 \vee M^3$, as shown in Figure 6.15(e). We see that there do exist 1-walks or 2-walks or 3-walks among vertices $A$, $B$, and $D$. There are no walks connecting $C$ with any of the other vertices. Thus the graph is disconnected.

## EXERCISES

1. Show that the order of a cycle must be $\geq 3$.
2. State a case for or against the following:
   (a) Every path is a trail.
   (b) Every trail is a path.
3. Identify all paths and all trails of length 5 starting at vertex $v_1$ in the graph shown.

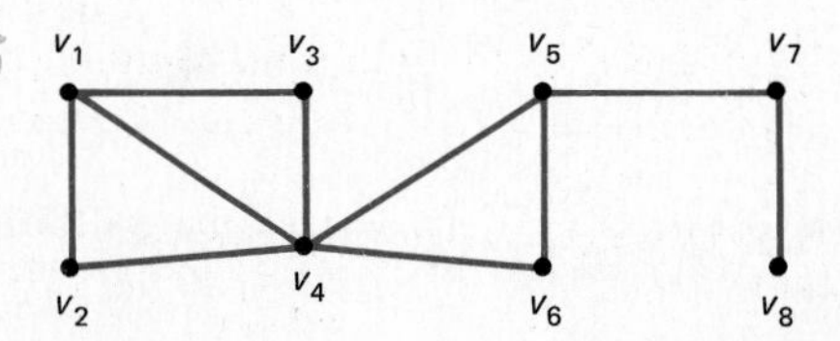


4. Which of the vertices of the graph in Exercise 3 are cut points? Which are bridges?
5. An *acyclic* graph is a graph without cycles. Draw several acyclic graphs. Try to determine the relationship between the size and the order of the graphs you drew.
6. Give an example of a graph with four components, each being acyclic (see Exercise 5).
7. Is it possible for a graph to contain more components than vertices?
8. Give an example of a connected graph having more cut points than bridges.
9. Give an example of a connected graph having more bridges than cut points.
10. Does every connected graph contain at least one cut point? At least one bridge?
11. Find $M$, $M^2$, $M^3$, $M^4$, and $M^5$ for the graph shown.

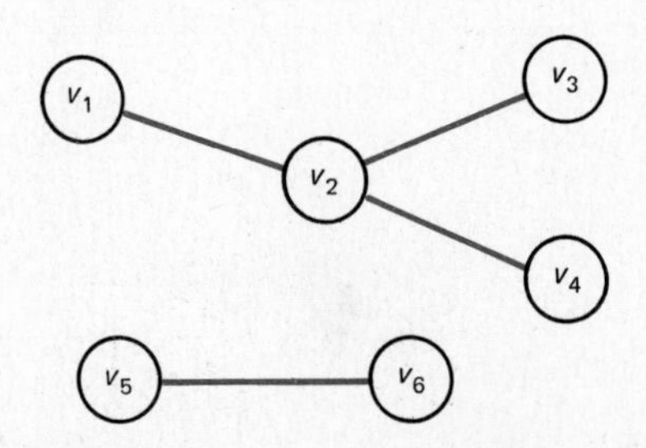


12. Find the connection matrix for the graph in Exercise 11.

## 6.4 Special Graphs

There are several ways that graphs and relations can be classified. We will look at several of them in this section. The theory is extensive and we will only tap the highlights in the text. Several of the exercises will explore the theory further.

**Complete Graphs** If a symmetric relation $R$ among the elements of a set is such that every element is related to every other element in the set (with the possible exclusion of itself) we say that the relation is *complete*. The graph of such a relation is a *complete graph*. A complete graph of order $p$ is denoted by $K_p$.

The complete relation between two elements $a$ and $b$ is shown in Figure 6.16. In the figure we depict the cross product, the adjacency matrix, and the graph $K_2$. Figure 6.16(c) is the only way the graph of a *complete* two-element relation can be drawn (other than by changing orientation).

Figure 6.17 shows the complete relation among three elements. The adjacency matrix is shown in lieu of the cross product. Note here again that Figure 6.17(b) is the only way this complete relation can be displayed as a graph. Of course, the vertices can be oriented differently. For obvious reasons $K_3$ is called a triangle.

Complete graphs of orders 4 and 5 are shown in Figure 6.18. Figure 6.18(c) is known also as the star graph. Note that the two $K_4$ graphs are isomorphic, as are the two $K_5$ graphs.

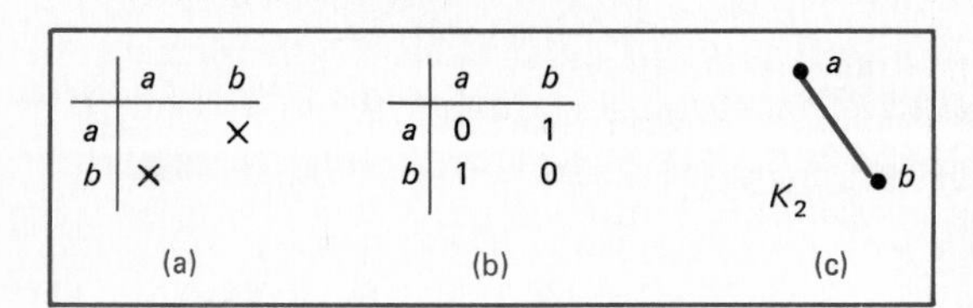


FIGURE 6.16 Complete relation of order 2.

| | a | b | c |
|---|---|---|---|
| a | 0 | 1 | 1 |
| b | 1 | 0 | 1 |
| c | 1 | 1 | 0 |

(a)

a b c
$K_3$
(b)

FIGURE 6.17 Complete relation of order 3.

**Planarity** The graphs of Figure 6.18 give rise to the question of whether or not certain graphs can be illustrated in a plane so that edges that are not supposed to intersect do not cross. These undesired crossings are called crossovers. [For example, edge $(e, c)$ crosses edge $(a, d)$ in Figure 6.18(d) at a point at which there is no vertex.] If it is possible to so construct such a graph $G$ without crossovers, then $G$ is termed *planar*. To be precise, a *planar graph* is one that can be drawn in a plane in such a way that no two edges intersect except at a vertex.

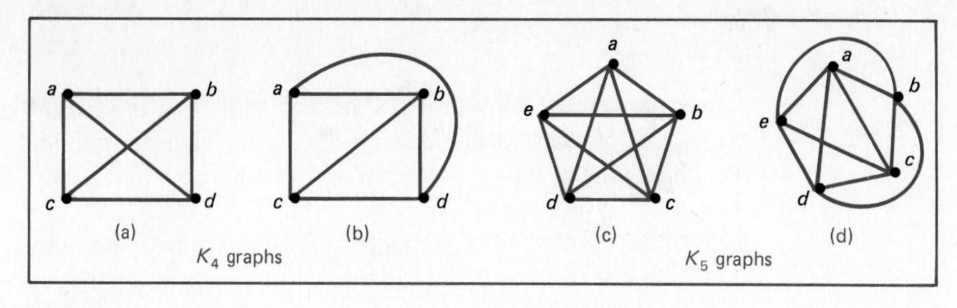


**FIGURE 6.18**

From Figure 6.18(b) it is seen that $K_4$ is planar. It has been proved that the star graph, $K_5$, is *nonplanar*, and that any $K_p$ for $p \geq 5$ is also nonplanar.

A relationship between planar graphs and regions in the plane is explored in Exercise 7, and an age-old problem of planarity and coloration is explored in Chapter 7.

Regular Graphs  A graph $G$ is *regular* if each vertex has the same degree. If this degree is $r$, then we say that $G$ is $r$-regular. The connected 0-regular and 1-regular graphs are shown in Figure 6.19. Several 2-regular and 3-regular graphs are shown in Figures 6.20 and 6.21.

Bipartite Graphs  It is at times instructive to partition the vertices of a graph $G(V, E)$ into two sets, $V_1$ and $V_2$, such that every vertex in $V_1$ is adjacent to one or more vertices in $V_2$, every vertex in $V_2$ is adjacent to one or more vertices in $V_1$, and no vertex is adjacent to another vertex in its own set. If a graph can be so partitioned into two parts, it is said to be *bipartite* (two parts).

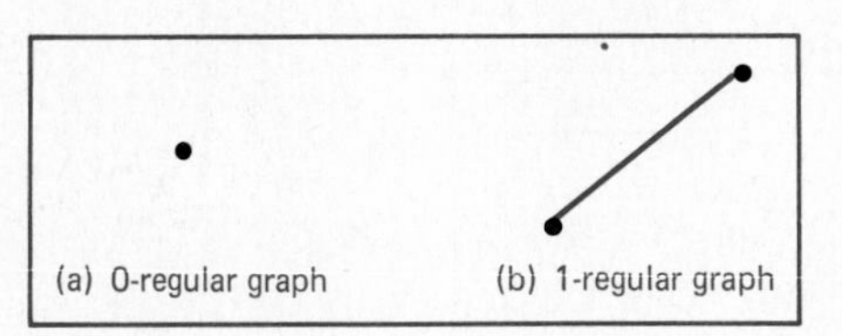


**FIGURE 6.19**

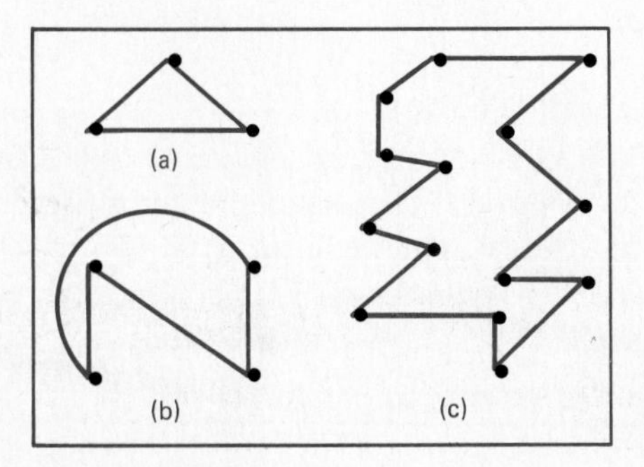


**FIGURE 6.20** 2-regular graphs.

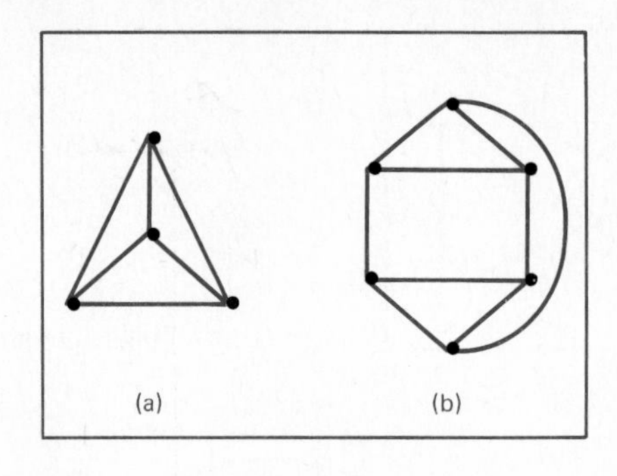


**FIGURE 6.21** 3-regular graphs.

It is not always evident from looking at the diagram of a graph that this is possible. For example, the graph in Figure 6.22(a) can be partitioned as shown in Figure 6.22(b). The set $V_1 = \{v_1, v_7, v_6, v_8\}$ and $V_2 = \{v_5, v_4, v_3, v_2, v_9\}$ in this bipartite graph.

*Complete bipartite graphs* are partitioned as above into two sets of vertices, $V_1$ and $V_2$, but additionally there is the restriction that *every* vertex in either set is adjacent to *every* vertex of the other set. When the orders of the two sets are $m$ and $n$, the complete bipartite graph is denoted by $K_{m,n}$. The graphs $K_{2,3}$ and $K_{4,4}$ are shown in Figure 6.23.

As an example of complete bipartite graphs, suppose that we have three houses, $H_1$, $H_2$, and $H_3$, each to be connected to three utilities—water ($W$), gas ($G$), and electricity ($E$)—by pipelines. The graph for the situation may be illustrated by Figure 6.24, in which the pipelines are the edges of the graph. This

**FIGURE 6.22** Bipartite graph.

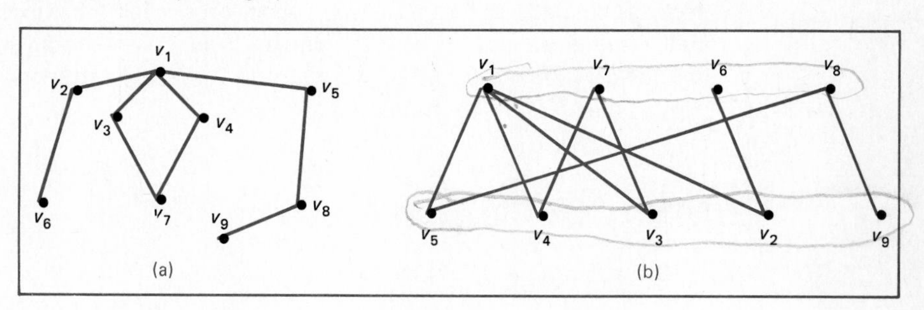


**FIGURE 6.23** Complete bipartite graphs.

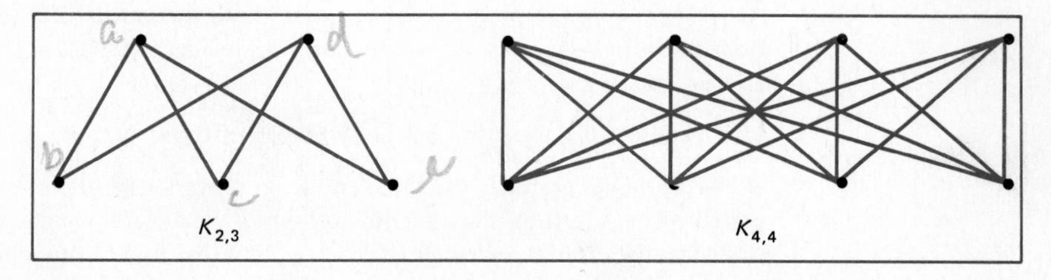

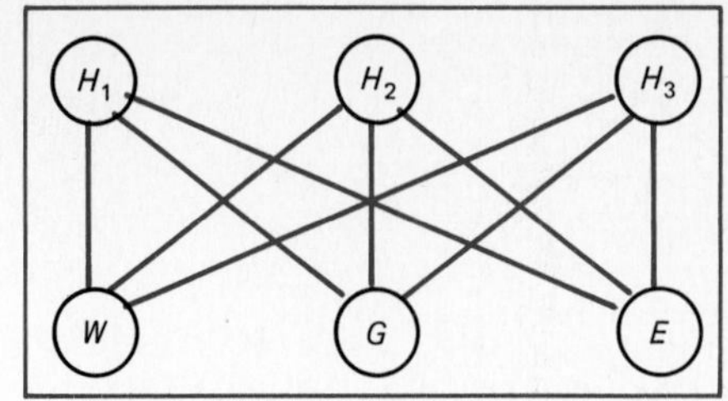


**FIGURE 2.24** Utility graph: $K_{3,3}$.

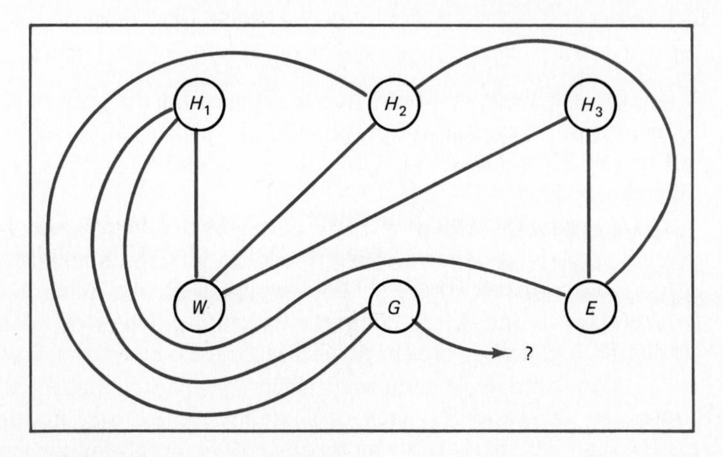


**FIGURE 6.25** $K_{3,3}$ is not planar.

is the classic example of the $K_{3,3}$ graph. A question that arises is: Can the connections be made so that no two pipelines cross each other except at the initial and terminal points? Or, is $K_{3,3}$ planar? We can start to answer the question as shown in Figure 6.25, but we will find our search fruitless, for it has been proven that $K_{3,3}$ is not planar.

## EXERCISES

1. Show that $K_p$ is $(p - 1)$ regular.
2. How many edges are there in a $K_{m,n}$ graph?
3. Prove that 3-regular graphs must have an even number of vertices (see Theorem 6.1).
4. A *tripartite* graph is similar to a bipartite graph except that the vertex set is partitioned into *three* subsets, so that no edge has incident vertices in the same set. Give an example of a tripartite graph.
5. Use Theorem 6.1 to show that the number of edges of a $K_n$ graph is $n(n - 1)/2$.
6. When a planar graph is drawn without crossovers, the plane is partitioned into regions. The $K_4$ graph shown below is seen to partition the plane into *four* regions ($R_4$ is on the exterior of the graph). This graph has four vertices and six edges.

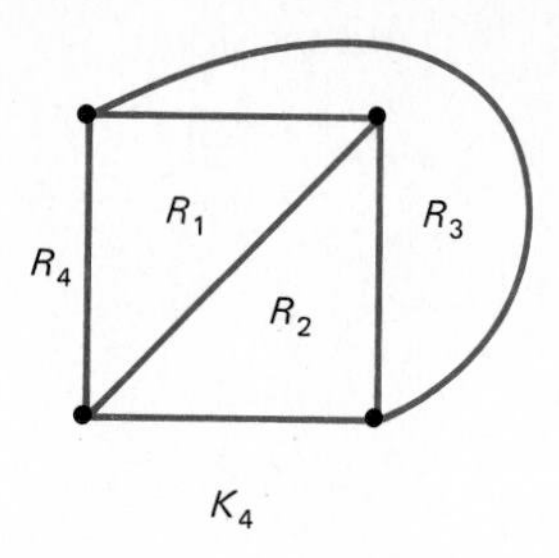


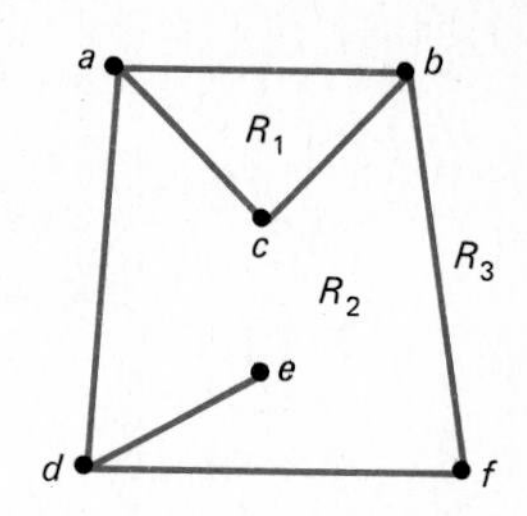


The other graph pictured has six vertices, seven edges, and only three regions ($R_3$ is on the exterior). An interesting observation emerges from this visualization. If we let $n$ be the order of the graph, $e$ the size, and $r$ the number of regions in the plane, then the following equation (Euler's formula) is true:

$$n - e + r = 2$$

For the two graphs above, we have:

1. $4 - 6 + 4 = 2$ for $K_4$
2. $6 - 7 + 3 = 2$ for the other graph

This formula can be proved by induction on $e$, the number of edges.

**Basis Step.** Let $e = 1$. Then there are two vertices ($n = 2$) and only one region. Thus $2 - 1 + 1 = 2$ and the result is true. *Prove* the induction step assuming the IH: $n - k + r = 2$ for $k$ edges.

7. Subgraphs of a graph $G$ that are complete and contained in no other complete subgraph of $G$ are called *maximal complete subgraphs.* Find all maximal complete subgraphs in graphs $G_1$ and $G_2$.

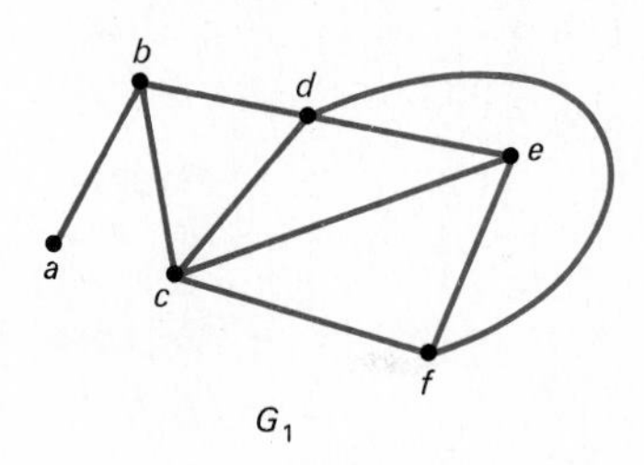


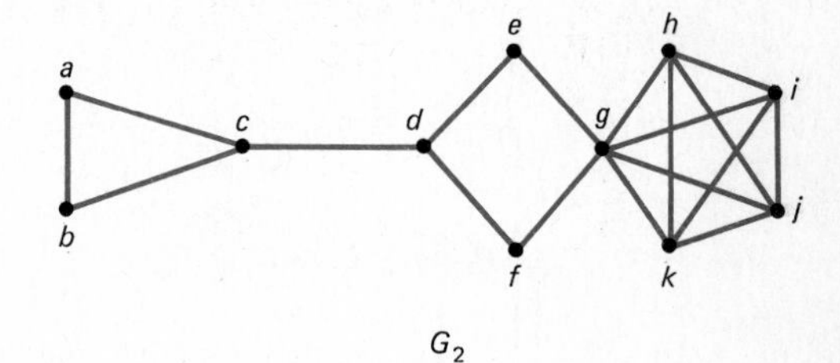


8. Due to the sociological applications of maximal complete subgraphs (Exercise 7), they are often referred to as *cliques.* When vertices are interpreted as individuals grouped together (e.g., committees, clubs, etc.), the edges can represent a relation of mutual friendship, interest, and so on. A measure of homogeneity of the group can be determined from the number of cliques present. (The fewer the cliques, the more homogeneous the group.)

   To determine the number of cliques of order 3 (triangular relations), one forms the adjacency matrix, $M$, and computes $\frac{1}{6} \times$ (sum of elements along the main diagonal of $M^3$). Find the number of 3-cliques present in the graphs $H_1$ and $H_2$ below by using this formula. Which graph represents the more homogeneous group?

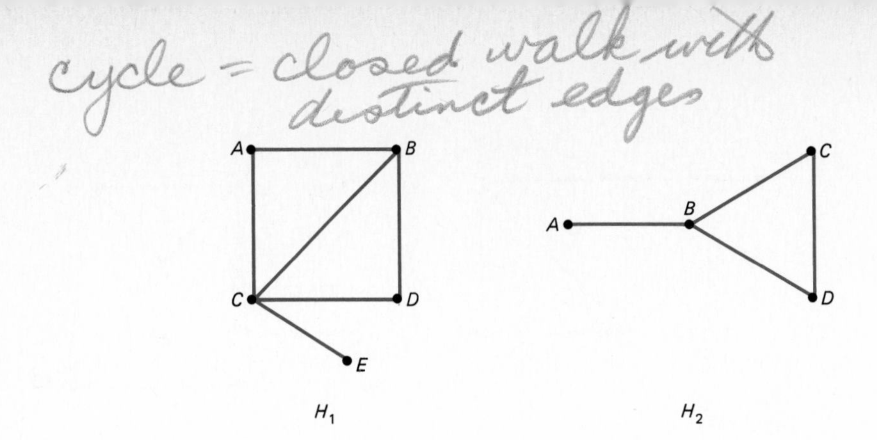


## 6.5 Trees

One of the special classifications of graphs that is extremely important to the field of computer science is that of a tree. A *tree* is defined to be a connected graph that has *no cycles* (acyclic) (see Exercise 5 of Section 6.3). This type of graph models many computer science applications and some of these will be discussed at length in Chapter 7. For now, let us examine some of the graph-theoretic topics of trees.

All trees with one, two, or three vertices are pictured in Figure 6.26. It should be evident that there is only *one* of each. There are different orientations of trees with two or three vertices, but they are, in fact, the same graphs, as pictured in Figure 6.26.

Where there are more than three vertices, there are several representations. There are two trees with four vertices and three with five vertices (see Figure 6.27). There are six trees with six vertices (Figure 6.28), 11 with seven vertices, 23 with eight vertices, 47 with nine vertices, and 106 with ten vertices [see Harary (1969, pp. 233–234)].

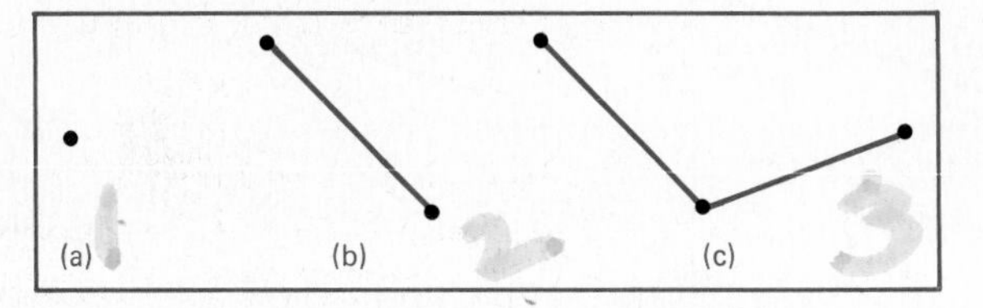


**FIGURE 6.26** Trees with one, two, or three vertices.

**FIGURE 6.27** Trees with four and five vertices.

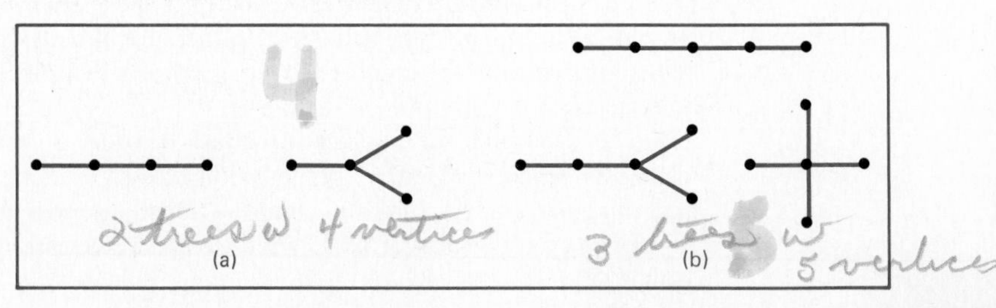

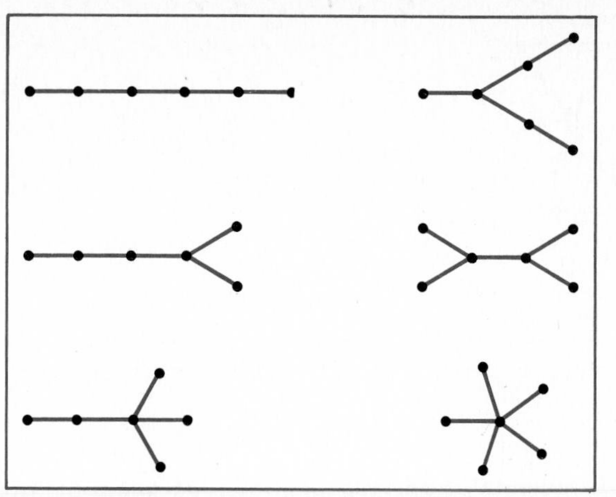

**FIGURE 6.28** Trees with six vertices.

More than one tree together in one graph structure is called a *forest* (Figure 6.29). By definition, a forest is a (disconnected) acyclic graph.

**Spanning Trees** A *spanning tree* of an arbitrary connected graph $G$ is a tree that contains all the vertices (the vertex set) of $G$. In order to create a spanning tree from a graph, one can merely delete edges from the graph that are not bridges until there are no more cycles. (A formal algorithm for this will be given in Chapter 7.) Spanning trees for a graph are not unique. See the two spanning trees for the graph of order 7 in Figure 6.30. Notice that in this figure each spanning tree contains six edges. Will this always be the case? Will there always be the same number of edges no matter how we delete cycles from a graph to form a spanning tree?

The question above is really more general in nature. We may ask if an arbitrary tree with $n$ vertices always contains the same number of edges. We answer this by proving the following two theorems and corollary.

□ **Theorem 6.4** Every pair of vertices in a finite tree $T$ is connected by exactly one path.

**Proof** We know that $T$ is connected; thus there is *at least one* path between each pair of vertices, say $u$ and $v$. We will show that there is only one such path.

Suppose that there are two distinct paths, $P_1$ and $P_2$, joining the vertices $u$ and $v$. As we travel from $u$ to $v$, there will be some vertex $w$ (possibly $w = u$) whose successor on $P_1$ is not on $P_2$. If we let vertex $x$ (possibly $x = v$) be the next vertex on $P_1$ which is also on $P_2$, then the portions of $P_1$ and $P_2$ that are

**FIGURE 6.29** Forest.

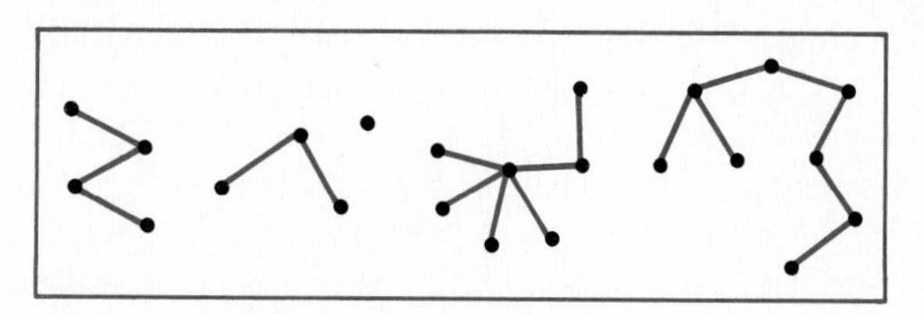

FIGURE 6.30 Two spanning trees for a graph.

between $w$ and $x$ form a cycle in $T$. But this contradicts the fact that $T$ is acyclic. Therefore, there is *only one* such path.

□ Corollary Every edge in a finite tree is a bridge.

The proof is left for the exercises.

□ Theorem 6.5 A finite tree of order $n$ (has $n$ vertices) is of size $n - 1$ (has exactly $n - 1$ edges).

**Proof** We prove this by mathematical induction (see Section 5.7).

**Basis Step.** Let $n = 1$. There is only one such tree of order 1 [see Figure 6.26(a)] and this tree has $1 - 1 = 0$ edges. Thus the basis is true. This is also evident in a tree of order 2, where we see that the size is 1 [see Figure 6.26(b)].

**Induction Hypothesis.** A tree of order $k > 1$ has $k - 1$ edges.]

**Induction Step.** We are to show that the addition of one other vertex, $w$ to $T$ (yielding a new tree $T'$ of order $k + 1$), will change the size to $(k - 1) + 1 = k$. Now in order for $T'$ to be connected, there will have to be at least one edge joining $w$ to $T$. Suppose that there are two edges joining $w$ to distinct vertices, $u$ and $v$ in $T$. There is already exactly one path between $u$ and $v$ (Theorem 6.4). The addition of these two edges creates another path, via $w$, thus forming a cycle. But since $T'$ can contain no cycles, no such two distinct edges from $w$ to $T$ can be introduced. Thus there is exactly one new edge, giving the size of $T'$ to be $(k - 1) + 1 = k$.

Rooted Trees Perhaps the most interesting and useful representation of tree structures for computer science are those known as rooted trees. The applications occur in many settings, and in the literature when trees are "rooted" the vertices are commonly referred to as nodes.

In these structures *one* of the nodes of the tree is distinguished as the *root*, and all other nodes eminate from it in a branching pattern much as limbs and branches grow from the root of a botanical tree. Such trees are usually pictured

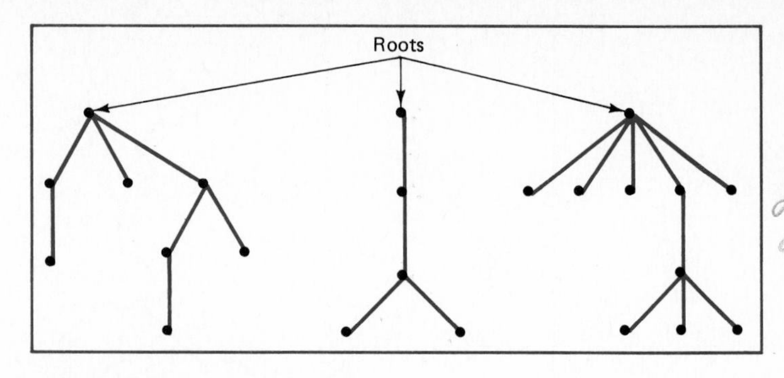


**FIGURE 6.31** Some rooted trees.

"upside down," with the root at the top. Several rooted trees are shown in Figure 6.31.

To use genealogical terms, the root of a tree is considered the *ancestor* of all the other nodes, which are themselves called *descendants* of the root. In general, the ancestors of any node of a rooted tree are all the nodes along the (unique) path from the root to that node. The nodes adjacent to any node $v$, which are pictured one level "lower" than node $v$, are the *children* of $v$, and $v$ is termed the *parent* of these adjacent nodes. Children of the same parent node are said to be *siblings*.

Notice that every path from the root "down" this tree has a *last* or *terminal* node. These terminal nodes are termed *leaves* of the tree and every leaf has degree 1. The set of all leaves of a rooted tree is sometimes called the *frontier* of the tree. The *height* of the tree is the length of the longest path from the root. The nodes that are not leaves are called *internal* nodes.

In many computer science applications there is the necessity to impose some order on the children of nodes in a tree. This is usually done in a left-to-right fashion. Thus the leftmost child pictured of a node is termed the first child, the next to the right is the second child, and so on.

As can be seen from Figure 6.32 of the next example, a rooted tree can be pictured in *levels* with the root at level 1, its children at level 2, its grandchildren at level 3, and so on.

**EXAMPLE 1** Consider the rooted tree $T$, shown in Figure 6.32. We make the following observations:

1. Node $A$ is the root. It is at level 1.
2. The children of node $A$ are nodes $B$, $C$, $D$, and $E$.
3. Nodes $J$, $K$, and $L$ are siblings whose parent is node $H$.
4. The leaves are nodes $F$, $G$, $C$, $J$, $K$, $M$, and $I$.
5. The first child of node $H$ is node $J$.
6. Nodes $F$, $G$, $H$, and $I$ are at level 3.
7. The height of the tree is 4.

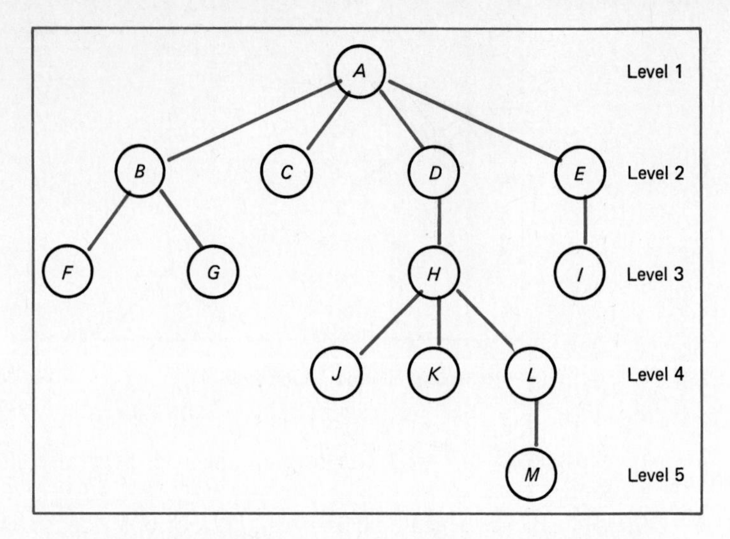


**FIGURE 6.32** Rooted tree.

Every node of a rooted tree, $T$, can be viewed, together with its descendants, as a rooted tree itself. This substructure is called a *subtree* of $T$. Thus in Figure 6.32, node $H$, together with its descendants, $J$, $K$, $L$, and $M$, is a subtree of the original tree, $T$, with node $H$ being the root of this subtree. Additionally, node $G$ is a subtree of $T$ with no descendants. It is also possible for a rooted tree to have zero nodes, in which case it is referred to as the empty tree. These should be allowed for purposes of applications.

If every node of a rooted tree, $T$, has $n$ or fewer children, then $T$ is referred to as an $n$-ary tree. The tree of Figure 6.32 is a 4-ary tree since every node has four or fewer children. For most applications $n$-ary trees are thought of as ordered. For example, the two (ordered) 3-ary (ternary) trees of Figure 6.33 are different because the children of node $A$ are listed in a different order.

Most computer science applications involve ordered 2-ary trees, known as *binary trees*. These may be empty or nonempty. Formally, nonempty binary

3 or fewer children

**FIGURE 6.33** Two ternary trees with different ordering.

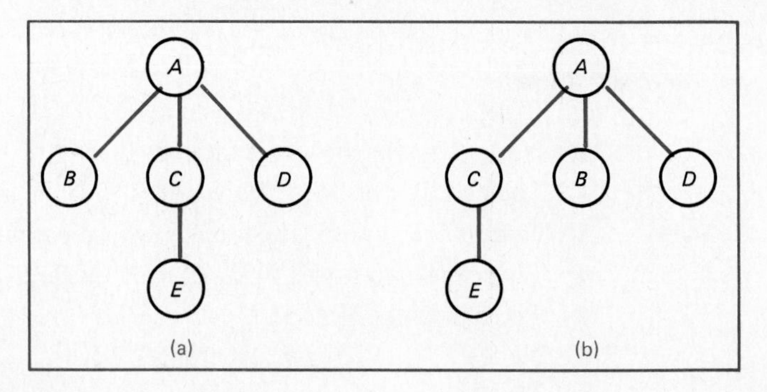

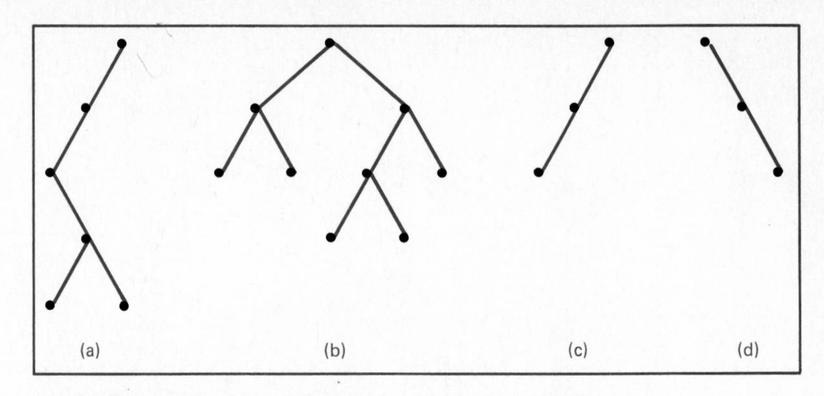


**FIGURE 6.34** Sample binary trees. Notice that the trees at (c) and (d) are different because of the different ordering. There are no right children in (c) and no left children in (d).

trees are ordered rooted trees where each node has *at most* two children (or two subtrees) (see Figure 6.34). If both children of a node $V$ are present, one is called the left child of $V$ and the other the right child of $V$. There are many applications of binary trees in computer science and some of these will be explored in Chapter 7.

□ Theorem 6.6 There are at most $2^{n-1}$ nodes at the $n$th level of a binary tree.

**Proof** We proceed by mathematical induction.

**Basis Step.** Let $n = 1$. We are at level 1. This is the root level, and there is $2^{1-1} = 2^0 = 1$ node, namely the root.

**Induction Hypothesis.** Assume that the maximum number of nodes at level $k$ is $2^{k-1}$.

**Induction Step.** Since the maximum number of nodes at level $k$ is $2^{k-1}$ (from the IH), and each node at level $k$ can have no more than two children, there will be a maximum of $2 \cdot 2^{k-1} = 2^k$ nodes at level $k + 1$.

## EXERCISES

1. Display four spanning trees of the following graph:

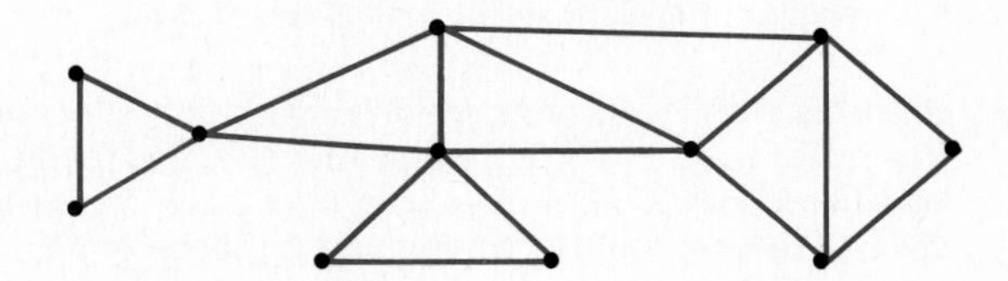

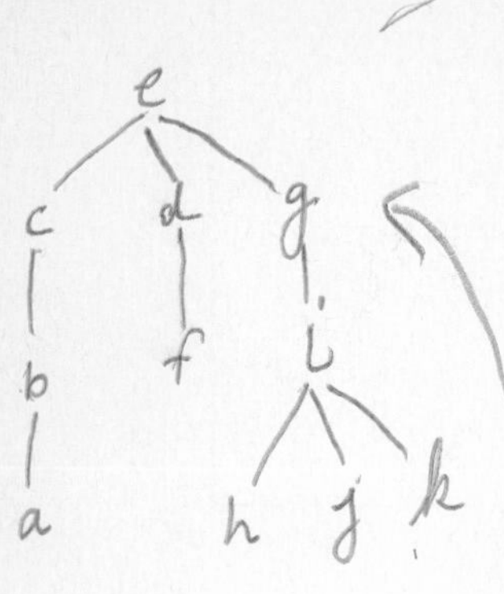

2. Given the following tree:

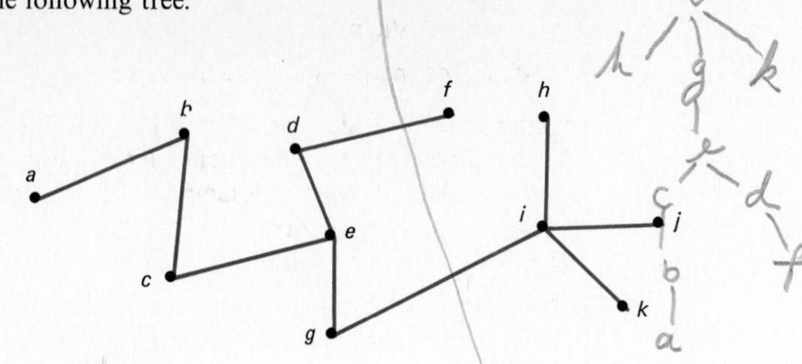


(a) Draw the tree as a rooted tree with vertex $e$ chosen as the root.

(b) What is the height of this rooted tree?

(c) At what level is vertex $k$?

(d) Draw the tree as rooted with vertex $j$ as the root.

(e) What is the height of this rooted tree (with the root being vertex $j$)?

(f) From part (d), at what level is vertex $a$ (with the root being vertex $j$)?

(g) If these rooted trees are ordered, does it matter how you drew them?

3. Show that the sum of the degrees of an $n$-order tree is $2n - 2$.

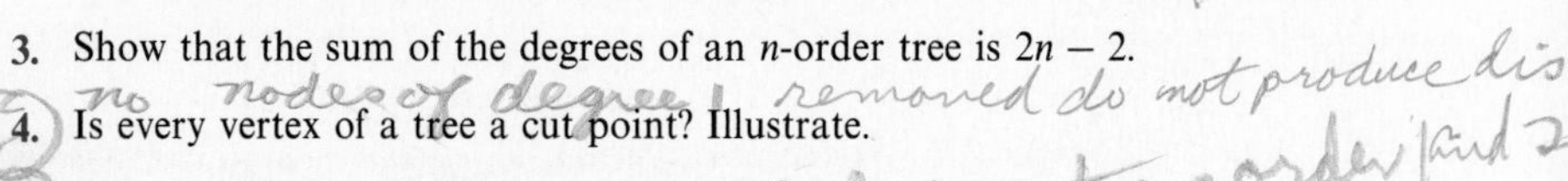

4. Is every vertex of a tree a cut point? Illustrate.

5. Is it possible for a tree to be complete?

6. Is the complement (see Exercise 8 of Section 6.2) of a spanning tree of a graph $G$ also a spanning tree?

7. Prove: In a tree every edge is a bridge.

8. Prove: Trees are bipartite graphs.

9. Prove that a forest with $k$ components and $n$ nodes has $n - k$ edges.

10. In Exercise 5(d) of Section 5.7 you were asked to show by mathematical induction that

$$\sum_{k=0}^{n} 2^k = 2^0 + 2^1 + 2^2 + \cdots + 2^n = 1 + 2 + 4 + 8 + \cdots + 2^n = 2^{n+1} - 1$$

By using this equality (prove it again if necessary) and by drawing some binary trees, demonstrate the validity of the following *theorem* (see also Theorem 6.6):

The maximum number of nodes of a binary tree of *height* $n$ is $2^{n+1} - 1$.

11. Explain how the following recursive definition of rooted trees describes the explanation of these in the text of this section.

**Definition** A rooted tree $T$ is a finite set of $n$ nodes ($n \geq 1$), $V$, such that (a) there is a specially designated node, $v_1 \in V$, called the *root* of $T$, and (b) the remaining nodes, $V - \{v_1\}$, are partitioned into $m > 0$ *disjoint* sets: $T_1, T_2, \ldots, T_m$, each of which is *itself* a rooted tree. Each of these $m$ rooted trees is called a subtree of $v_1$.

12. Show that there are at least two vertices of a tree of order $n \geq 2$ whose degree is 1.

13. Given $G$ is a finite graph of order $n > 1$. Prove the following statements:

(a) If $G$ is acyclic with size $n - 1$, then $G$ is a tree.

(b) If $G$ is connected with size $n - 1$, then $G$ is a tree.

14. A binary tree in which each node has either zero or *exactly two* children is known as a *strictly binary tree* (for short, *s.b.t.*). In the literature these are also known as *2-trees*. Figure 6.34(b) is an example of an s.b.t., whereas the other trees pictured in that figure are not. Notice that there will always be an *odd number* of nodes in an s.b.t. Draw a few to verify this observation.

(a) It can be proved by induction that there are exactly $\frac{1}{2}(n + 1)$ leaf nodes in an s.b.t. with $n$ nodes. The *basis* of the induction is a tree with one node and it is a leaf: $\frac{1}{2}(1 + 1) = 1$ leaf.

Carry out the induction step noting that the I.H. is that a $k$ order s.b.t. (with $k$ odd) has exactly $\frac{1}{2}(k + 1)$ leaves. The "next" tree for the induction step will have $k + 2$ nodes, since the order is to remain an odd number.

(b) Show that there are $\frac{1}{2}(n - 1)$ internal (non-leaf) nodes in an $n$ order s.b.t.

(c) If a strictly binary tree has a total of 49 nodes, how many of these are leaves? Internal nodes?

(d) Show that in a strictly $m$-ary tree (every node has either zero or exactly $m$ children) with $n$ nodes that there are exactly $(n(m - 1) + 1)/m$ leaves and $(n - 1)/m$ internal nodes.

15. A *full binary tree* is a binary tree in which every level has the maximum number of nodes possible. Full binary trees with two, three, and four levels are shown below.

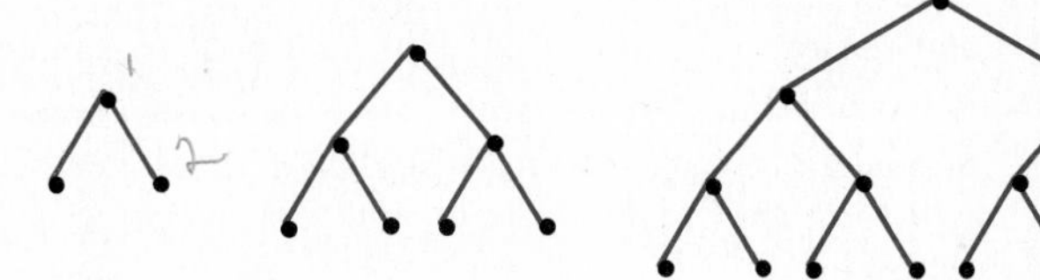

(a) Show that a full binary tree with $k$ levels (with the root at level one) has exactly

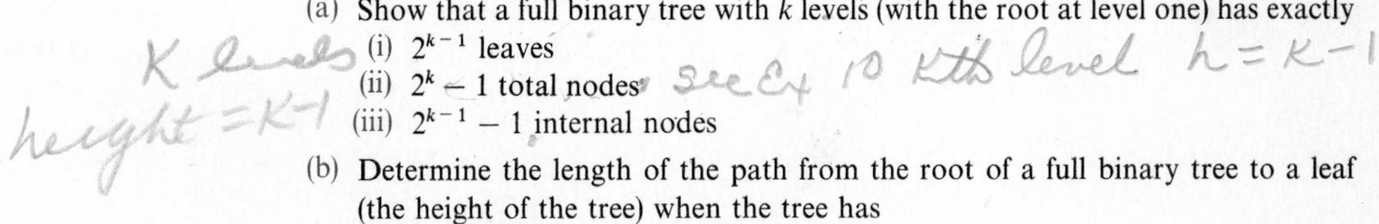

(i) $2^{k-1}$ leaves

(ii) $2^k - 1$ total nodes

(iii) $2^{k-1} - 1$ internal nodes

(b) Determine the length of the path from the root of a full binary tree to a leaf (the height of the tree) when the tree has

(i) 8 levels

(ii) 15 total nodes

(iii) 127 total nodes

16. (*Base 2 logarithms*) As you have learned in other courses, logarithms are defined in terms of a real number $b > 0$, which is called the base. The exponential equation, $y = b^x$, has as its inverse the logarithmic function, $x = \log_b y$. In words: "if the logarithm (log) of $y$ to the base $b$ is $x$, then $x$ is the *exponent* to which the base $b$ is raised to obtain $y$." Thus the log(arithm) is an exponent. Logs are not defined for a base $\leq 0$.

Any positive real number can be chosen as the base. The base with which most are familiar is $b = 10$. Logs to base 10 are known as "common" logs and before the advent of modern-day calculators and computers were used extensively in numerical hand-calculations. We have the following numerical equivalencies.

$$10^4 = 10000 \quad \text{is equivalent to } \log_{10}10000 = 4$$

$$10^{-1} = 0.1 \quad \text{is equivalent to } \log_{10}0.1 = -1$$

$$10^0 = 1 \quad \text{is equivalent to } \log_{10}1 = 0$$

In certain physical situations the naturally occurring base is the irrational number $e = 2.71828\ldots$, with which you may be familiar from the study of calculus. Logs to base $e$ are called "natural" logs and are abbreviated by ln. Thus $\log_e x$ is written as $\ln x$.

Many computer algorithms are based on extensions and applications of binary trees and full binary trees (see Exercise 15 above), and much of the analysis is in terms of powers of two. Consequently logarithms with base $b = 2$ enter the picture frequently. Base 2 logs are abbreviated by *lg*. Thus, for a real number $x > 0$, $\log_2 x$ is written as $\lg x$. Some numerical equivalencies are:

$$2^3 = 8 \quad \text{is equivalent to } \lg 8 = 3$$

$$2^6 = 64 \quad \text{is equivalent to } \lg 64 = 6$$

$$2^{10} = 1024 \quad \text{is equivalent to } \lg 1024 = 10$$

We will also be interested in the integer portion of base 2 logs, indicated by INT($\lg x$). Thus INT($\lg 10$) = 3, INT($\lg 70$) = 6, INT($\lg 2000$) = 10, INT($\lg 31$) = 4, and so forth.

To see the utility of base 2 logarithms, consider a full binary tree (see Exercise 15) with five levels. The distance from the root to any leaf is 4 (the height of the tree). From Exercises 10 and 15(a) above we see that there are $2^5 - 1 = 31$ total nodes in the tree. Notice that INT($\lg 31$) = 4, the height of the tree. Now suppose there are 63 total nodes in a full binary tree. At what level are the leaves? We get:

$$2^k - 1 = 63$$

$$2^k = 64$$

$$k = \lg 64 = 6 \text{ levels;}$$

height of tree

and the distance from the root to a leaf is 5. *Note:* INT($\lg 63$) = 5.

If the nodes of a binary tree are well balanced about the root and about each internal node, the above formula gives an approximation to the longest path from the root to a leaf. If there are $n$ nodes in such a tree, then the distance from the root to the leaf furthest from the root should be no more than INT($\lg n$).

Thus in a well-balanced tree with 70 nodes the maximum distance from the root to a leaf will be no more than INT($\lg 70$) = 6.

(a) Determine the value of each of the following base 2 logarithms:
  (i) $\lg 32$
  (ii) $\lg 4096$
  (iii) INT($\lg 15$)
  (iv) INT($\lg 150$)
  (v) INT($\lg 300$)

(b) Suppose a well-balanced rooted binary tree has 500 nodes. What is the approximate longest path from the root to a leaf?

(c) Suppose the longest path from the root to any leaf in a well balanced binary tree is 9. Give an approximate number of nodes in the tree.

## 6.6 Directed Graphs

Directed graphs (*digraphs* for short) are graphs in which the edges (usually referred to as arcs) imply an order relation among the vertices. Digraphs are depicted as ordinary (undirected) graphs but the arcs are drawn with arrows indicating the underlying order relations among the vertices. The digraph pictured in Figure 6.35 with six vertices and six arcs has as its vertices, the set $\{a, b, c, d, e, f\}$, and its arcs are $\{\langle a, b\rangle, \langle a, d\rangle, \langle b, d\rangle, \langle b, c\rangle, \langle e, c\rangle, \langle f, e\rangle\}$ listed as ordered pairs. It should be noted that even though the arc $\langle a, b\rangle$ is in the digraph of Figure 6.35, the arc $\langle b, a\rangle$ is not. Thus we see that the relation represented here is *not* symmetric as was true with the previous undirected graphs in this chapter.

Digraphs are useful models in a variety of situations. Computer programming flowcharts are digraphs with the vertices being the various instructions and the arcs representing the flow of control. Maps of one-way streets are digraphs. The switching circuit diagrams of Chapter 4 are digraphs with the various gates being the vertices.

The theory involved with digraphs roughly parallels that of undirected graphs, but the nonsymmetric nature adds a few minor complications not present with undirected graphs.

An arc $\langle u, v\rangle$ joining the vertices $u$ and $v$ has as its *initial* point, $u$ and as its *terminal* point $v$. The indegree of a vertex $u$, *indeg*($u$), is the number of arcs *terminating* at $u$. The outdegree, *outdeg*($u$) is the number of arcs *beginning* at $u$. Obviously, the sum of all the indegrees of a digraph equals the sum of all the outdegrees, which equals the number of arcs. A vertex with zero indegree is termed a *source*. One with zero outdegree is a *sink*.

source
sink

In the digraph of Figure 6.35 we see that

$$\text{indeg}(a) = \text{indeg}(f) = 0$$

$$\text{indeg}(b) = \text{indeg}(e) = 1$$

$$\text{indeg}(c) = \text{indeg}(d) = 2$$

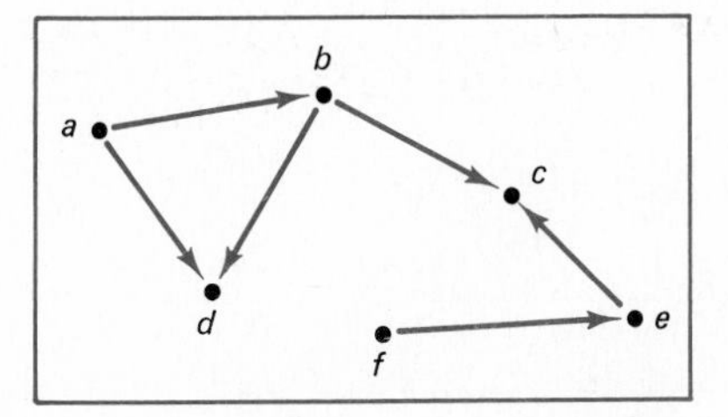


**FIGURE 6.35** Digraph.

and

$$\text{outdeg}(c) = \text{outdeg}(d) = 0$$

$$\text{outdeg}(e) = \text{outdeg}(f) = 1$$

$$\text{outdeg}(a) = \text{outdeg}(b) = 2$$

We also note that vertices $a$ and $f$ are sources and vertices $d$ and $c$ are sinks.

The concepts of walks, paths, trails, and cycles in directed graphs closely parallel those for undirected graphs studied previously in this chapter. Attention has to be paid to the direction established by the arcs involved, however. Specifically, a directed walk is an alternating sequence of vertices and (correctly directed) arcs; a directed path is a directed walk with *distinct* vertices, and so on.

There is a further concept of a *semiwalk* (semipath, semitrail) in which the direction of the arcs is disregarded. A semiwalk in a digraph is identically the same as a (undirected) walk in the underlying undirected graph. Connection in digraphs will be explored in the exercises (see Exercise 7).

As with undirected graphs, adjacency matrices are useful visualizations of digraphs. The definition of these matrices is exactly parallel to that used earlier. Note that they are not symmetric matrices. This is evident from the adjacency matrix in Figure 6.36 of the digraph at Figure 6.35.

The *row sums* of the adjacency matrix of a digraph indicate the outdegrees of the vertices, while the column sums show the indegrees. Sources and sinks can be observed directly when these column sums and row sums, respectively, sum to zero, as shown in Figure 6.36.

Directed graphs can have parallel arcs, multiple arcs, and loops. These are depicted in the digraph of Figure 6.37(a). The adjacency matrix of this digraph is shown in Figure 6.37(b). Note the parallel arcs joining vertices $a$ and $c$, the multiple arcs between $a$ and $b$, and the loop at vertex $d$.

As with undirected graphs, adjacency matrices of digraphs are useful for answering questions concerning (directed) walks and reachability within the di-

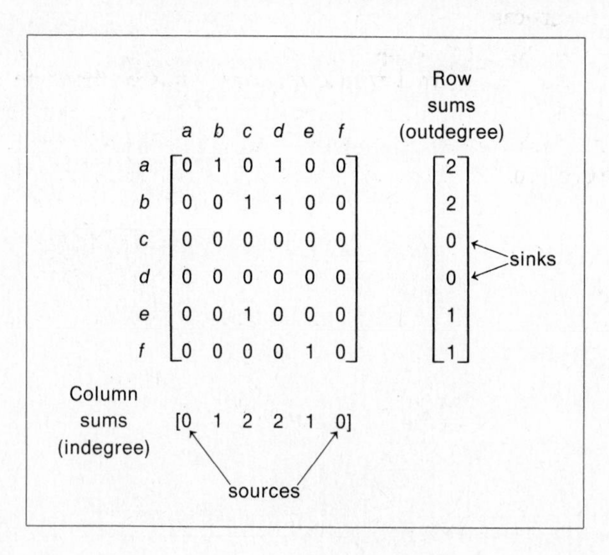


FIGURE 6.36 Adjacency matrix of the digraph of Figure 6.35.

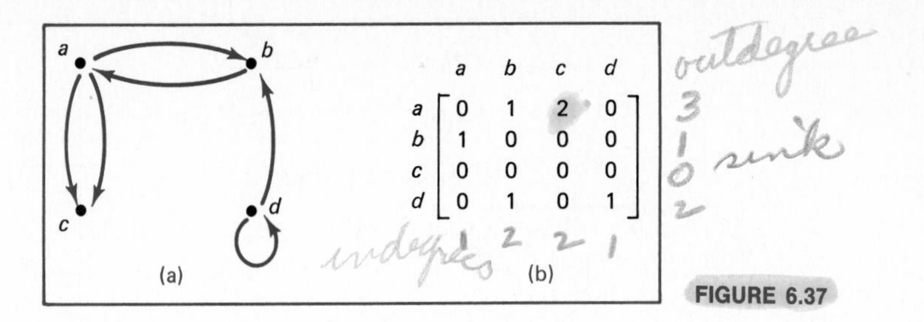


FIGURE 6.37

graph. Recalling the associated discussion regarding undirected graphs and Theorem 6.3, we present an accompanying theorem concerning digraphs:

□ Theorem 6.7 Let $M$ be the adjacency matrix of a digraph $D$ of order $n > 1$. Then the $(i, j)$ entry of $M^p$ gives the number of (directed) $p$-walks from vertex $v_i$ to vertex $v_j$.

**EXAMPLE 1** To illustrate this theorem, consider the digraph, its adjacency matrix, $M$, and the powers of this matrix, $M^2$ and $M^3$, as shown in Figure 6.38. In $M^3$ the 1 in the (1, 2) position represents the one 3-walk from vertex $a$ to vertex $b$, namely $a$–$b$–$a$–$b$; the 1 in the (3, 3) position is the walk $c$–$a$–$b$–$c$; the 2 in position (4, 2) represents the two 3-walks from $d$ to $b$: $d$–$b$–$a$–$b$ and $d$–$c$–$a$–$b$.

A *complete directed graph* is a complete (see Section 6.4) undirected graph with an arrow attached to each edge. Such a graph easily pictures a round-robin *tournament* in which each team or player (represented by vertices) plays each of the other teams or players. The winner of the match played between vertices $v_i$ and $v_j$ (no draws allowed) is indicated by the arc $\langle v_i, v_j \rangle$. The relation represented is: $v_i$ "beats" $v_j$. A five-vertex tournament is pictured in Figure 6.39. The winner of the tournament can be declared to be the vertex with the largest outdegree. Vertices $v_1$ and $v_4$ tied in this respect, but $v_4$ is the eventual winner since the arc $\langle v_4, v_1 \rangle$ is in the digraph.

Trees can be pictured with directions. Thus we have no directed cycles in these digraphs. Usually, these directions indicate the parent-child relation in the tree in which arrows point from a parent to its children, as shown in Figure 6.40(a), (b), and (c). These directed acyclic (connected) graphs are sometimes called *dags* for short. Family trees and the NCAA basketball tournament can

FIGURE 6.38

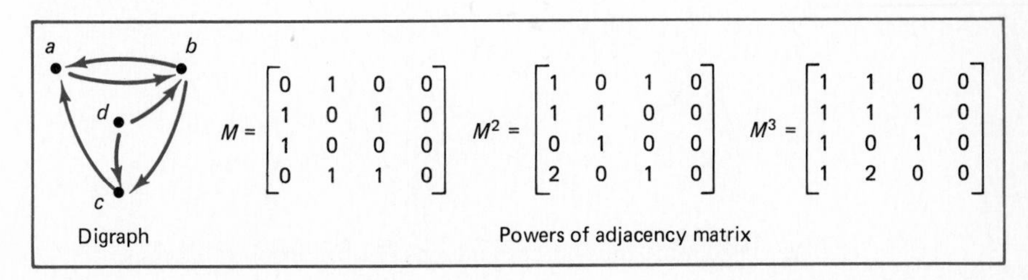

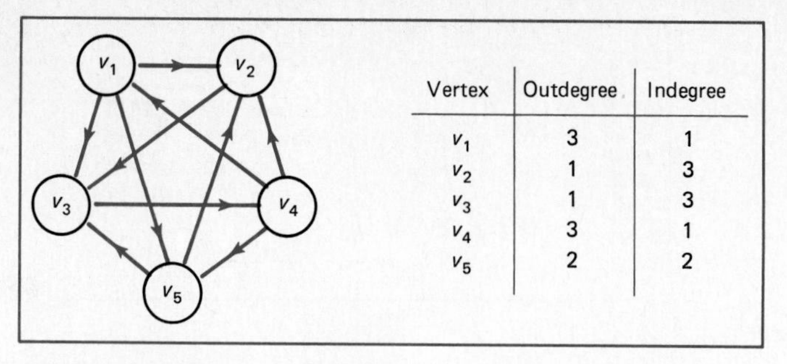



| Vertex | Outdegree | Indegree |
|---|---|---|
| $v_1$ | 3 | 1 |
| $v_2$ | 1 | 3 |
| $v_3$ | 1 | 3 |
| $v_4$ | 3 | 1 |
| $v_5$ | 2 | 2 |

**FIGURE 6.39** Five-vertex tournament.

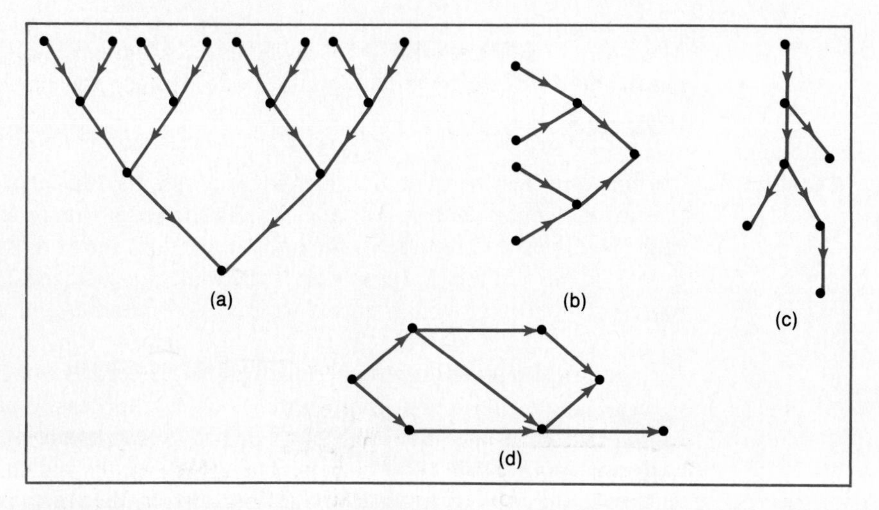



**FIGURE 6.40** Some directed acyclic graphs (dags).

be depicted as dags, as can the administrative structure of a business. Also, as we shall see in Chapter 7, computer searches are often modeled by dags.

Partial order relations can also be depicted as dags. We will explore these diagrams further in Chapter 9, but for now we should observe that these are relations with no directed cycles.

## EXERCISES

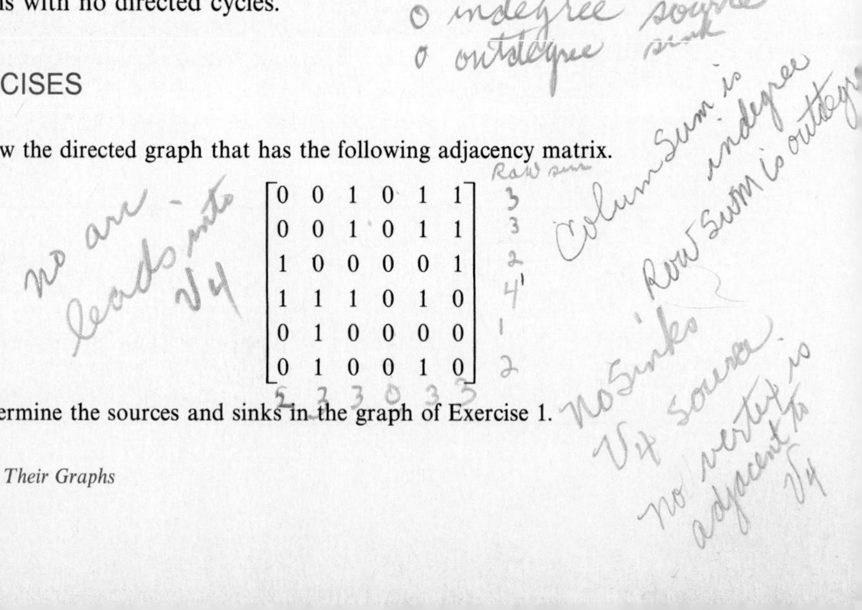

1. Draw the directed graph that has the following adjacency matrix.

$$\begin{bmatrix} 0 & 0 & 1 & 0 & 1 & 1 \\ 0 & 0 & 1 & 0 & 1 & 1 \\ 1 & 0 & 0 & 0 & 0 & 1 \\ 1 & 1 & 1 & 0 & 1 & 0 \\ 0 & 1 & 0 & 0 & 0 & 0 \\ 0 & 1 & 0 & 0 & 1 & 0 \end{bmatrix}$$

2. Determine the sources and sinks in the graph of Exercise 1.

3. (a) Construct the adjacency matrix for the directed multigraph shown.

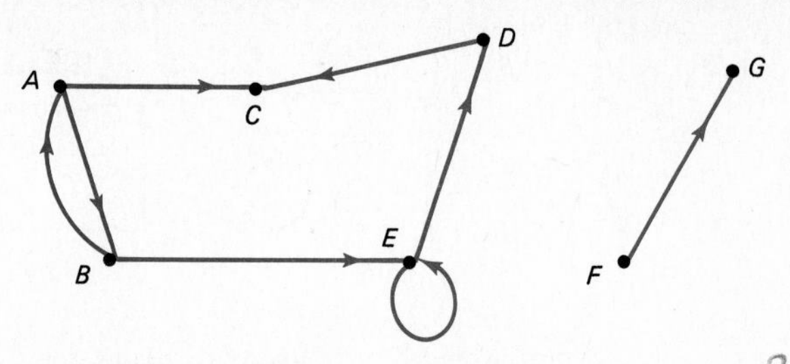


(b) Give the indegrees and outdegrees of the vertices.

(c) Identify all sources and sinks.

4. Give the indegree and outdegree of vertices of the digraph of Figure 6.37.

5. Identify sources and sinks of the tournament digraph of Figure 6.39.

6. By using the adjacency matrix and its powers, identify all directed walks of length 2, of length 3, and of length 4 in the digraph pictured.

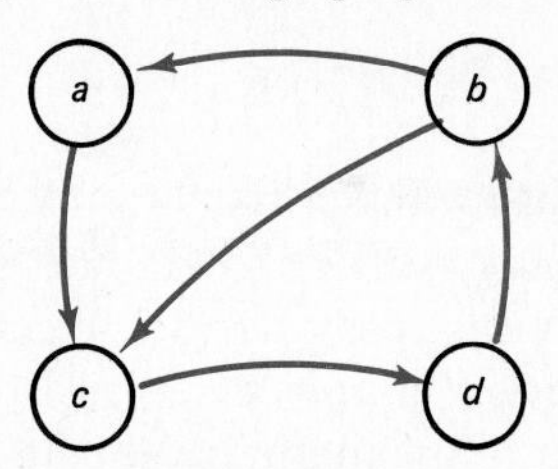


7. The concept of connection in digraphs takes three forms. A digraph $D$ is said to be *strongly* connected if for *each* pair of vertices $v_1$ and $v_2$ in $D$ there is a (directed) path from $v_1$ to $v_2$ *and* there is also a (directed) path from $v_2$ to $v_1$. The digraph of Exercise 6 is strongly connected because there is a directed path between every pair of vertices.

A digraph $D$ is said to be *weakly* connected if there is a *semipath* between each pair of vertices $v_1$ and $v_2$. If we disregard the direction arrows of the digraph, then weak connection is identical to ordinary connection in the underlying non-directed graph.

The third form of connection is known as *unilateral* connection. A digraph is unilaterally connected if for every pair of vertices $v_1$ and $v_2$ there is a directed path either from $v_1$ to $v_2$ or from $v_2$ to $v_1$.

Digraph (a) below is unilaterally connected since there is a path from node $A$ to node $C$ but the reverse path does not exist. It is also weakly connected. Digraph (b) displays a strongly connected digraph (hence also weakly and unilaterally connected).

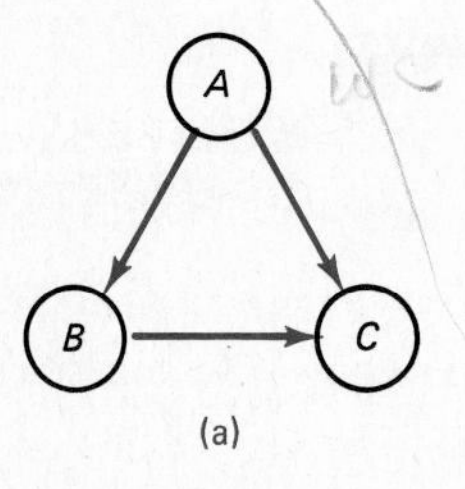


(a)

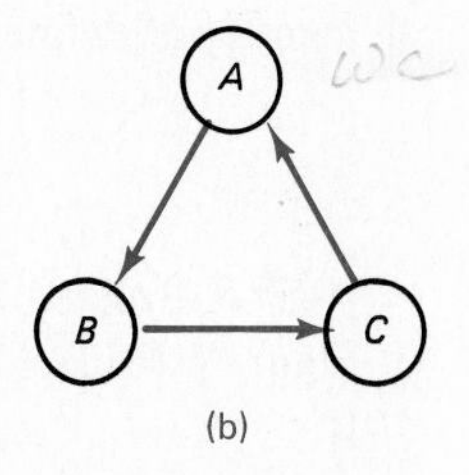


(b)

Determine whether each of the following digraphs is strongly, weakly, or unilaterally connected.

(a) (b) (c) (d)

(e) (f) (g)

## 6.7 Weighted Graphs and Networks

As is the case with so many terms in graph theory, the word *network* has been used by many authors to denote many graph situations. The term itself originated from the study of electric currents and electrical "networks," and much of the theory of graphs emerged from these beginnings.

Much of the literature equates networks with our graphs, but this should not confuse the alert reader. For the purposes of this book we use the term "network" as synonymous with graphs (multigraphs, digraphs) that have numbers assigned to the edges. These numbers are usually referred to as weights: thus *weighted graphs.*

Weighted graphs (i.e., networks) have a variety of applications. Maps showing distances between cities are examples of networks; the distances on the roads (edges) between the cities (vertices) are the weights. Similarly, edges can be assigned costs and lengths of time between events and activities.

Several of these applications will be explored in Chapter 7, such as finding the shortest path from one vertex in a network to another and finding the minimal spanning (weighted) tree (sometimes called the economy tree) of a network.

When vertices are connected by edges having weights assigned to them, the usual matrix representation is known as the "cost matrix," $C$. The cost matrix of a network resembles the adjacency matrix of graphs in that the elements are defined as follows:

$$C_{ij} = \begin{cases} \text{weight assigned to arc } \langle i, j \rangle \\ 0 \qquad \text{if there is no } \langle i, j \rangle \text{ arc} \end{cases}$$

The directed network of Figure 6.41(a) has as its cost matrix the matrix of Figure 6.41(b). Quite obviously, cost matrices of networks need not be symmetric.

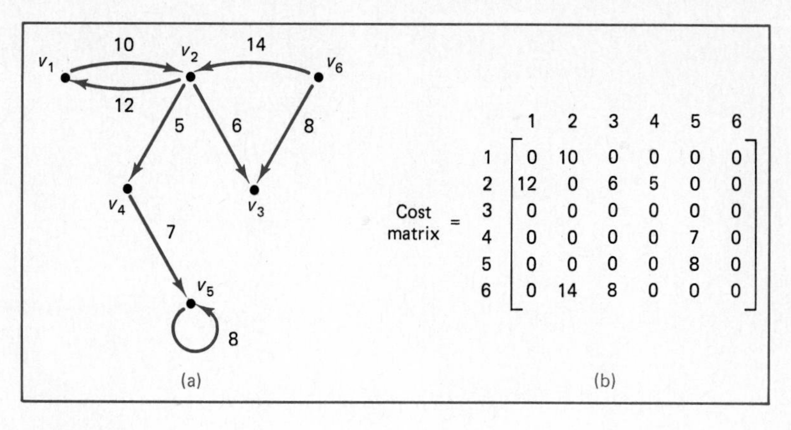


FIGURE 6.41

## Summary and Selected References

As noted, the theory of graphs has developed from a number of sources and can be studied from a variety of approaches. The definitive mathematical treatise on graph theory is Harary (1969). Other theoretical approaches can be found in Behzad and Chartrand (1971) and Christofides (1975). On the more applied side are Deo (1974), Bondy and Murty (1976), and Busacker and Saaty (1965).

Graph theory books from an intuitive view without an abundance of rigor include Ore (1963), Malkevitch and Meyer (1974), and the delightful presentation by Chartrand (1977). As applied to computer science, the reader will find that the following have substantive sections: Berztiss (1975), Korfhage (1974), Prather (1976), Preparata and Yeh (1973), Liu (1977), Gersting (1982), and Sahni (1981).

# 7 Applications of Graph Theory

## 7.1 Introduction

The theory of graphs is extensive and in Chapter 6 we only hit the highlights and explored enough definitions to make the reader conversant with graphs as an important structure of a discrete nature. In like manner, the applications of the theory are numerous. In fact, there are volumes devoted to this aspect of graph theory alone. In this chapter we make no attempt to recount them all. We will try, however, to give some of the flavor of these applications as they relate to computer science.

As a discipline, computer science relies heavily on the practical, applied side of graph theory. Graphs are used as tools to develop further theories in computer science, and, as mentioned previously, to model many situations that can be solved effectively by computers. Thus some knowledge of graph theory and its applications acts as a double-edged sword in your attack on further computer science courses.

In the recent computer science literature, graph-theoretic applications have become almost synonymous with the study of the development of associated algorithms. The representation of graph structures in a form easily manipulated by computers in an extremely important topic, and these forms are numerous. Perhaps the most useful and intuitive way for us to think of this storage is as a two-dimensional array to represent the adjacency or cost matrix of a graph or network. Most of the relevant information can be obtained by examining the entries of this array. Once stored in the memory of a machine, how best to manipulate these data is a proper topic of study in a course on the design and analysis of algorithms. For the purposes of this book we omit such development. We explore how graphs are useful and give some of the flavor of their applications.

## 7.2 Path Problems

Eulerian Graphs The birth of graph theory is usually pinpointed to the year 1736, when an amusing problem was solved by the Swiss mathematician Leonhard Euler (1707–1783). The problem had been posed as a puzzle by the inhabitants of the then Prussian city of Königsberg, now Kaliningrad in Russia. The city is located on the banks and on two islands of the river Pregel. The land masses are connected by seven bridges as shown in Figure 7.1. The problem posed to Euler was this: As the city dwellers strolled about town, was it possible to return to the starting point after having crossed each of the seven river bridges once and only once?

Euler converted the problem to a graphic representation. Each land mass would be a vertex and the various bridge connections would be the edges joining the vertices. The resulting multigraph is pictured in Figure 7.2, and the puzzle thus becomes (in graph-theoretic terms): Can one trace the figure starting at one of the vertices in a continuous fashion, travel each edge only once, and return to the starting vertex (a closed walk)? You should try this without lifting your pencil from the paper as you trace the figure.

**FIGURE 7.1** City of Königsberg with its seven bridges.

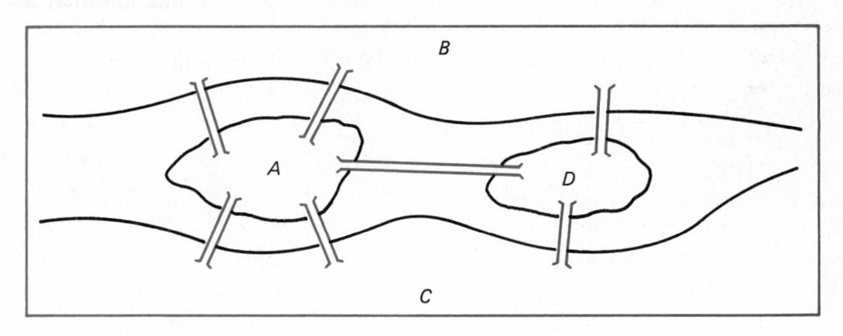


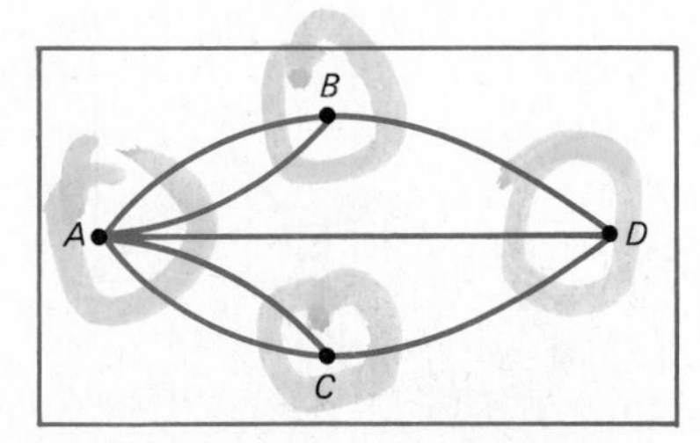


**FIGURE 7.2** City of Königsberg as a multigraph.

Such a traversal would be a closed *trail* (a circuit) since the *edges* are to be distinct in this walk. The number of times each vertex (land mass) is encountered is not under consideration. Euler proved that *no* such closed traversal is possible in the multigraph of Figure 7.2. When a (multi) graph is so traversable it is said to be *Eulerian* and possesses an *Eulerian circuit.* An *Eulerian trail* is a walk that contains all the edges (exactly once) but is *not* closed. A graph having such a trail is said to be *traversable.*

The graph of Figure 7.3(a) is easily seen to be Eulerian. A closed trail is the sequence of vertices: $A$, $B$, $F$, $A$, $E$, $C$, $D$, $F$, $C$, $A$. This trail contains each of the nine edges once and only once. On the other hand, the graph of Figure 7.3(b) does not contain a *closed* trail covering each edge. It does, however, possess an Eulerian trail: $U$, $V$, $Z$, $X$, $V$, $Y$, $U$, $W$, $Y$, $X$, $W$, thus is traversable.

What is it about a graph that allows it to have an Eulerian trail or have a closed Eulerian trail? The answer lies in the following theorems.

□ Theorem 7.1 A connected (multi) graph $G$ is Eulerian (has a closed traversable trail or circuit) iff every vertex of $G$ is even.

First, let us show that if $G$ is Eulerian (and hence a circuit $C$ exists), then every vertex must be even.

**Proof** Each occurrence of a given vertex in the circuit $C$ contributes 2 to the degree of the vertex since the circuit must enter and leave the vertex by different edges. Since each edge appears exactly once in $C$, then every vertex must have even degree.

Now let us show the *converse;* that is, if each vertex of $G$ has even degree, then there will be an Eulerian circuit. The proof of this part is relatively easy to conceptualize but lengthy to explain. We approach the proof from a constructive sense; that is, we present an *algorithm* for finding a circuit $C$ and will substantiate each claim made, as we go. We start by assuming that each vertex of $G$ is even.

**FIGURE 7.3**

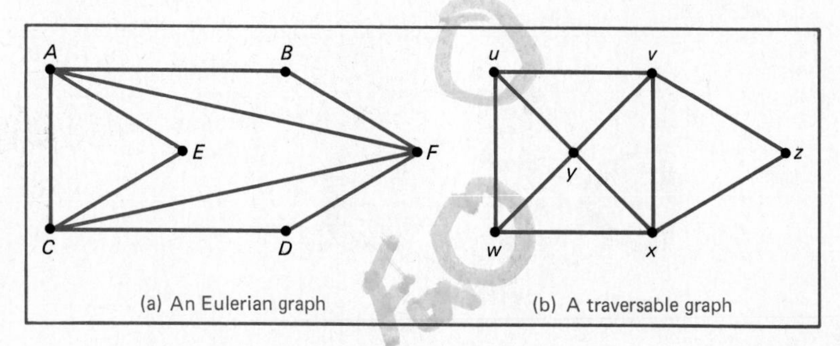

**Proof**

1. Start the construction of $C$ at any arbitrary vertex $v$ of $G$.
2. From $v$, choose any edge (unused at this point) and continue to wander among the vertices of $G$ using only *unused* edges. When a vertex $w$ is reached out of which there are no unused edges, we claim that a *cycle* has been constructed; that is, vertex $w$ = vertex $v$. (Not all the edges of $G$ have necessarily been used. However, if they have all been used, the required Eulerian circuit has been found.)

To substantiate the claim that $w = v$, we assume to the contrary that $w \neq v$. Now each previous occurrence of vertex $w$ in $C$ used one edge to enter and a different edge to exit. On this last entry to vertex $w$ only one edge has been used. Since $w$ has even degree, then there must be at least one edge incident with $w$ not on circuit $C$, which means that $C$ can be continued, and thus does not terminate at vertex $w$. Thus, by contradiction, vertex $w$ = vertex $v$. If, however, the circuit, $C$, thus constructed contains all the edges of $G$, then $C$ is Eulerian, thus proving the theorem, and the construction halts.

3. We have constructed a circuit $C$ that does not contain all the edges of $G$. Since $G$ is connected there is at least one vertex $u$ on $C$ that is incident with edges not on $C$. Remove all edges of $C$ from $G$ obtaining a sub-(multi) graph $G'$ of $G$. Every vertex of $G'$ is even since every vertex of $C$ was even. Choose the vertex $u$ to begin a new circuit which will be determined as in step 2.
4. Continue steps 2 and 3 obtaining circuits which can be inserted in the original circuit $C$ at any one instance of vertex $u$ in that circuit. Since there are a finite number of vertices and edges in $G$, this procedure can be accomplished in a finite number of times; thus an Eulerian circuit can be obtained.

We *illustrate* this construction by using the graph of Figure 7.4(a). It has an Eulerian circuit since each vertex is even. Let us proceed with its construction.

At step 1 of the construction we start with vertex $v_1$ and by step 2 produce the circuit $v_1, v_4, v_7, v_8, v_4, v_3, v_5, v_1$ of length 7, as shown by the arrows of Figure 7.4(a).

Figure 7.4(b) shows the resulting graph with circuit $C$ deleted as outlined in step 3. In this new graph we can start at vertex $v_4$ and produce the circuit $v_4, v_5, v_6, v_{10}, v_9, v_4$ of length 5 as shown by the arrows in Figure 7.4(b). This new circuit can be inserted at $v_4$ in the first circuit, yielding the circuit

$$v_1, v_4, \underbrace{v_5, v_6, v_{10}, v_9, v_4,}_{\text{second circuit}} v_7, v_8, v_4, v_3, v_5, v_1$$

of length 12.

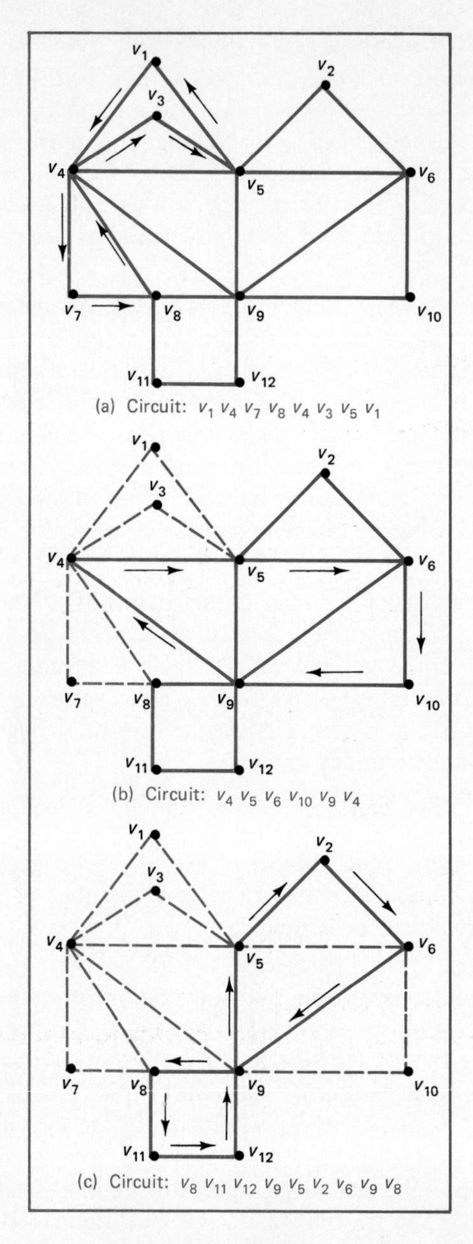

(a) Circuit: $v_1\, v_4\, v_7\, v_8\, v_4\, v_3\, v_5\, v_1$
(b) Circuit: $v_4\, v_5\, v_6\, v_{10}\, v_9\, v_4$
(c) Circuit: $v_8\, v_{11}\, v_{12}\, v_9\, v_5\, v_2\, v_6\, v_9\, v_8$


FIGURE 7.4

Now delete this second circuit [Figure 7.4(c)], yielding another subgraph with every vertex being even. To complete matters we may now start at vertex $v_8$ and produce the circuit $v_8v_{11}v_{12}v_9v_5v_2v_6v_9v_8$ of length 8. Inserting this new circuit at $v_8$ of the last one we get the circuit

$$v_1\underbrace{v_4v_5v_6v_{10}v_9v_4}_{\text{second circuit}}v_7\underbrace{v_8v_{11}v_{12}v_9v_5v_2v_6v_9v_8}_{\text{third circuit}}v_4v_3v_5v_1$$

Thus ends the construction of the desired circuit.

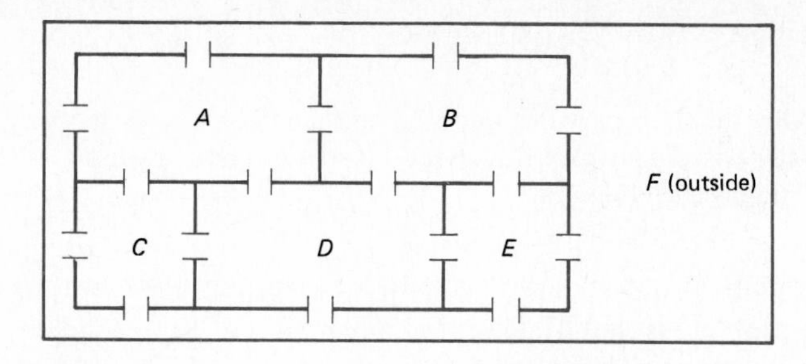


**FIGURE 7.5** Floor plan.

□ Theorem 7.2 A connected (multi) graph $G$ is traversable (a noncircuit trail including all the edges) iff $G$ has exactly two odd vertices. The trail will begin at one of these odd vertices and end at the other.

**Proof** This theorem follows immediately from Theorem 7.1. Let $v$ and $u$ be the two odd vertices. Then attach an edge joining vertex $v$ to vertex $u$ (an additional edge if they are already adjacent). Now every vertex is even and therefore an Eulerian circuit exists from Theorem 7.1. Construct this circuit. Now delete the new $(v, u)$ edge. Then a (nonclosed) trail will exist starting at either vertex $v$ or vertex $u$ and ending at the other vertex.

We see now why the people of Königsberg were having trouble with their Sunday strolls. The land masses were all "vertices" with odd degree!

**EXAMPLE 1** One often encounters Eulerian graphs as puzzles. Consider the famous floor plan (Figure 7.5) that consists of five rooms interconnected with themselves and the outside by doors on every wall. The puzzle is to start in one room or the outside, walk through every doorway exactly once, and return to the starting point. Of course, one tries this with pencil and paper and attempts the trace without lifting the pencil from the paper.

We redraw the plan as a multigraph in Figure 7.6. Note the number of odd vertices. There are four of them. Thus the trace cannot be accomplished. (Since Euler solved this problem for us, maybe it won't appear in the puzzle section of the papers anymore!)

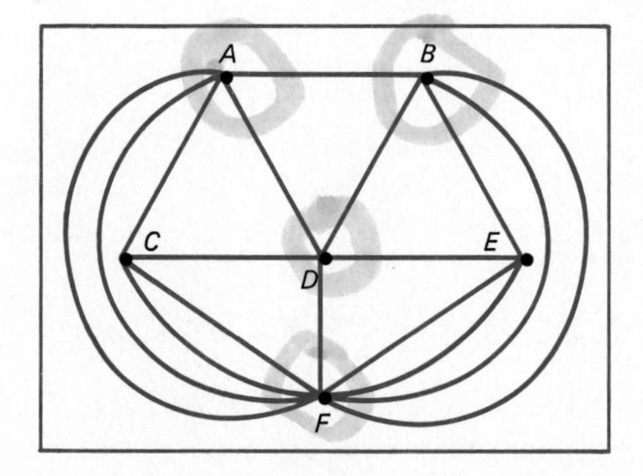


**FIGURE 7.6** Multigraph of the floor plan.

## EXERCISE

1. Modify the floor plan in Figure 7.5 so that there are no doorways to the outside. Is this revised floor plan traversable? If so, can one start in any of the rooms? Draw the modified multigraph.

Hamiltonian Graphs: The Traveling Salesman Problem Another closely related problem is that of determining whether a graph can be traversed by visiting each *vertex* of the graph once and only once. This problem was first examined by the Irish mathematician Sir William Rowan Hamilton (1805–1865). The problem posed is to find a *cycle* that passes through each vertex of the graph. Such a cycle is called a *Hamiltonian cycle* and a graph that possesses one is called a *Hamiltonian graph.* A path (not closed) that passes through every vertex of a graph is said to be a *Hamiltonian path.* A graph may possess a Hamiltonian path without possessing a Hamiltonian cycle. This can be seen from the graphs of Figure 7.7. The graph of Figure 7.7(c) is a planar representation of Hamilton's "traveler's dodecahedron" (1859), in which he labeled each vertex as a well-known city. The object of his "game" was for the player to travel "around the world" by visiting each city exactly once by moving according to the indicated edges. The heavy lines in the figure indicate one such path.

There is a certain similarity between the graph properties of being Eulerian and Hamiltonian. In the former, one passes through each edge exactly once, and in the latter, each vertex exactly once. The property of being Eulerian was determined by the degrees of the vertices as shown in Theorem 7.1. In spite of the analogy there is, unfortunately, no known effective way to determine whether a graph is Hamiltonian. There are several sufficient conditions, one of them being that the graph be complete, but there are no known necessary conditions.

A special case of the problem of finding a Hamiltonian cycle is the classic *traveling salesman problem.* This is one of the best known problems of a class of those that are easy to state but very difficult to solve. In this problem a

**FIGURE 7.7**

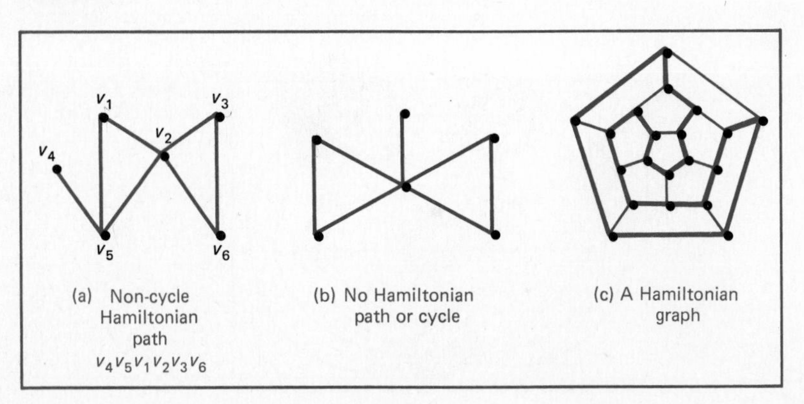


(a) Non-cycle Hamiltonian path $v_4v_5v_1v_2v_3v_6$

(b) No Hamiltonian path or cycle

(c) A Hamiltonian graph

salesman is to visit a number of cities in a round trip from home and would like to do this in the most inexpensive manner possible. The problem can be viewed as a network with the vertices being labeled as cities and the weighted edges being distances between the cities. In graph-theoretic terms, the problem is to find the *shortest Hamiltonian circuit* of the network.

The most straightforward way to attempt a solution is to consider one of the cities (vertices) as the home base of the salesman. Then, of the remaining $n - 1$ vertices of the network of order $n$, list all $(n - 1)!$ cycles and their total distances, and choose the shortest of these computed distances. This exhaustive inspection may be effective for small $n$. The practicality for large $n$ is questionable. Each of the $(n - 1)!$ cycles to be inspected involves $n$ additions. Thus we have $n(n - 1)! = n!$ additions to compute.

Suppose that there are 20 cities in the route (probably not unreasonable for the average salesman). This would involve 20! additions. This value is approximately $2.4 \times 10^{18}$. Now assuming a computer that can perform 1,000,000 additions per second, it would take more than 76,000 years just to complete the additions in this combinatorial explosion!

Many attempts have been made to solve the traveling salesman problem, most with limited success. Many approaches are through heuristic algorithms, or *heuristics*. These are methods that are usually good, but not necessarily optimal, and they are faster and easier to implement than any known exact algorithm.

One such heuristic for the traveling salesman problem, known as the *greedy* method, is presented here to give an example of one that is intuitively easy to implement but can be far from optimum. The algorithm produces partial solutions that "look good" at any particular time (locally) but can be globally "costly" in that the cheapest tour is not necessarily found.

□ Algorithm: Greedy Traveling Salesman

1. Set the total cost to 0.
2. Select a home vertex. Designate it as $v$.
3. Choose the *least costly unused* edge from $v$, $(v, u)$. Mark this edge used; add its cost to the total cost.
4. Let $v \leftarrow u$. (Designate $u$ as the new vertex, $v$.)
5. Repeat steps 3 and 4 until all vertices are visited and the tour is back at "home."

To illustrate this "greedy algorithm," consider the network of Figure 7.8(a). We choose the least costly edge from "home," this being to vertex $C$, and the cost of the tour is 10 [see Figure 7.8(b)]. From vertex $C$ we travel to vertex $A$, giving the total cost at this time to be 25 [Figure 7.8(c)]. From vertex $A$ we go to vertex $B$, making the total cost 50 [Figure 7.8(d)]. Going from $B$ to the home vertex, we find that the total cost is 110 [Figure 7.8(e)]. It is obviously *not* a minimum tour. The tour $H$–$C$–$B$–$A$–$H$ gives a cost of 75.

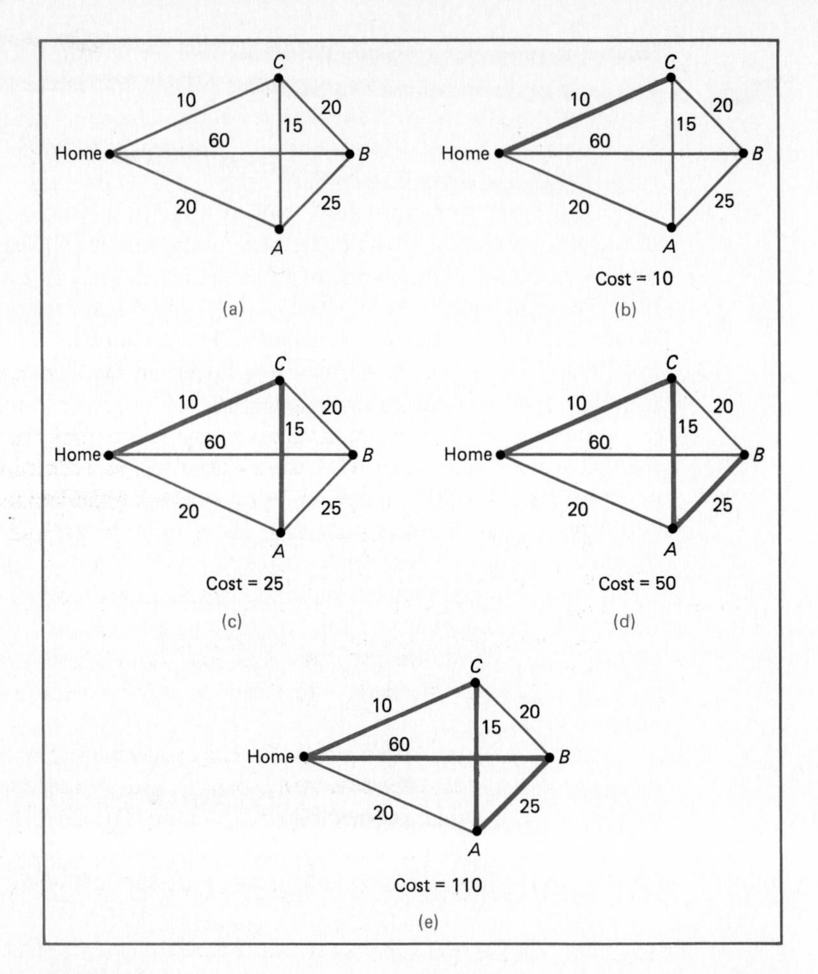


**FIGURE 7.8**

Shortest Paths in Networks One of the useful applications of networks is that of finding a shortest path between two vertices of the network. Usually, the network is given as directed, but this is not a necessity. If we consider the set $V$ in our network $G(V, E)$ as a set of cities and the set $E$ as the interconnecting highways, then the usual assignment of weights would be the distances between the cities.

As with the traveling salesman problem, the length of a path in $G$ will be defined as the sum of the weights (distances) along the edges of the path. Our problem will be to determine a (not necessarily unique) path of shortest length between a beginning vertex (the source of the path) and a terminal one (the destination).

For purposes of illustration, consider the network $G$ of Figure 7.9. We will show a portion of a scheme that will find a shortest route from the source, vertex $A$, to the destination, vertex $K$. The complete algorithm will not be given. Enough will be shown so that the reader can understand its nature. Notice in

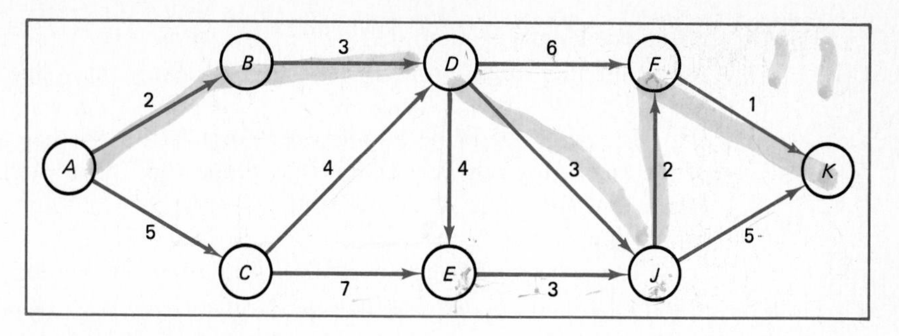


FIGURE 7.9 Network.

the network that even though there is a direct path from vertex $D$ to vertex $F$ of length 6, it is more economical to go by way of vertex $J$, in which case the length is 5.

One way to find the shortest path between two vertices is to enumerate *all* possible paths. In a "small" network this may be entirely feasible, but for reasons presented earlier, this becomes untenable in terms of computer usage. For example, one can assign the source vertex, $A$, as the root of a *tree* and construct branches indicating the various path lengths. This scheme is partially shown in Figure 7.10. Then one merely sums all the branch lengths to terminal vertices and takes the shortest one.

In our procedure to follow, due to Dijkstra, we use the idea from this tree enumeration. We find a shortest path from the source vertex, $A$, to vertices adjacent to it, then a shortest path to other vertices is found, until eventually we stop when a shortest path to the destination vertex is found.

After the costs from $A$ to $B$ and from $A$ to $C$ are determined to be 2 and 5, respectively, it is seen that the cheapest way to get to vertex $D$ is through $B$,

FIGURE 7.10 Enumeration of all paths from vertex $A$ of network of Figure 7.9.

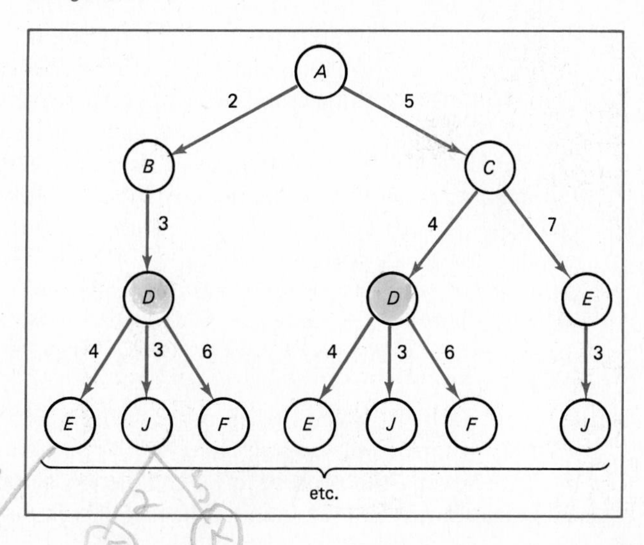

where the length is computed to be 5. (If we go via $C$, the length is 9.) Thus we retain the length from $A$ to $D$ by assigning the value 5 to a variable, len($D$).

The length of the path to vertex $E$ from vertex $A$ is seen to be the *minimum* of the paths through $D$, with len($D$) = 5 and through $C$, with len($C$) = 5. It is found that the value to be assigned to len($E$) then is 9 via vertices $B$ and $D$ rather than vertex $C$.

The procedure is continued by calculating the lengths to vertices $F$ and $J$ and eventually $K$, the destination. It can be seen that the shortest path from vertex $A$ to vertex $K$ is 11 and the path is $A$–$B$–$D$–$J$–$F$–$K$.

The algorithm is given here in abbreviated form only, with the idea being that the reader will sense the nature of the problem as an application of graphs. Most books devoted to data structures, algorithm analysis, and operations research will present detailed algorithms and storage management of data. The thrust there will be on the development of algorithms. The purpose here is to present a survey of such applications.

Critical Path Method and Scheduling Directed networks often represent models for problems involving scheduling and planning. The vertices usually represent activities or events of a project and the directed edges portray the time it takes to complete the individual event. The directions also indicate prerequisite requirements in that if the edge $\langle v_i, v_j \rangle$ is present, it is understood that activity $v_i$ has to occur before $v_j$ can be started. For this reason, such structures are called activity digraphs.

In such projects where cost is related to productivity it is essential to optimize the total time spent. The method to accomplish this is to find the *longest* path from a source vertex to a destination vertex. There will be no way the project can be completed until all the prerequisites are fulfilled with their associated times for completion. This scheduling is critical to the timing of the overall project. The method is known as the *critical path method* (CPM). Another name for a digraph structure representing the project is *PERT* chart, an acronym for "performance evaluation and review techniques."

To illustrate the method, we can again use the network pictured in Figure 7.9. We can say that at the start of the project we are at vertex $A$ and the completion will be at vertex $K$. It is seen that it will take 2 units of time to complete activity $A$ before activity $B$ can be begun, and 5 units of time to complete the activity before vertex $C$. Both activities $B$ and $C$ are required to be finished before activity $D$ can be started. Thus instead of looking for a shortest path from vertex $A$ to vertex $D$, we need to know the amount of time required before activity $D$ can be begun in order to properly allocate resources. This *longest* path is seen to be 9 (because of the 5 units from $A$ to $C$ and the 4 units from $C$ to $D$).

Similarly, it is seen that in order to begin activity $E$, we need to allocate 13 units of time from the start at vertex $A$. This path is represented by $A$–$C$–$D$–$E$. Following the process to its completion at vertex $K$, the reader should show that it will take 21 units of time from start to finish.

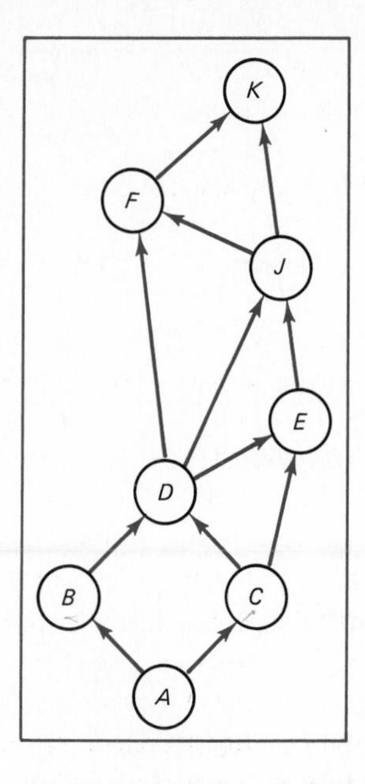


**FIGURE 7.11** Activity graph as a partial order (see Figure 7.9).

An important point to note about activity digraphs is that there must be no (directed) cycles present. Thus these graphs represent an application of *dags* (directed acyclic graphs) mentioned in Section 6.6. If a cycle were in the structure, then an activity could become a prerequisite of its own prerequisite and the project could never be completed!

Another interesting point to be made concerning activity digraphs is that they are *partially ordered* with respect to the relation "is a prerequisite for." The partial order is incompletely specified since reflexivity does not hold (event $X$ is *not* a prerequisite for itself). You should show that this relation is antisymmetric and transitive. Also notice that no order is specified between activity $B$ and activity $C$. Figure 7.11 shows this partial order for Figure 7.9 in a different orientation.

To here Wed

Planarity and Coloration As we encountered earlier with the traveling salesman problem, many far-reaching problems are so simple to state and yet so difficult to solve. One of the most famous is one that stems from a map-coloring problem: Given a map representing various regions (states, countries, etc.), what is the minimum number of colors required to color the regions so that no two adjacent regions have the same color? This minimum number for a map, $M$, is known as the *chromatic* number of $M$, designated by $\chi(M)$. ($\chi$ is the lowercase Greek letter chi.)

As usual, as an aid in visualizing the situation, the map can be represented graphically. One labels the vertices of the graph as the regions of the map and the edges reflect the relation of adjacency. Only those regions that have borders in common will be joined by edges. For example, the map of Figure 7.12(a) can be drawn as the graph of Figure 7.12(b). Note that regions $A$

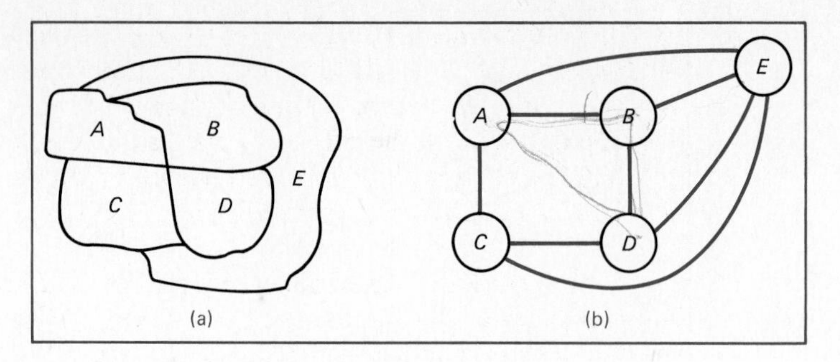


**FIGURE 7.12** Map with its graphic representation.

and $D$ are not considered adjacent, nor are $B$ and $C$. They have no common boundary.

One may attempt to color a map as follows. For the map in Figure 7.12, we may color region $A$ with color 1 (this merely represents the first color: red, blue, green, or whatever). Quite obviously, regions $B$, $C$, and $E$ cannot be assigned color 1, for they are adjacent to region $A$. So, arbitrarily, we assign color 2 to region $B$. Since region $C$ is not adjacent to region $B$, we may also color it with color 2. There is no need to use a third color on region $D$; it is not adjacent to region $A$, so we may use color 1 again. This coloring, however, forces us to use color 3 on region $E$. Thus the chromatic number, $\chi(M)$, is 3. A coloring of this map is shown in Figure 7.13(a). A map that requires four colors is shown in Figure 7.13(b) and (c).

From these examples, two questions can be observed. (1) What is the minimum number of colors required to color any given map; or how many crayons should a cartographer have at a minimum? (2) What kind of graphs arise from maps? We explore question (2) first.

It should be evident that the graphs that arise from maps have to be drawn so that no edges cross in the plane except at vertices. There should be no crossovers. This is exactly the definition of *planar* graphs given in Section 6.4. We won't prove this fact here, but the reader should draw enough maps and associated graphs to be convinced.

**FIGURE 7.13**

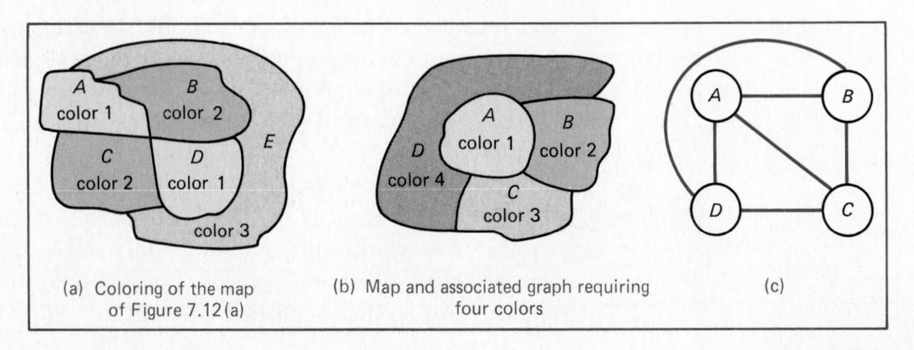


(a) Coloring of the map of Figure 7.12(a)

(b) Map and associated graph requiring four colors

(c)

Question 1 has plagued some of the best mathematical minds since it was (supposedly) first posed in 1852 in a letter from Augustus De Morgan to Sir William Rowan Hamilton. We saw in Figure 7.13 that it took three colors on one map and four on the other. No one has been able to demonstrate a map that took more than four; thus the famous conjecture: the minimum number of colors required to color a map (or color the vertices of a planar graph) is four.

This four-color conjecture remained unsolved until 1976 when a computer-aided proof was announced by Apel and Haken showing that the minimal number of colors required is, indeed, four. The proof took over 1200 hours of computer calculations, analyzing about 2000 graphs involving millions of cases. To this date, though, no "elegant" mathematical solution has been found for this baffling conjecture. The interested reader should not expend too much effort to find one.

Earlier we showed briefly how one might find a coloring for an arbitrary map or graph. Let us now present an algorithm that has been found useful, but not perfect:

□ Welsh-Powell Algorithm First represent the map as a graph and order the vertices according to decreasing degree. (This ordering may not be unique if some vertices have the same degree.) Assign color number 1 to the vertex of largest degree. Proceeding sequentially through the ordered list of vertices, assign this color to any vertex not adjacent to a vertex having the same color. Repeat this process with the second color, and so on, until all vertices have been colored.

Let us illustrate this algorithm on the map and associated graph in Figure 7.14. First we color vertex $E$ with color 1. Proceeding down the list, the first vertex not adjacent to $E$ is vertex $C$. Color it 1. The rest of the vertices are either adjacent to $E$ or $C$. Going back to the top of the list, assign color 2 to vertex $F$, and also to vertices $A$ and $H$. The third pass through the list will assign color 3 to vertex $B$ and also to the rest of the vertices, $I$, $D$, and $G$. Thus $\chi(M) = 3$. We get the coloring shown in Table 7.1.

Graph coloring has useful application to problems of scheduling. For example, let us schedule classes for a school. Denote the vertices of the graph as

**FIGURE 7.14**

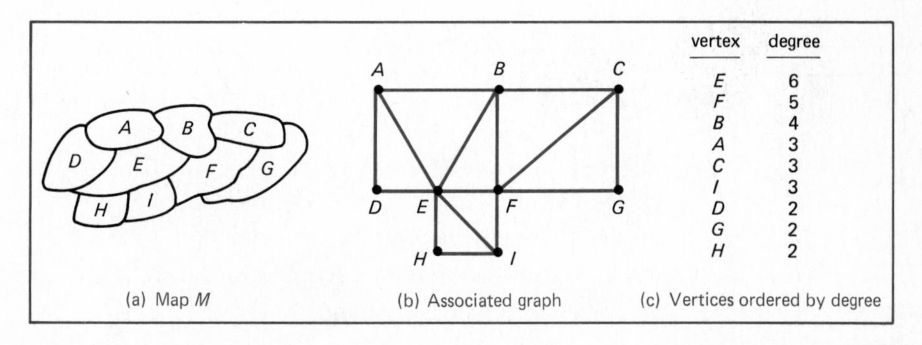


| vertex | degree |
|---|---|
| E | 6 |
| F | 5 |
| B | 4 |
| A | 3 |
| C | 3 |
| I | 3 |
| D | 2 |
| G | 2 |
| H | 2 |

(a) Map $M$ (b) Associated graph (c) Vertices ordered by degree

TABLE 7.1

| Vertex | Color: First Pass | Second Pass | Third Pass |
|---|---|---|---|
| *E* | 1 | | |
| *F* | | 2 | |
| *B* | | | 3 |
| *A* | | 2 | |
| *C* | 1 | | |
| *I* | | | 3 |
| *D* | | | 3 |
| *G* | | | 3 |
| *H* | | 2 | |

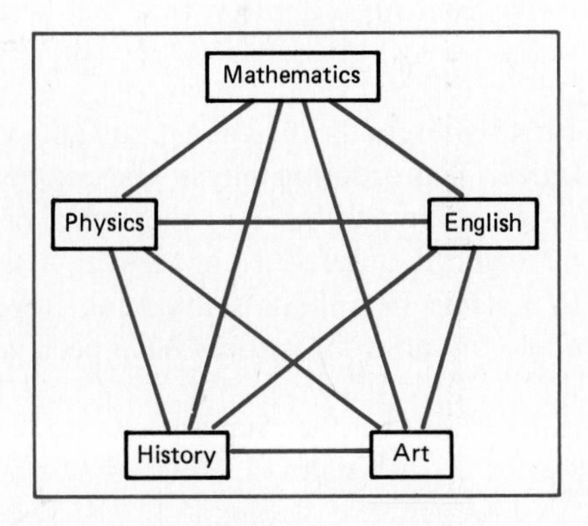


**FIGURE 7.15** Nonplanar scheduling (coloring) graph.

classes and join two vertices by an edge iff at least one student has to take the associated classes simultaneously. One must schedule these classes at different times, and this is equivalent to assigning the vertices different colors.

We do not necessarily get *planar* graphs from this scheduling problem. The $K_5$ graph (the star graph) represents a student who takes five classes (see Figure 7.15). We know from Chapter 6 that this graph is not planar, so we need at least five different times (colors) in the schedule.

## EXERCISES

2. Which of the graphs below
   (a) Are Eulerian (possess an Eulerian circuit)?
   (b) Are traversable (possess an Eulerian trail)?
   (c) Are Hamiltonian (possess a Hamiltonian cycle)?
   (d) Possess a Hamiltonian path?

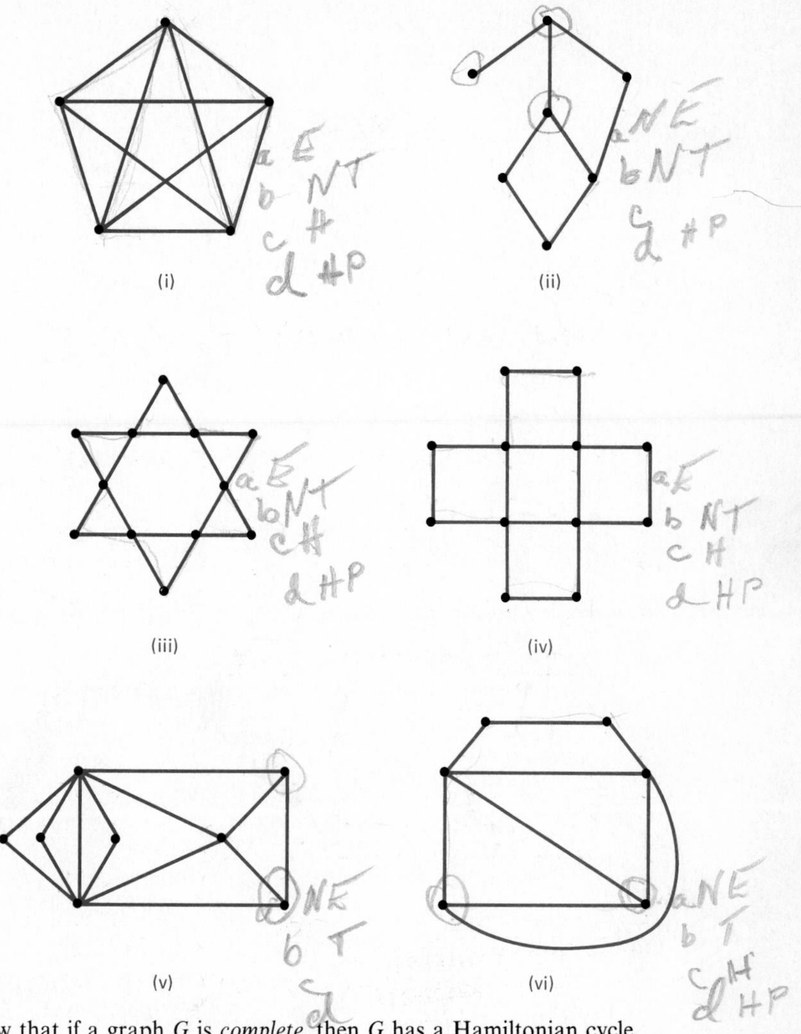


3. Show that if a graph $G$ is *complete*, then $G$ has a Hamiltonian cycle.

4. Find a least costly tour covering all the vertices of the network below. Use vertex $A$ as the home base of the "salesman." Show that the greedy method does not produce an optimum tour.

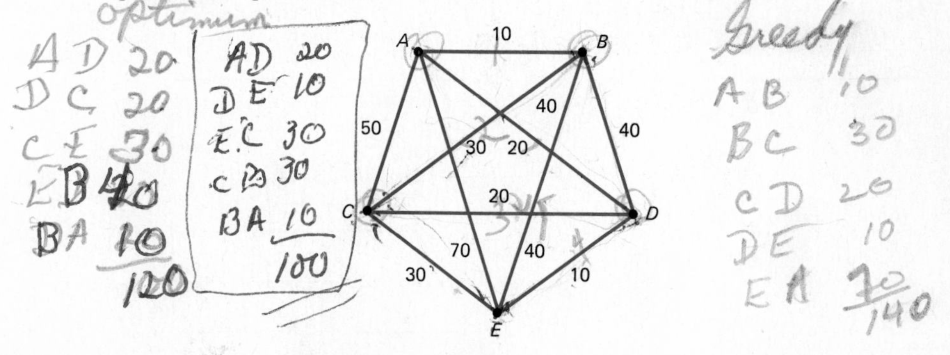

5. In the digraph below find a shortest path from vertex $v_1$ to vertex $v_{11}$.

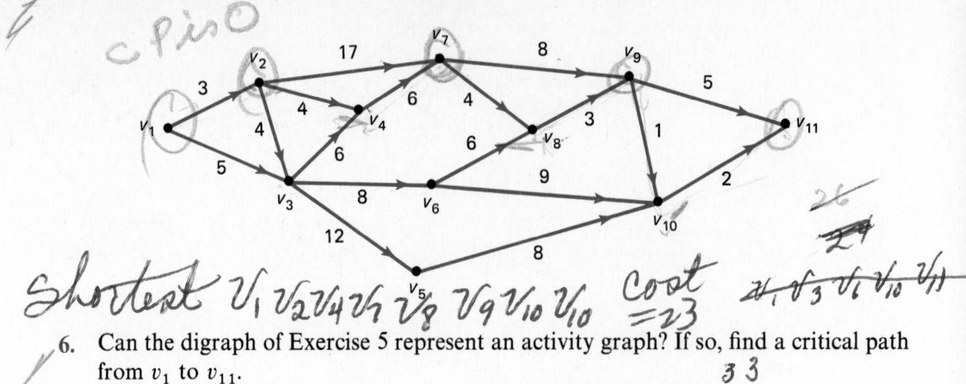


6. Can the digraph of Exercise 5 represent an activity graph? If so, find a critical path from $v_1$ to $v_{11}$.

7. Can the digraph below represent an activity graph? Why or why not?

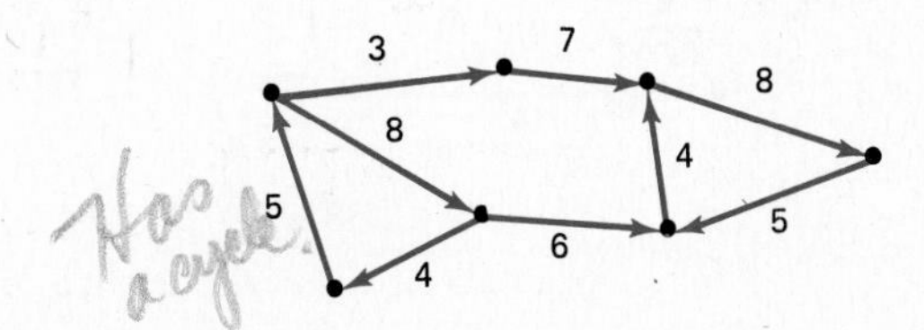

8. Can the graphs below be the graph of maps (i.e., do they have planar representation)?

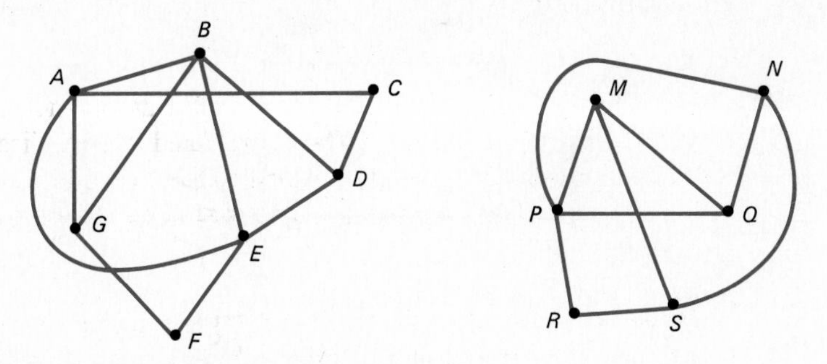


9. Show that all trees are 2-colorable.

10. Show that all bipartite graphs are 2-colorable.

11. Find the chromatic number of the map below. Use the Welsh-Powell algorithm to display a coloring of the regions. Show the associated graph.

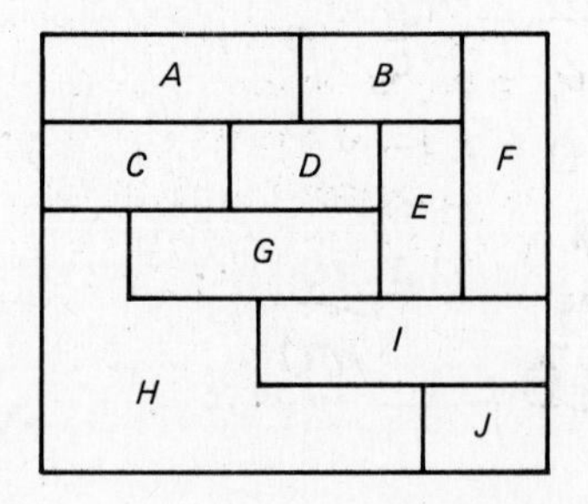

12. (a) For the map below use the Welsh-Powell algorithm on the vertices ordered as

$$F \quad K \quad A \quad B \quad C \quad G \quad I \quad D \quad E \quad H \quad J$$

and show that the coloring produced uses more than four colors.

(b) Revise the ordering (or revise the algorithm) to display a coloring using $\leq$ four colors.

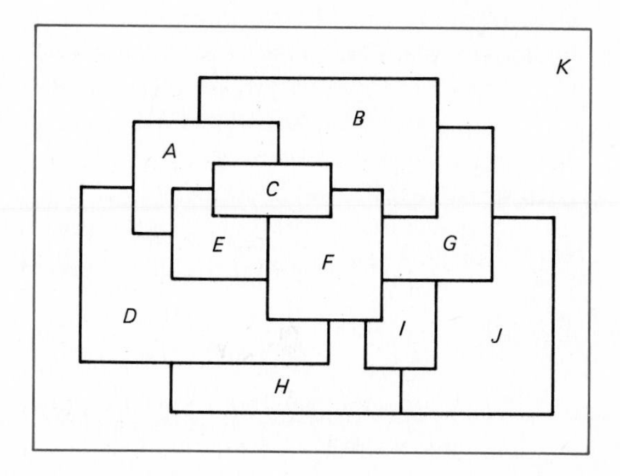


## 7.3 Applications of Trees

Trees are probably the single most useful graph structure as applied to computer science. In Section 7.2 we examined one application, the enumeration tree, in an attempt to locate shortest paths in graphs. Other varied applications will be presented in this section. As can be seen, their applicability is wide and diverse.

Minimum Spanning Trees As a first application of trees, let us review briefly the procedure mentioned in Section 6.5 for creating spanning trees from a graph. There we stated: "One can merely delete edges that are not bridges from a graph until there are no more cycles." This is very easy to do visually since a human being can "see" cycles.

Asking a computer to do this is another matter. First, the graph structure has to be stored, possibly as an adjacency matrix, in such a manner that it becomes meaningful to ask: (1) Is this edge a bridge (that is, upon deletion, will the resulting graph be disconnected?) In other words, you will need an algorithm to determine if a particular graph is connected. (2) Are there more cycles in the graph? Here again an algorithm will have to be written to check to see if the graph is acyclic.

Such questions as these can become horrendous to answer unless the data representing the graph are structured in such a manner as to make such decisions fairly routine. This structuring of data is again best left to other courses.

Let us suppose that our graph in question has weighted edges; thus is a network, with the edges representing costs, time spent, or distances or some other measure between vertices. There may be many spanning trees of ordinary (nonweighted) graphs, but when one adds weights, a meaningful question becomes: What edges should be left in the network in order that the spanning tree have a total minimum "cost"? The job, then, is to construct a *minimum spanning tree* (sometimes called an *economy tree*).

It turns out that there are *two* relatively (at least conceptually) easy algorithms for this. The key point to knowing whether or not we have actually built a tree and thus to terminate the algorithm is to make use of Theorem 6.5, which states that a finite tree of order $n$ has exactly $n - 1$ edges.

The first algorithm, REMOVE, due to Prim, starts with a network of order $n$ and size $m$. It successively deletes high-cost edges that are not bridges until $n - 1$ edges remain. The second algorithm, BUILD-UP, due to Kruskal, starts with a network of $n$ vertices (and no edges initially). It successively adds low-cost edges that do not form cycles, until the count reaches $n - 1$.

□ Algorithm: REMOVE

1. Start initially with a connected network $G(V, E)$ of order $n$ and size $m$.
2. Repeat the following (a)–(c) below until size $= n - 1$.
   (a) Select the edge $e \in E$ of greatest weight.
   (b) Let $G' = G - \{e\}$.
   (c) If $G'$ is connected, let
      (i) $G = G'$
      (ii) $E = E - \{e\}$
      (iii) Size = size $-$ 1
   Otherwise, let $G$ remain "as is."

To illustrate the algorithm REMOVE on the network of Figure 7.16, we first order the edges according to weight, from high to low, as shown in Table 7.2. We first select edge $\langle A, C\rangle$ of greatest weight and determine that $G$ is still connected. Thus our new network has $\langle A, C\rangle$ deleted and the size is now 8. This is illustrated in Figure 7.17(a).

**FIGURE 7.16** Network of order 6 and size 9.

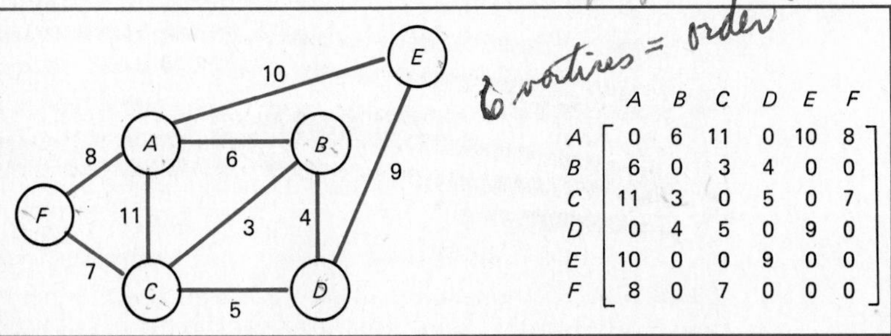

**TABLE 7.2**

| *Ordering of Edges* | *Weight* |
|---|---|
| $\langle A, C \rangle$ | 11 |
| $\langle A, E \rangle$ | 10 |
| $\langle D, E \rangle$ | 9 |
| $\langle A, F \rangle$ | 8 |
| $\langle F, C \rangle$ | 7 |
| $\langle A, B \rangle$ | 6 |
| $\langle C, D \rangle$ | 5 |
| $\langle B, D \rangle$ | 4 |
| $\langle B, C \rangle$ | 3 |

Next we find that the next most expensive edge $\langle A, E \rangle$ does not disconnect $G$ either. This edge is deleted and size = 7 [see Figure 7.17(b)].

The next edge to be deleted, $\langle D, E \rangle$, is observed to be a bridge, so it is left in. The next highest, however, $\langle A, F \rangle$, can be deleted, reducing the size to 6 [see Figure 7.17(c)].

The next edge $\langle F, C \rangle$ is a bridge. So is the next one, $\langle A, B \rangle$. We finally delete $\langle C, D \rangle$ and observe that the size is now 5, which equals 6 − 1, thus a tree [Figure 7.17(d)]. The total cost is 29.

**FIGURE 7.17** Producing a minimum spanning tree.

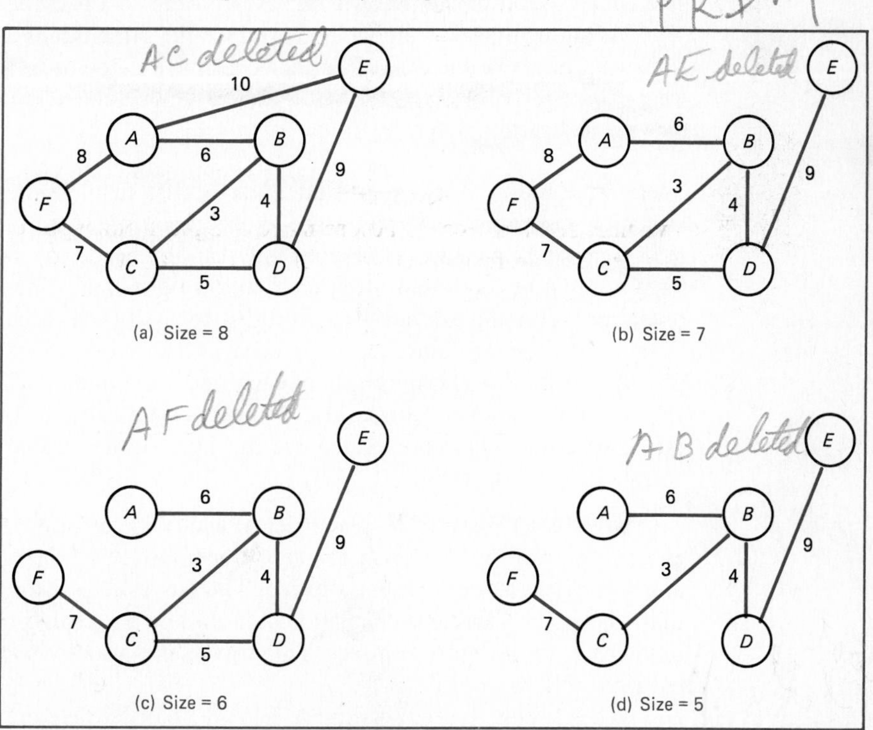

□ Algorithm: BUILD-UP

1. Start initially with the vertices only of the network $G(V, E)$ of order $n$ and size zero.
2. Repeat (a)–(c) below until size $= n - 1$.
   (a) Select the edge $e \in E$ of lowest cost
   (b) Let $G' = G \cup \{e\}$
   (c) If $G'$ has no cycles let
      (i) $G = G'$
      (ii) size = size + 1
   Otherwise, let $G$ remain "as is."

To illustrate algorithm BUILD-UP on the network of Figure 7.16 we first order the edges from low cost to high. This is obviously the reverse ordering of that seen in algorithm REMOVE. The network initially is as in Figure 7.18(a) with no edges. The first edge to be added is $\langle B, C\rangle$ with weight 3 [see Figure 7.18(b)]. The next edge $\langle B, D\rangle$ introduces no cycles, so it is added next [see Figure 7.18(c)].

Edge $\langle C, D\rangle$ of weight 5 introduces a cycle, so it is skipped and edge $\langle A, B\rangle$ is added, giving the size now to be 3, as in Figure 7.18(d).

Edge $\langle F, C\rangle$ of weight 7 is next. No cycles are introduced; thus it is added (see Figure 7.18(e)]. Edge $\langle A, F\rangle$ introduces a cycle, so is skipped. Edge $\langle D, E\rangle$ adds no cycle, so it is added, giving the size to be 5, as shown in Figure 7.18(f). The construction of the tree is complete and the cost is 29, as before.

As shown, the two algorithms produce the same minimum spanning tree. This will always be the case when the edges have different weights. For equally weighted edges, one chooses the one that is convenient, and thus the trees will not always be unique.

Tree Traversals A problem that often occurs in computer applications is that of systematically traversing the nodes of a rooted tree (usually binary) so that every node is visited at least once. Recall that nodes of rooted binary trees have at most two children, a left child and a right child. There are three usual orders of traversing a binary tree from left to right (and three more from right to left, but these are not commonly used). All three traversal schemes involve processing the three components of each node: the node itself as the root of a subtree, the node's left subtree, and the node's right subtree. The *order* in which the three components are visited are the important steps to take in the traversal. They are as follows:

**PREORDER TRAVERSAL** The order in which the components are visited for any subtree will be: *first* visit the root node; *next* visit the entire left subtree; and *last* visit the entire right subtree. The traversal is defined recursively in that when each subtree is encountered, it is to be traversed in preorder fashion until there are no more subtrees. This traversal is usually referred to as node-left-right (NLR) order.

To traverse the tree of Figure 7.19 in preorder fashion, we first visit the

Kruskal.

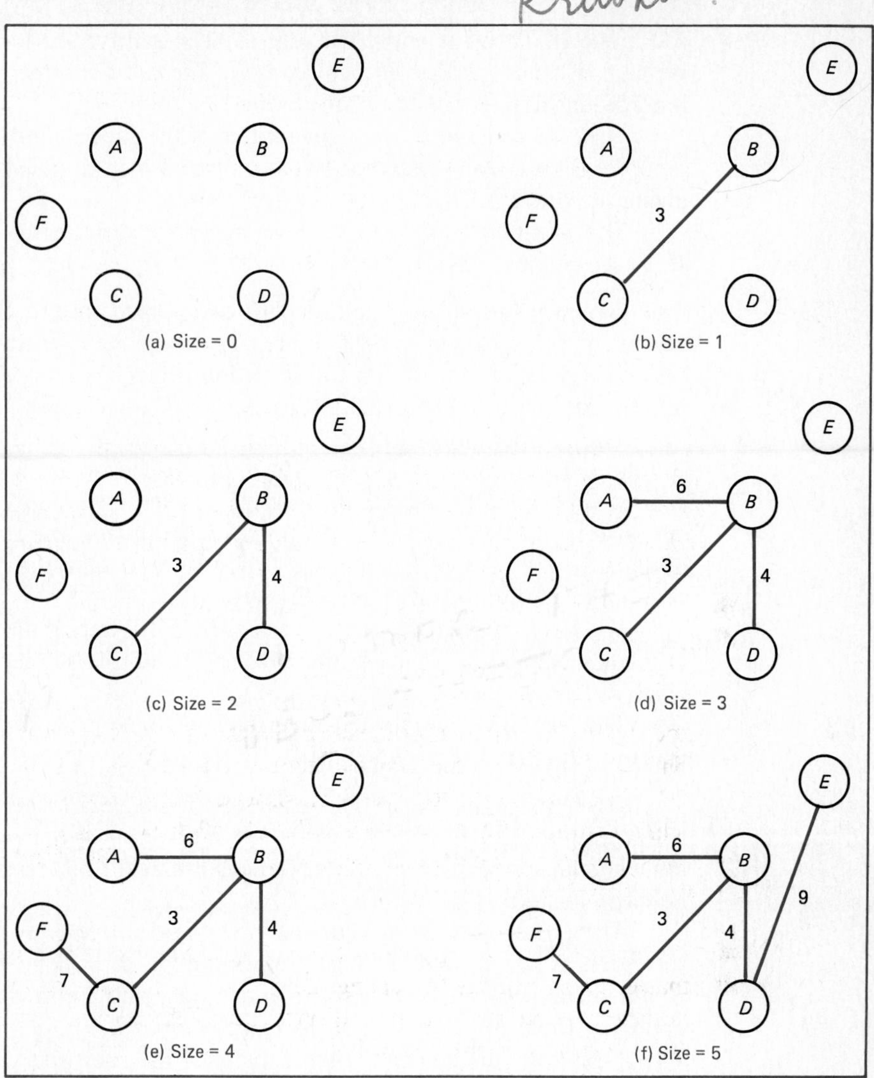


**FIGURE 7.18**

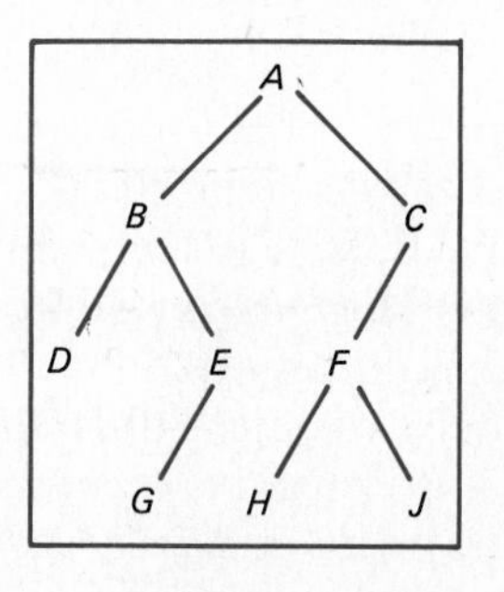


**FIGURE 7.19** Binary tree.

root node $A$. Then the entire left subtree of $A$ is traversed in preorder. There, we first visit node $B$. Its left subtree is $D$. The right subtree of $B$ is visited in preorder fashion, giving the order $E$ and $G$.

Now we go back to the right subtree of the root $A$, and visit its nodes in preorder. This gives $C$ first, then its left subtree, which is visited in NLR fashion, giving the order $F$, $H$, $J$.

The preorder traversal thus gives the node order as: $A\ B\ D\ E\ G\ C\ F\ H\ J$ [see Figure 7.20(a)].

**Inorder Traversal** The order the components are to be visited for any subtree: *first* visit the entire left subtree; *next* visit the root node; and *last* visit the entire right subtree. Again the definition is given recursively and is usually referred to as left-node-right (LNR) order.

The inorder traversal for the tree of Figure 7.19 is as follows: We encounter the root node, $A$. Do *not* visit it, but visit its entire left subtree. In this subtree we encounter the root $B$, traverse its left subtree, and since there is no left subtree of $D$ we visit it first. Thus the first node of the traversal is $D$. We go back to the root of the subtree, $B$, visit it, and take the right subtree. Its order of visitation is first $G$, then $E$.

We have finished the visitation of the left subtree of the root $A$; thus we visit it and head right. The visitation of the right subtree is in the following order: $H$, $F$, $J$, $C$.

Thus we have the inorder traversal as $D\ B\ G\ E\ A\ H\ F\ J\ C$. This traversal is illustrated in Figure 7.20(b).

**Postorder Traversal** The order of components to be visited here will be for any node: *first* visit the entire left subtree; *next*, the right subtree; and *last*, the root node. The subtree traversals are again recursive, and the order is: left-right-node (LRN).

The postorder traversal for our now-famous tree is as follows. First we travel left as far as possible, finally visiting node $D$, back to the subtree node $B$, but before visiting it we visit its entire right subtree in postorder fashion, obtaining the order $G$ and $E$. Now we go back to $B$.

Next we visit the entire right subtree of the root node, $A$, obtaining the order: $H$, $J$, $F$, and then $C$. Finally, we visit the root of the tree, $A$. The postorder traversal is, then, $D\ G\ E\ B\ H\ J\ F\ C\ A$. This traversal is illustrated in Figure 7.20(c).

Some uses of these traversals will be seen in the next few applications.

**Trees of Expressions** Any expression with binary operators, such as the usual ones of arithmetic or the logical operators of propositional calculus, can be represented as ordered rooted binary trees. The representation takes the form: the operator is the label of the root of a subtree and its two arguments or operands as the labels of the left and right subtrees.

Thus the expression $a + b$ could be represented as shown in Figure 7.21(a). The logical expression $p \wedge (q \rightarrow r)$ would be expressed as shown in Figure 7.21(b). If these two trees are traversed in *inorder* fashion (LNR) we

FIGURE 7.20

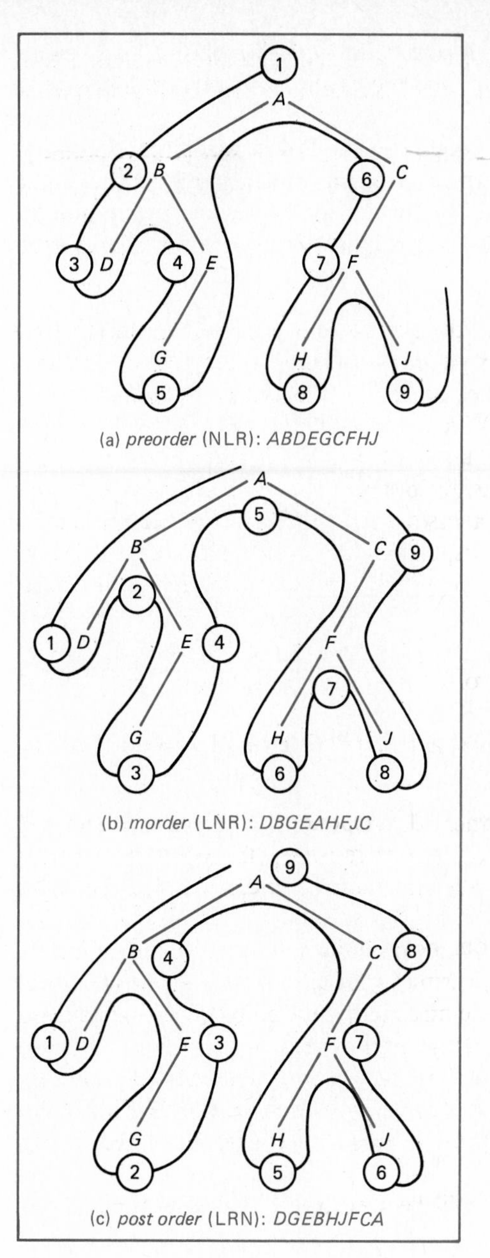


(a) *preorder* (NLR): *ABDEGCFHJ*

(b) *inorder* (LNR): *DBGEAHFJC*

(c) *post order* (LRN): *DGEBHJFCA*

**FIGURE 7.21** Binary expression trees.

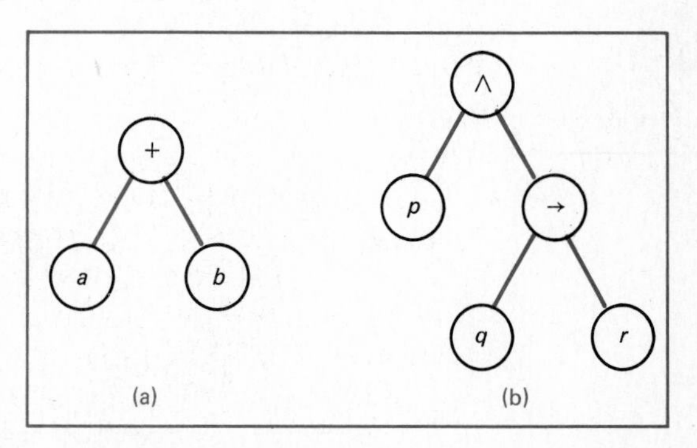


(a) (b)

would obtain the expressions: $a + b$ and $p \wedge q \rightarrow r$. The *preorder* traversals (NLR) would yield $+ab$ and $\wedge p \rightarrow qr$; and the *postorder* traversals (LRN) yield $ab+$ and $pqr \rightarrow \wedge$.

Observe that the preorder traversal gives the Polish prefix representation (Section 5.3) and the postorder traversal gives the Polish postfix form. These two types of traversals are extremely important for expression evaluation by compilers and much has been studied regarding efficient schemes of traversal.

**EXAMPLE 1** Consider the expression

$$((a + b) * (c/d)) + (d - (e \uparrow (f * g)))$$

It can be represented by the (ordered, rooted) binary tree shown in Figure 7.22(a).

A traversal in preorder (prefix) would yield

$$+ \; * \; + \; a \; b \; / \; c \; d \; - \; d \uparrow e \; * \; f \; g$$

FIGURE 7.22

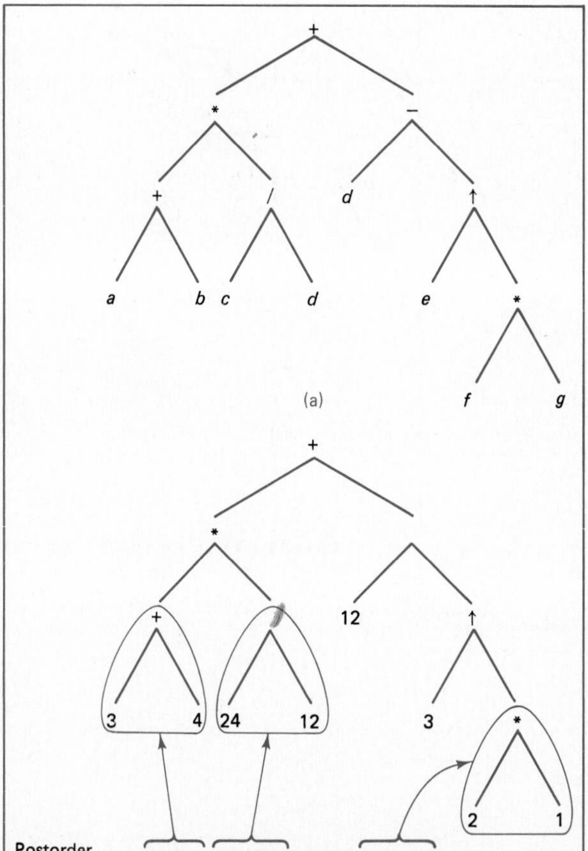

(a)

Postorder traversal (3)(4) + (24)(12)/ * 12 (3) (2) (1) * ↑ − +

(b)

The postorder (postfix) traversal is

$$a \; b \; + \; c \; d \; / \; * \; d \; e \; f \; g \; * \; \uparrow \; - \; +$$

A reason for studying such traversals becomes apparent when one wants to *evaluate* the expression. For this purpose, let us assign values to the letters (variables):

$$a = 3, \quad b = 4, \quad c = 24, \quad d = 12, \quad e = 3, \quad f = 2, \quad g = 1$$

Thus our tree becomes as shown in Figure 7.22(b), and the evaluation becomes apparent. The terminal subtrees (circled) correspond directly to the order in which the binary operations can be evaluated in the postfix expression.

Each of these subtrees can now be evaluated and they can be replaced in the tree, *pruned* as shown in Figure 7.23. Further reduction and pruning of Figure 7.23(a) yields Figure 7.23(b), and finally, Figure 7.23(c).

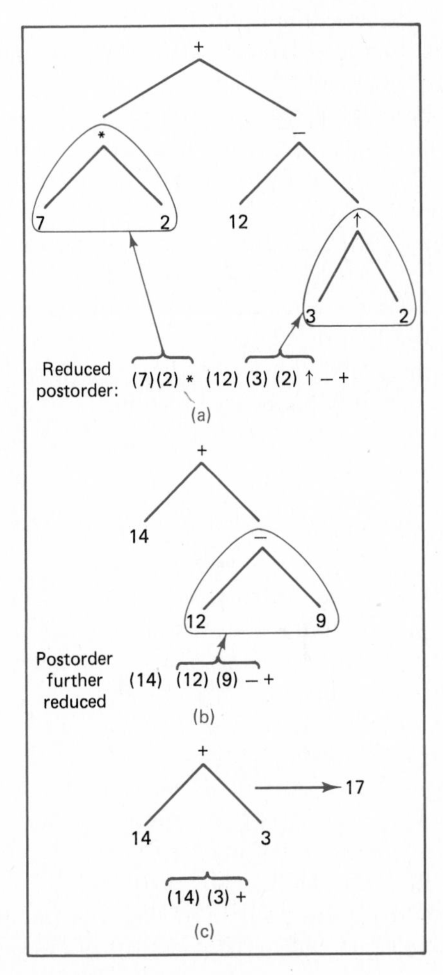


**FIGURE 7.23**

**Sort Trees** Rooted trees give an efficient way to *sort* data into order. If you are presented with a set of random data, either numerical or alphabetic, this procedure to sort them is to assign the first item encountered to the root of a tree. Then the next item will either be "less than" or "greater than" the value of the new root. This next item can then be placed as the left child or right child of the root, depending on whether it is "less than" or "greater than" the root value. This item then becomes a root of its own subtree and the process continues until all the data are encountered.

The procedure is basically very simple and can be explained by an example. Suppose that we are given the sentence

"Ordered Rooted Trees Present An Efficient Way To Sort Data"

and we wish to place the words in alphabetical (lexicographic) order.

The first item encountered is "Ordered"; it becomes the root of the tree [Figure 7.24(a)]. The next item value is "Rooted"; it is lexicographically greater than "Ordered," so it is placed as the right child of the root [Figure 7.24(b)]. The value "Trees" is next and it is "greater than" the root and its right child, thus is placed as the next right child [Figure 7.24(c)].

The fourth item, "Present," is "greater than" the root but "less than" the node "Rooted," so it is placed as this second node's left child [Figure 7.24(d)].

The next item, "An," is found to be "less than" the root, so is placed as the root's left child [Figure 7.24(e)].

The next two items, "Efficient" and "Way," will be placed as shown in Figure 7.24(f).

After the last three items are read and stored on the sort tree, we have the final result, indicated in Figure 7.24(g).

Once the tree is completely stored with data, one traverses it in *inorder* fashion to produce the ordered (alphabetized) listing:

The *ordered* list:

An Data Efficient Ordered Present Rooted Sort To Trees Way.

**Searching** Not only is this an efficient sort algorithm but once the data are stored in the tree in this fashion, one can easily "search" the tree to locate particular items. This search turns out to be equivalent to the "binary search" you may have learned in other courses.

Suppose that we search the tree for the item "To." We first encounter the root and note that "To" is "greater than" the value there, which is "Ordered," so we next compare with the root of the right subtree. In this fashion we will ignore the whole left subtree, thus cutting the number of items to be searched by *about* one-half. (The actual fractional part of the tree that is ignored varies depending on the order in which the data were read in.)

Each time we encounter a root of a subtree in our search we either find the item or we ignore about one-half of the items left in the subtree.

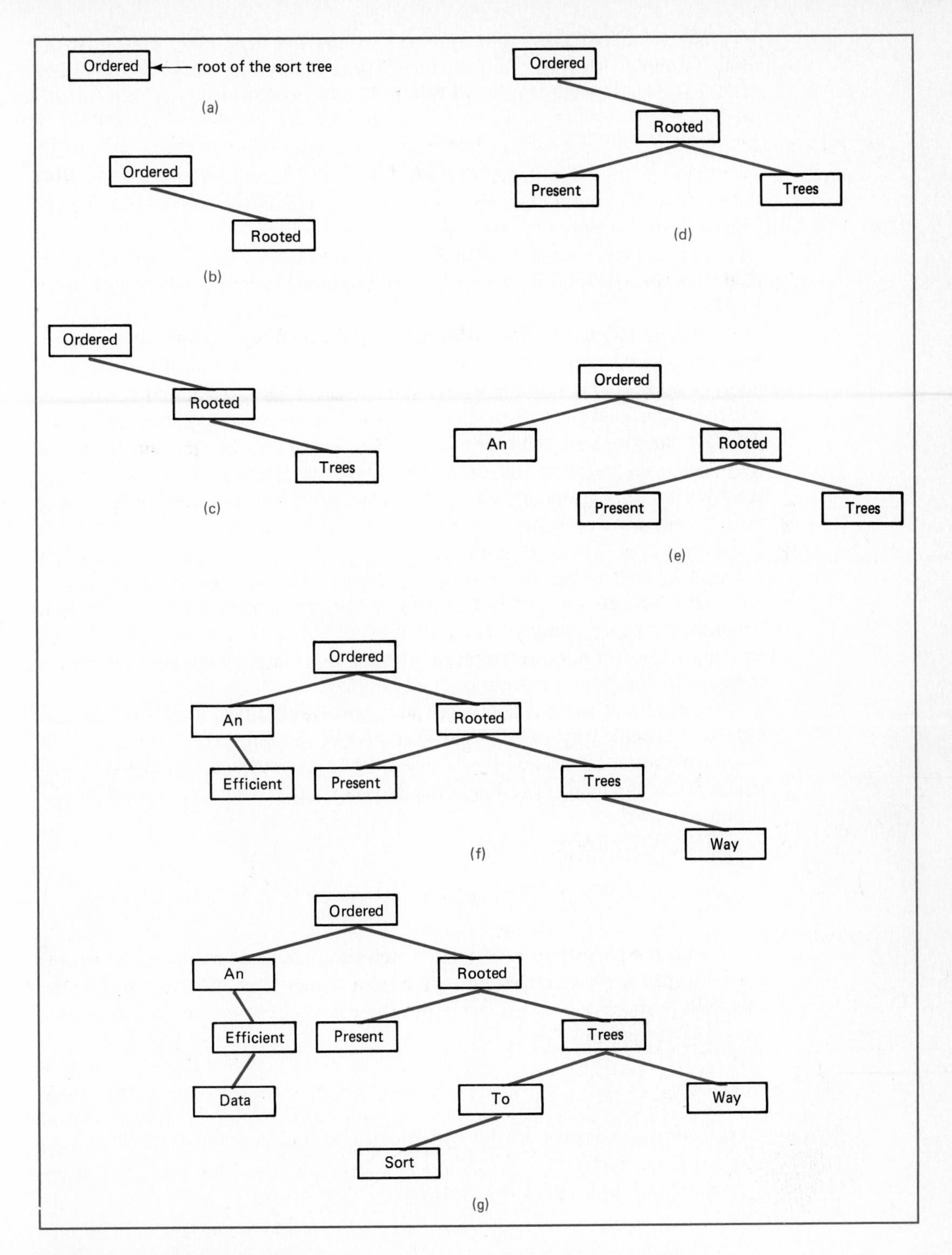


**FIGURE 7.24** Forming a sort tree.

Notice that it takes four "probes" to find the item "To." First was the root, "Ordered," then the value "Rooted," and finally the value "Trees" before "To" was found. See Exercise 11 for further discussion of searches in binary sort trees.

Derivation (Parse) Trees A topic of major concern to all computer scientists is the study of programming languages. These are languages designed by human beings for the purpose of communicating with machines, namely computers. As such, they are classified as artificial or formal languages as opposed to natural languages, such as English or French, used by people to communicate with other people.

Formal languages differ from natural languages in a very important respect. A natural language is one that has developed through daily usage by many people. It was not deliberately constructed; there was neither structure nor conscious grammar; it was just born and it grew according to social needs. A formal language, on the other hand, is one based entirely on a preconceived set of rules. Its grammar was devised first, and the language was generated instantly from this grammar. By contrast, the grammar of a natural language evolves only after centuries of use in an attempt to categorize the rules for the language.

As a result of the differences in development, the rules and grammar of a formal language are usually relatively simple, whereas the rules of a natural language are rather complex. After all, most of you have spent at least 12 years learning your own natural language, whereas it probably took less than a year to be fairly fluent in one programming language.

Formal and natural languages share two similarities. They both have a *syntax* as specified by rules of the grammar of the language. And they both share the notion of *semantics*—the meaning to be attached to strings or sentences of the language. The semantics of a language is not determined by the grammar alone.

The FORTRAN statement

$$X = X + Y ** 2$$

is syntactically correct; it fits the rules of assignment and arithmetic operators. Semantically, it is meaningless if no previous values have been assigned to the variables X and Y.

The statement

$$X = X(3) + Y ** 2$$

is syntactically incorrect, for no variable can be both a scalar variable and an array at the same time.

Similarly, in English, the sentence

"Colorful numbers dream quickly"

has no semantic content but is a syntactically correct sentence, since the various parts of speech fit together according to the rules of the grammar.

One of the useful techniques in the study of the syntax of a grammar of both natural and formal languages is the *derivation tree* (also known as the *parse tree*). A grammar is basically a set of rewrite rules. For example, in English, every sentence, S, can be divided (parsed) into two components: a noun phrase, ⟨NP⟩, usually followed by a verb phrase ⟨VP⟩. This can be expressed by the rewrite rule: S → ⟨NP⟩ ⟨VP⟩.

This rule can be conveniently expressed as a rooted tree (the beginnings of the parse tree) [Figure 7.25(a)]. The three components S, ⟨NP⟩, and ⟨VP⟩ are known as syntactic elements and the graphic structure is known as *phrase structure*.

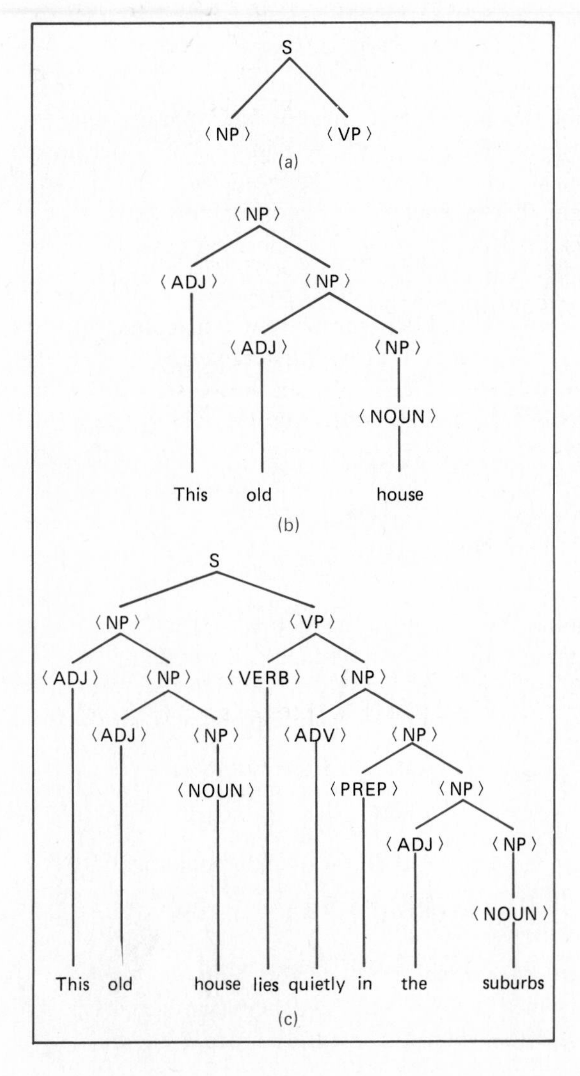


**FIGURE 7.25** Forming a derivation (parse) tree of an English sentence.

A noun phrase, ⟨NP⟩, can be rewritten variously (among other ways) as

$$\langle \text{NP} \rangle \rightarrow \langle \text{ADJ} \rangle \langle \text{NP} \rangle$$

$$\langle \text{NP} \rangle \rightarrow \langle \text{NOUN} \rangle$$

$$\langle \text{NP} \rangle \rightarrow \langle \text{PREP} \rangle \langle \text{NP} \rangle$$

So that a phrase such as "this old house" is the adjective, ⟨ADJ⟩, "this," followed by a noun phrase, ⟨NP⟩, "old house," and this phrase can further be parsed as shown in Figure 7.25(b).

Verb phrases can similarly be parsed into their component parts.

$$\langle \text{VP} \rangle \rightarrow \langle \text{VERB} \rangle \langle \text{NP} \rangle$$

with

$$\langle \text{NP} \rangle \rightarrow \langle \text{ADV} \rangle \langle \text{NP} \rangle$$

By using the foregoing rewrite rules for noun phrases and verb phrases, let us show in Figure 7.25(c) a complete derivation of an English sentence: "This old house lies quietly in the suburbs." It should not be too hard to see from this parse tree that the sentence is correct syntactically, because it follows the rewrite rules of English grammar. The meaning changes drastically when some of the syntactic categories are replaced by other terminal values (words). For example, in the above, we used

$$\langle \text{ADJ} \rangle \rightarrow \text{this, old, the}$$

$$\langle \text{NOUN} \rangle \rightarrow \text{house, suburbs}$$

$$\langle \text{VERB} \rangle \rightarrow \text{lies}$$

$$\langle \text{ADV} \rangle \rightarrow \text{quietly}$$

$$\langle \text{PREP} \rangle \rightarrow \text{in}$$

If we add other words (and the problem in English is that there are so many of them) as values to the syntactic elements:

$$\langle \text{ADJ} \rangle \rightarrow \text{brick, old, new, all}$$

$$\langle \text{NOUN} \rangle \rightarrow \text{fire, numbers}$$

$$\langle \text{VERB} \rangle \rightarrow \text{burns, work}$$

$$\langle \text{ADV} \rangle \rightarrow \text{quickly, majestically}$$

$$\langle \text{PREP} \rangle \rightarrow \text{on}$$

we can obtain other syntactically correct sentences, such as: "The brick house burns quickly in this fire", and "Old new numbers work majestically on all suburbs": one with semantic content, and one without.

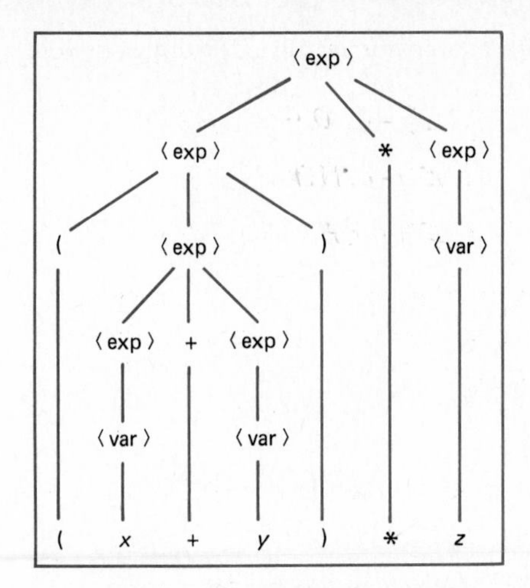


**FIGURE 7.26** Parse tree of an expression.

As an example of a parse tree for the expressions, $(x + y) * z$, we can use the following rewrite rules:

$$\langle \text{exp} \rangle \rightarrow (\langle \text{exp} \rangle)$$

$$\langle \text{exp} \rangle \rightarrow \langle \text{exp} \rangle * \langle \text{exp} \rangle$$

$$\langle \text{exp} \rangle \rightarrow \langle \text{exp} \rangle + \langle \text{exp} \rangle$$

$$\langle \text{exp} \rangle \rightarrow \langle \text{var} \rangle$$

$$\langle \text{var} \rangle \rightarrow x, y, z$$

to obtain the parse tree shown in Figure 7.26.

The procedure used for determining parse trees for expressions and other statements of programming languages is indispensible in the construction of compilers and translators of these formal languages.

## EXERCISES

1. Find minimum spanning trees for the networks below. Use both methods presented in this section. Explain any discrepancies in the spanning trees found by the two methods.

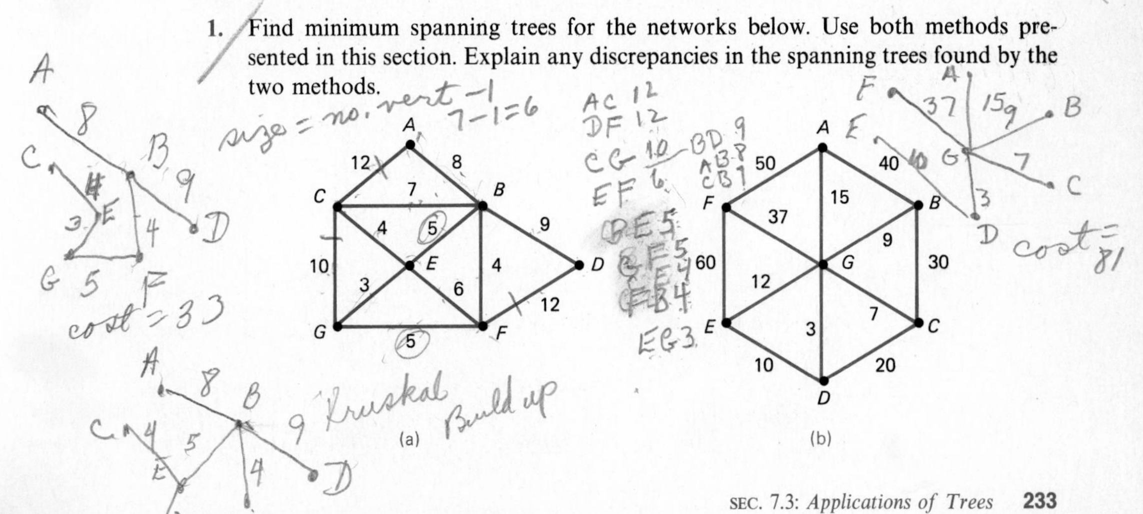

2. Give preorder, inorder, and postorder traversals of the trees below.

(a) (b) (c)

3. Draw the rooted trees of the expressions
   (a) $(((a + b) * c) - d) + (e \uparrow (f - (g * h)))$
   (b) $(p \wedge r) \vee ((q \rightarrow p) \leftrightarrow (r \vee (p \wedge q)))$

4. Traverse each tree of Exercise 3 in preorder and postorder.

5. Evaluate the tree of Exercise 3(a) by showing the pruning process for the values $a = 5$, $b = 4$, $c = 3$, $d = 15$, $e = 4$, $f = 21$, $g = 6$, and $h = 3$.

6. Show sort trees for the following data items.
   (a) The words read in order:

   "All dogs eat fried grits instead of popcorn sitting together under vans."

   (b) The months of the year input in chronological order (i.e., "Jan.," "Feb.," . . . , "Dec.").

7. Explain why a binary search for the word "popcorn" would not be very efficient on your sort tree of Exercise 6(a).

8. Construct parse trees for the following sentences.
   (a) "The old cat slept under the tree."
   (b) "Blue telephones think quickly."

9. Languages are termed as *ambiguous* if some of the sentences can have more

⟨S⟩ ⟨NP⟩ ⟨VP⟩ ⟨pronoun⟩ ⟨V⟩ ⟨NP⟩ ⟨N⟩ They are flying planes

⟨S⟩ ⟨NP⟩ ⟨VP⟩ ⟨pronoun⟩ ⟨V⟩ ⟨NP⟩ ⟨ADJ⟩ ⟨NP⟩ ⟨N⟩ They are flying planes

than one parse tree. The sentence "They are flying planes" can create two images depending on how it is parsed:

Construct two different parse trees for the following sentences.

(a) "Flying planes can be dangerous."

(b) "Her classes have satisfied students."

(c) "I saw them both dancing and singing last night."

(d) "They are failing students."

(e) "Time flies." (Let "time" be the verb with an understood subject.)

**10.** Expression trees discussed in this section are examples of *strictly binary trees* in which each internal node has exactly *two* children. The variables of the expression are the leaf nodes and the operators, the internal nodes. Use the discussion and formulas of Exercise 14 in Section 6.5 to answer the following questions:

(a) If an expression has eight instances of variables (leaves), how many total nodes will there be in the expression tree?

(b) How many operators (internal nodes) must there be?

(c) Suppose an expression has a total of 23 variables and operators (thus 23 nodes in the expression tree). How many instances of variables are there? How many operators are there?

**11.** In the sort tree of Figure 7.24(g), we notice that there are 10 nodes. We saw that the distance from the root "Ordered" to the node "To" was 3. When we consider base 2 logarithms from Exercise 16 in Section 6.5, we see that lg 10 = 3, which should represent the maximum distance from the root to any node. Explain the discrepancy occurring from the fact that the distance from the root to the node "Sort" is 4.

## 7.4 Applications to Puzzles and Games

To close this chapter we will discuss some of the varied applications of graphs to puzzles and games. Some of the techniques shown are also applicable to situations outside the realm of games.

**EXAMPLE 1** (*The Cannibal-Missionary Problem*) Consider the well-known puzzle involving the three cannibals and the three edible missionaries. They are all initially on the left bank of a river and wish to get to the right side. The only transportation is a two-person rowboat that can be rowed by a single person—either a cannibal or a missionary. For obvious reasons, at no time can the cannibals outnumber the missionaries. How do they manage to cross the river with no one getting eaten?

**Solution** It is convenient to label the vertices of the graph as "states" or positions representing the numbers of cannibals and missionaries on either side and some indication of which bank the boat is on. There are 16 possible "states," but because of the outnumbering condition, some of them are clearly impossible.

These states are shown in Figure 7.27 with the number of cannibals being the first element of each ordered pair and the number of missionaries, the second element.

We will indicate the vertices of our directed graph with the state number and either L or R to indicate the bank on which the boat is. The start state is state 1 with the boat on the left, which we write as shown in Figure 7.28(a).

It is possible to move from this start state to five others with the boat now on the right, but only three are acceptable [Figure 7.28(b)].

From state 5, R we can only move to 1, L, which seems to be a fruitless venture, so we abandon this path. From state 6, R we can move to 2, L, which is impossible, and to 5, L and, of course, back to 1, L. From 9, R we can go back to 1, L and to 5, L [Figure 7.28(c)].

Advancing from 5, L, we can, of course, retrace our steps to 6, R and 9, R. It turns out that there is one other legal move, 13, R. Moves 10, R and 7, R are illegal. Thus we have the situation shown in Figure 7.28(d).

From 13, R we show the next three legal steps [Figure 7.28(e)].

It can be noted that at this point, the finish of the trip can be carried out as the first part only in reverse order to complete our state diagram [Figure 7.28(f)]. There are 11 crossings needed, as shown in Figure 7.29.

**EXAMPLE 2** (*The Decanting Problem*) Suppose that you are given three jugs, A, B, and C, with capacities 8, 5, and 3 liters, respectively, but none are calibrated. Jug A is filled full with 8 liters of wine. The objective of this game is, by a series of pourings back and forth among the three jugs, to divide the 8 liters into exactly two equal parts: 4 liters in jug A and 4 liters in jug B.

**Solution** Here again we can represent the vertices of our digraph as "states" of activity, each one being an ordered triple. The triples are to represent the amounts in jugs A, B, and C, respectively.

**FIGURE 7.27** Sixteen "states" for the cannibal-missionary problem.

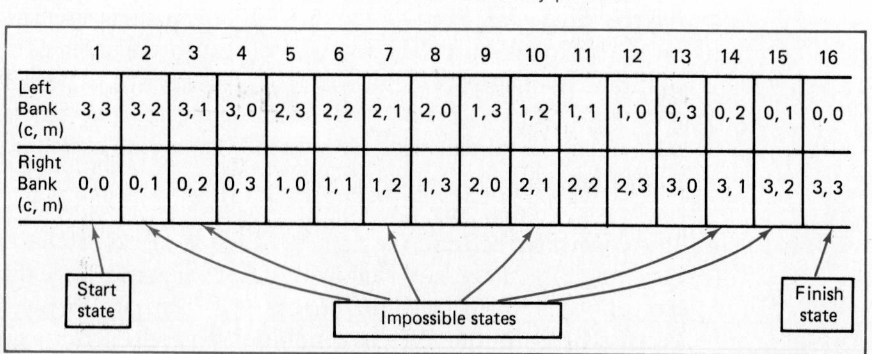

| | 1 | 2 | 3 | 4 | 5 | 6 | 7 | 8 | 9 | 10 | 11 | 12 | 13 | 14 | 15 | 16 |
|---|---|---|---|---|---|---|---|---|---|---|---|---|---|---|---|---|
| Left Bank (c, m) | 3, 3 | 3, 2 | 3, 1 | 3, 0 | 2, 3 | 2, 2 | 2, 1 | 2, 0 | 1, 3 | 1, 2 | 1, 1 | 1, 0 | 0, 3 | 0, 2 | 0, 1 | 0, 0 |
| Right Bank (c, m) | 0, 0 | 0, 1 | 0, 2 | 0, 3 | 1, 0 | 1, 1 | 1, 2 | 1, 3 | 2, 0 | 2, 1 | 2, 2 | 2, 3 | 3, 0 | 3, 1 | 3, 2 | 3, 3 |

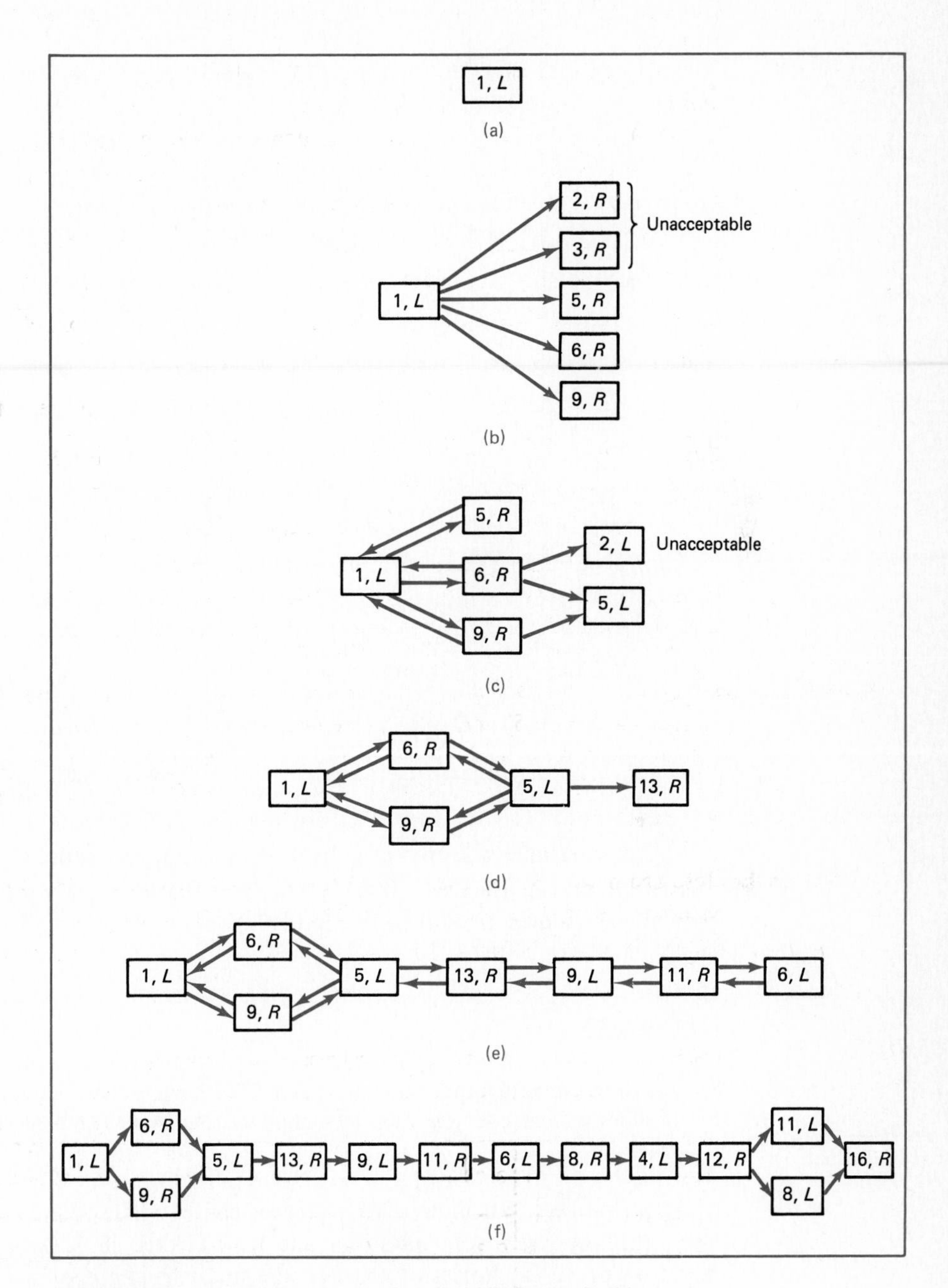


**FIGURE 7.28** State Diagram for the Cannibal-Missionary Problem.

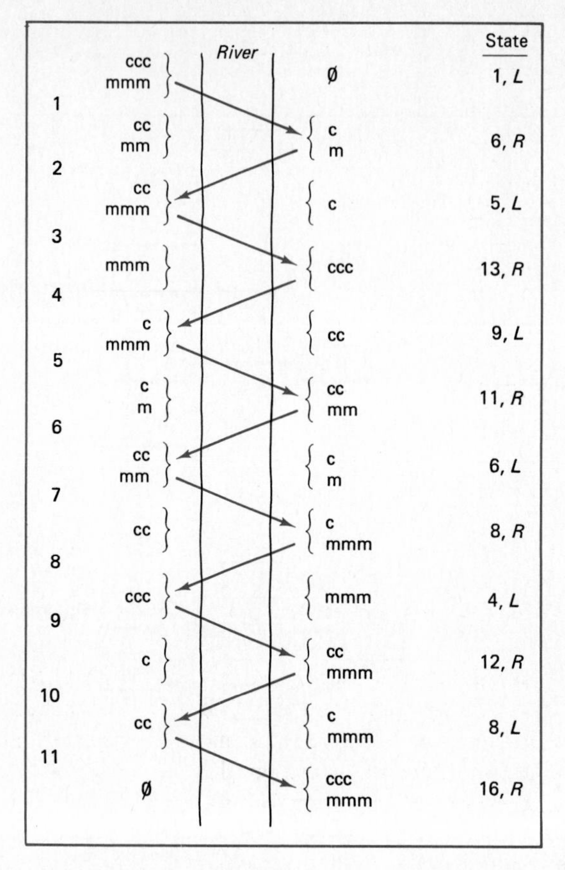


FIGURE 7.29

The start state will obviously be $\boxed{8, 0, 0}$. From this initial state we see the possible moves in Figure 7.30(a). We wish to finish at the state $\boxed{4, 4, 0}$. The possible next states are shown in Figure 7.30(b).

## EXERCISE

1. Finish the state digraph shown in Figure 7.30 showing that the minimum number of states is *eight*, counting both the initial and final states. Thus *seven* pourings are required.

Decision Trees When alternative choices can be made based on discrete answers (usually, yes-no) to questions. and then further choices are to be made based on previous choices of answers, the situation can give rise to a *decision tree.* For some problems these trees enumerate all possible histories of a particular solution. These trees are rooted with the root representing the first decision or choice to be made. Each successive internal node of the tree represents a further decision and the leaves of the tree represent the ultimate solution. This

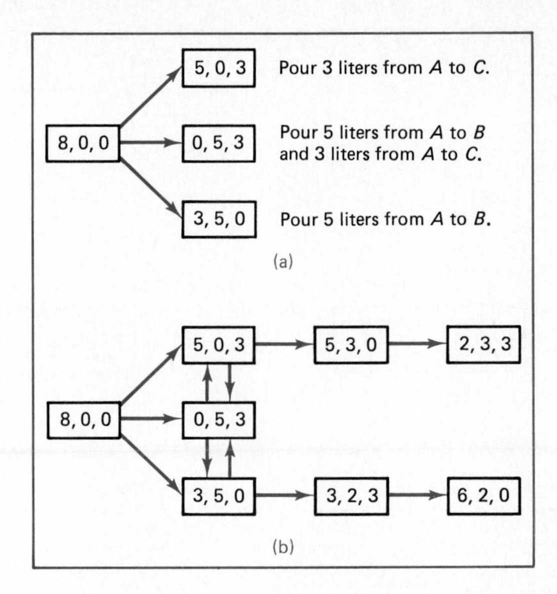


**FIGURE 7.30**

technique was used at various times throughout the book to illustrate some counting strategies.

Trees as useful tools of decision making should not technically be classified in the puzzle and game category. Thus one may wonder about their placement in this section. The reason is that the power of this technique can best be illustrated in the context of a puzzle.

As an illustration, consider the following puzzle:

**PUZZLE** You are given 12 seemingly identical coins. One of them, however, is counterfeit and weighs *less than* the other 11, all of which weigh the same. By using an equal-arm balance weight, determine in no more than *three* weighings which is the culprit coin.

To solve this problem, consider each weighing you perform as the node of a tree, and its possible discrete outcomes are the branches to subsequent weighings (decisions). Your first weighing will be the root of the tree. If you weigh six of the coins against the other six, then there are *two* possible outcomes: The left pair will be heavier than the right pair, or it will be lighter. If you weigh, say, four coins against four others, then there will be *three* branches, the two outcomes above and the possibility that they weigh the same and you have not chosen the culprit coin to weigh.

Let us begin by weighing six coins against the other six. For convenience, assign the first 12 letters of the alphabet as the coins and the weighings can be carried out as shown in the decision tree of Figure 7.31. We note that the first and second weighings have two outcomes, whereas the third weighing has three outcomes (solutions).

This decision tree can be programmed fairly easily. First, initialize all 12 variables with the value 1. Then randomly assign one of them the value 0. At

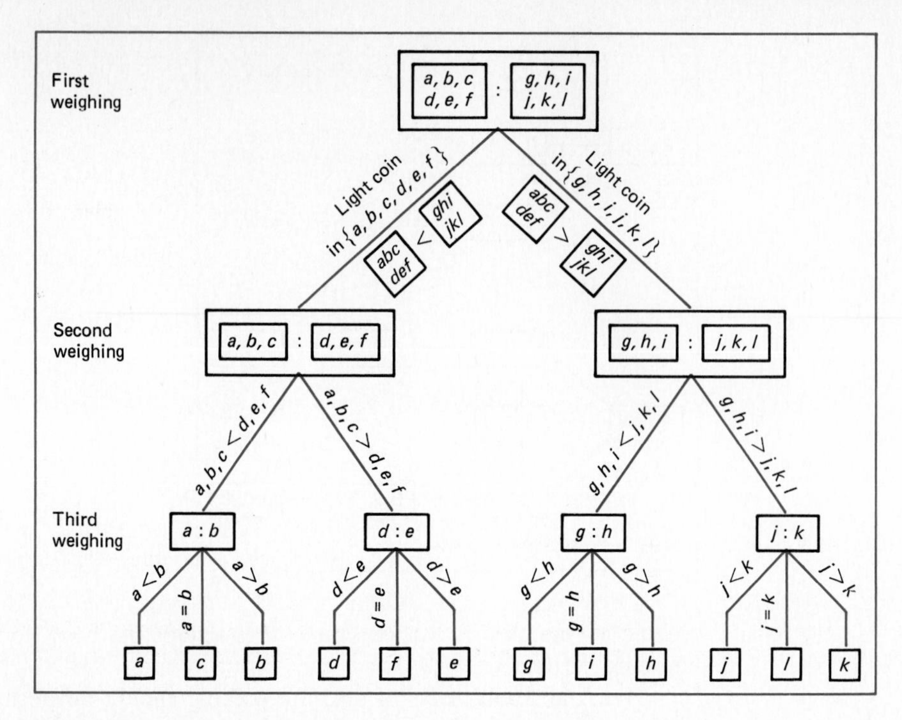


**FIGURE 7.31** Decision tree.

each decision point (weighing) merely sum the values of the variables and take the appropriate branch. Notice that some of the possible branches are not used, for they are impossible occurrences. In real-life decision making it is best to utilize as many branches as possible early. Thus further decisions can be made from these earlier decisions (see Exercise 5).

## EXERCISES

2. *The Ferryman's Dilemma* A ferryman is charged with carrying across a river a wolf, a sheep, and a bag of cabbage. His boat is too small to carry more than himself and one of the "passengers" at a time. He cannot leave the wolf alone with the sheep nor can he leave the sheep alone with the cabbage. Show how he can solve his dilemma with a graph-theoretic solution.

3. *The Problem of the Jealous Husbands* Three married couples are invited to a party on the tenth floor of an apartment building. The only access from the lobby is a small elevator that holds only two people, and one person has to be at the controls at all times or the elevator will not run. They all arrive in the lobby at the same time. The trip up is complicated by the fact that all the husbands are very jealous and will not permit their wives to be left without them in a company where other men are present. Construct a graph to show how the trip up can be accomplished.

4. *Another Decanting Problem* You have two uncalibrated jugs—one holding 5 liters and the other, 3 liters—and an unlimited supply of wine (maybe water is best for this).

By a sequence of filling and emptying the jugs and pouring back and forth between the two jugs, obtain exactly 4 liters in the first jug.

5. Suppose that you have eight coins and are told that one is counterfeit and has a *different* weight (heavier or lighter) than the other seven. Determine which coin is counterfeit by no more than three weighings. (*Hint:* To start, weigh three of the coins against three others.) Note that there will be 16 different outcomes.
6. Repeat Exercise 5 using 12 coins, with one being different. There will be 24 possible outcomes. (*Hint:* Start with four coins against four more.)
7. Repeat Exercise 5, this time with 13 coins, with one of them being *heavier* than the rest.
8. Repeat Exercise 5, this time with eight coins, with one of them being *heavier* than the rest, and employ only *two* weighings.

## Summary and Selected References

In this chapter we presented many of the applications of graph theory as problems that a computer scientist may encounter. The range of such application is broad. For further reading on the methods and techniques presented, one is directed first to the several books on data structures, such as Horowitz and Sahni (1982), Berztiss (1975), Aho, Hopcroft, and Ullman (1983), Augenstein and Tenenbaum (1979), Lewis and Smith (1982), Standish (1980), Tenenbaum and Augenstein (1981), Tremblay and Sorenson (1984), Wirth (1976), Baron and Shapiro (1980), Reingold and Hansen (1983), and Beidler (1982).

The books by Deo (1974), Horowitz and Sahni (1978), Goodman and Hedetniemi (1977), and Fisher (1977) give very clear introductions to the uses of graph techniques in the design of computer algorithms as do Aho, Hopcroft, and Ullman (1975), and Baase (1978) from a more advanced point of view.

Applications of graph theory as mathematical models of many real-world situations are presented in Roberts (1976), Maki and Thompson (1973), and Harary, Norman, and Cartwright (1965).

# Discrete Counting: An Introduction to Combinatorics

# 8

## 8.1 Introduction

At several places throughout this book we have made reference to counting the number of distinct elements of discrete systems. This topic of counting, or *enumeration*, comes under the general heading of *combinatorics* or combinatorial analysis.

The topic is important in many aspects of computing, and due to the impact that has been made by computer science, combinatorics is one of the fastest growing areas of modern mathematics as applied to the discipline. Such problems as scheduling, storage and size requirements, time and memory allocations, and algorithm design and analysis are of fundamental consideration to those engaged in the study of computer science. An understanding of each of these subjects requires some knowledge of counting techniques.

Since this book is meant to be a first-level survey of discrete mathematical structures, we make no attempt to cover the breadth of known enumeration techniques. Many of the topics overlap with other disciplines, notably probability, and undoubtedly the reader will encounter them there. We do include here some of the more common techniques as an introduction to the overall subject, and no attempt is made to be complete, since you will probably devote an entire course of study to this important topic. Although this chapter is short, the concepts introduced are powerful and far-reaching in both mathematics and computer science. A thorough understanding of the basic rules presented here will provide you with considerable insight when later confronted with a discrete counting situation.

In all of discrete counting there are two fundamental rules that apply. They are the addition rule and the multiplication rule. We state them without proof and accept them as axioms in our discussion.

□ Fundamental Addition Rule If the number of elements of a set $A_1$ is $n_1$, the number of elements of set $A_2$ is $n_2$, and the two sets are *disjoint* (i.e., $A_1 \cap A_2 = \emptyset$), then the total number of elements in the two sets is $n_1 + n_2$. This can be extended to any number of *disjoint* sets $A_1, A_2, \ldots, A_m$. If the numbers in these sets are $n_1, n_2, \ldots, n_m$, then the total number of elements in $A_1$ or $A_2$ or $\cdots$ or $A_m$ is $n_1 + n_2 + \cdots + n_m$.

This rule was examined in Section 2.7, where we explored the general subject of counting the number of elements of finite sets. There we noted that this rule for two sets becomes

$$\#(A_1 \cup A_2) = \#(A_1) + \#(A_2) - \#(A_1 \cap A_2)$$

to account for the possibility that sets $A_1$ and $A_2$ may not be disjoint. The reader is encouraged to review Section 2.7.

□ Fundamental Multiplication Rule If some procedure can be *ordered* into $n$ parts (sets) and there are $r_1$ outcomes (elements) in the first part, and *following* this first part there are $r_2$ outcomes for the second part and following this second part there are $r_3$ outcomes for the third part, and so on until we encounter $r_n$ outcomes for the $n$th part, then the number of total outcomes (elements) for the entire procedure is the product $r_1 \cdot r_2 \cdot r_3 \cdots r_n$.

We have encountered this rule previously also. In determining the total number of truth tables for two variable propositions in Section 1.9, we implicitly used this rule. There we noted that there are two distinct choices for the first row, either a T or an F; and *following* this, there are two distinct choices for the second row; and following this choice there are again two choices for the third and then again for the fourth. By using the fundamental multiplication rule we arrive at $2 \cdot 2 \cdot 2 \cdot 2 = 16$ possible combinations of T and F for the two propositions. These truth tables were illustrated in Table 1.30.

We demonstrate other cases of applying the multiplication rule in the following examples.

**EXAMPLE 1** Suppose that a chairman and a vice-chairman need to be selected from a committee consisting of five people. This represents two parts, as mentioned in the multiplication rule. If the selections are ordered so that the chairman is selected first and then following this selection, the vice-chairman is selected, there are *five* possible outcomes for the first part. Once the chairman has been selected, there are only *four* possible outcomes for the vice-chairman. Thus from the multiplication rule we get $5 \cdot 4 = 20$ possible selections for the two positions.

**EXAMPLE 2** Suppose that you want to compose an acronym for a new computer system you have developed. It should consist of three letters, with the stipulation that the first letter (first part) is to be a "C"; the second letter is to be a vowel (thus five possible elements: A, E, I, O, U); and the third letter can be any of the 26 letters of the alphabet. How many possible acronyms can you have? By using the multiplication rule, you will have

$$\underset{\substack{\uparrow \\ \text{first} \\ \text{part}}}{\underline{\;1\;}} \cdot \underset{\substack{\uparrow \\ \text{second} \\ \text{part}}}{\underline{\;5\;}} \cdot \underset{\substack{\uparrow \\ \text{third} \\ \text{part}}}{\underline{\;26\;}} = 130$$

possibilities.

**EXAMPLE 3** In Example 7 of Section 5.2 we explored the question of how many functions there are from one set to another. There we stated that if the domain had $n$ elements and the codomain had $m$ elements, then there are $m^n$ possible functions. This result is easily seen by the fundamental multiplication rule.

The first part is the mapping of $a_1$, the first domain element. There are $m$ possible outcomes. Following this mapping there are $m$ possible mappings for the second domain element, $a_2$, yielding $m \cdot m$ for a two-element domain. For an $n$-element domain we get

$$\underbrace{m \cdot m \cdots m}_{n \text{ times}} = m^n$$

possible functions.

## EXERCISES

1. License plates of a certain state are composed of two digits followed by three letters. A restriction is that the first digit cannot be a zero (0). There are no restrictions on the other digit and the three letters can be any of the 26 letters of the alphabet. How many possible license plates are there? Use the fundamental multiplication rule.
2. An 8-bit computer word is a string of eight digits, each a 0 or a 1.
   (a) How many 8-bit computer words are there?
   (b) How many such words are there if the first digit cannot be a 0?
   (c) How many such words are there if the first three digits cannot be a 0?
   (d) How many such words are there if each group of four digits must start with a 1?
3. A club has 20 members. Six of the members have agreed to serve as president if elected. All the members have agreed to serve as vice-president. If the vice-presidential election follows the presidential election, how many selections are possible for the two officers of the club?
4. There are several roads from the town of Ash to the town of Elk, all by way of Big Creek. There are three roads from Ash to Big Creek and two from Big Creek to Elk.

(a) How many ways can a person travel from Ash to Elk by way of Big Creek?

(b) How many possible round trips are there between Ash and Elk?

(c) How many such round trips are there if no road is traveled more than once?

(d) In addition to the route by way of Big Creek, one may travel from Ash to Elk via the town of Deer. There are two roads from Ash to Deer and two from Deer to Elk. How many routes are there now from Ash to Elk? [*Hint:* Draw a diagram (digraph) and use the fundamental addition rule in conjunction with the fundamental multiplication rule.]

5. A three-digit decimal number is composed of three digits such that the first digit is not a zero. Compute the number of

   (a) All three-digit numbers.

   (b) All three-digit even numbers.

   (c) All three-digit numbers divisible by 10.

   (d) All three-digit numbers divisible by 5.

6. In how many ways can five files be stored on three external storage devices? Think of this as a mapping from the set of files to the set of storage devices.

7. Suppose in Example 3 that the function specified is an injection (1–1); that is, no element in the codomain can be used as an image more than once. Under this restriction how many possible functions are there from a four-element domain to a seven-element codomain? From an $n$-element domain to an $m$-element codomain with $n \leq m$?

## 8.3 Arrangements: Permutations

Often one wants to know how many possible arrangements or orderings there are of a number of distinct elements. To motivate the concept, suppose that there is only one element to "arrange." Then there is obviously only one "arrangement." If we have two distinct objects, $a_1$ and $a_2$, to arrange, we know from the multiplication rule that in the arrangement, the first position (outcome) can occur in one of *two* ways, either with the $a_1$ element or with the $a_2$ element. *Following* this placement there is only *one* possibility for the second element to be placed. This then gives us $2 \cdot 1 = 2$ possible arrangements.

When we have a third element, $a_3$, to place in the arrangement, the first position can be filled by any one of *three* elements. Following this placement we have *two* possibilities for the second position, and finally only *one* left for the last position. This gives us $3 \cdot 2 \cdot 1 = 6$ arrangements. This arrangement of three elements can be observed from Figure 8.1.

The emerging pattern can be observed. If we have four distinct objects to arrange, there will be $4 \cdot 3 \cdot 2 \cdot 1 = 24$ possible arrangements. This can be carried on for $n$ elements, yielding $n(n-1)(n-2) \cdots 3 \cdot 2 \cdot 1$ arrangements. An exact proof of this can be given by mathematical induction to verify the phrase "carried on for $n$ elements" in the last sentence.

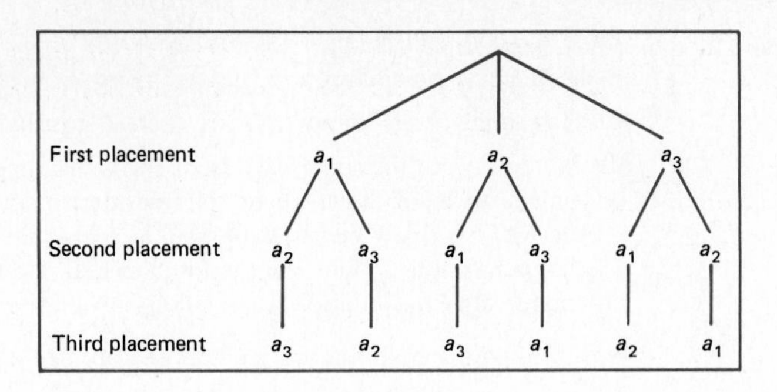


**FIGURE 8.1**

The product above, $n(n-1)(n-2)\cdots 3\cdot 2\cdot 1$, we have seen earlier in Chapter 5 as $n!$. Thus we have the following

□ Theorem 8.1 For $n$ distinct objects there are $n!$ possible arrangements. Each of the arrangements is known as a permutation of the objects. Thus there are $n!$ permutations of $n$ objects. We symbolize this as $P(n, n)$.

If of $n$ possible distinct objects we want to order only $r$ of them, how many arrangements are there? We let $r$ be less than $n$ and take on the cases listed in Table 8.1. So we see that the number of arrangements (permutations) of $r$ objects out of $n$ distinct objects is $n(n-1)(n-2)\cdots(n-r+1)$. This is denoted by $P(n, r)$, and we have

$$P(n, r) = n(n-1)(n-2)\cdots(n-r+1) \tag{1}$$

Equation (1) can be rewritten as follows:

$$P(n, r) = n(n-1)\cdots(n-r+1)\frac{[(n-r)(n-r-1)\cdots 3\cdot 2\cdot 1]}{[(n-r)(n-r-1)\cdots 3\cdot 2\cdot 1]}$$

$$= \frac{n!}{(n-r)!}$$

For the special case of this result in which $r = n$, we have

$$P(n, n) = \frac{n!}{(n-n)!} = \frac{n!}{0!}$$

**TABLE 8.1**

| $r$ | *Number of Arrangements from* n *Objects* |
|---|---|
| 1 | $n$ |
| 2 | $n(n-1)$ |
| 3 | $n(n-1)(n-2)$ |
| 4 | $n(n-1)(n-2)(n-3)$ |
| ⋮ | |
| $r$ | $n(n-1)(n-2)\cdots(n-(r-1)) = n(n-1)(n-2)\cdots(n-r+1)$ |

We know from above that $P(n, n) = n!$. Since $n!/0!$ should equal $n!$, we have the following definition: $0! = 1$, and this is the definition given throughout mathematics for $0!$.

**EXAMPLE 1** Suppose that we wish to construct a four-letter word by choosing any letter of the alphabet for each of four positions, with the restriction that once a letter has been used, it may not be used again.

This problem represents a permutation of 26 objects taken four at a time. Thus we have

$$P(26, 4) = \frac{26!}{(26-4)!} = \frac{26!}{22!} = \frac{26 \cdot 25 \cdot 24 \cdot 23 \cdot 22!}{22!} = 26 \cdot 25 \cdot 24 \cdot 23$$

Suppose that the restriction were lifted. That is, it makes no difference whether a letter has been used or not. A possible word then could be "BBBB." How many possible words of this type are there?

## EXERCISES

1. List the six permutations of the letters a, b, and c.
2. After a first course in BASIC a student can take any one of three programming courses: FORTRAN, COBOL, or Pascal. In how many ways can the student order a programming curriculum that includes all four languages such that no two will be taken simultaneously?
3. Compute the following.
   (a) $P(10, 6)$ (b) $P(5, 1)$
   (c) $P(6, 0)$ (d) $P(n, 2)$
4. In Exercise 3(d) you should have showed that $P(n, 2) = n(n-1)$. Find $n$ if $P(n, 2) = 42$. (*Note:* $n$ can only be positive.)
5. A computer program calls five procedures and no procedure can be called twice. In how many different orders can the procedures be called?
6. Show that $P(n+1, r) = (n+1) \cdot P(n, r-1)$

## 8.4 Combinations

If we want to pick $r$ objects from a set of $n$ distinct objects without regard to how they are arranged, we have what is known as a *combination.* The symbol used to denote a combination of $r$ objects from a set of $n$ is $C(n, r)$ or, alternatively, $\binom{n}{r}$. Both symbols are prevalent in the literature, and we will use the two interchangeably. Each symbol is read: "the number of combinations of $n$ things taken $r$ at a time."

To *count* the number of combinations there will be of $n$ things taken $r$ at a time is a relatively easy task when one recalls that there are $r!$ permutations (arrangements) of $r$ distinct objects.

For example, we showed above that for three elements, $a$, $b$, and $c$, there are $3! = 6$ ways to *arrange* them: *abc*, *acb*, *bac*, *bca*, *cab*, and *cba*. Even though each of these represents a different arrangement, they *all* constitute the *same combination* of the elements. Thus there is only *one* combination of $n$ elements taken $n$ at a time. Hence $C(n, n) = \binom{n}{n} = 1$.

For the same three elements there are $P(3, 2) = 3!/(3 - 2)! = 6$ arrangements of elements taken two at a time. They are *ab*, *ba*, *ac*, *ca*, *bc*, and *cb*. Obviously, there are only *three* distinct combinations: $\{a, b\}$, $\{a, c\}$, and $\{b, c\}$.

What this leads to is the observation that to obtain the number of *combinations* of $r$ objects taken from $n$ objects is to find the number of permutations of $n$ objects taken $r$ at a time *divided* by the total number of the $r$ permutations. Thus the formula

$$\binom{n}{r} = C(n, r) = \frac{P(n, r)}{r!}$$

This formula can be rewritten as

$$\begin{aligned}\binom{n}{r} = C(n, r) &= \frac{P(n, r)}{r!} \\ &= \frac{n!}{(n - r)!r!}\end{aligned}$$

We see the special case

$$C(n, n) = \frac{n!}{n!0!} = 1$$

mentioned above.

There are two equivalencies to be noted regarding combinations.

1. Since

$$\binom{n}{r} = \frac{n!}{(n - r)!r!} = \frac{n!}{r!(n - r)!}$$

we have

$$\binom{n}{r} = \binom{n}{n - r} \qquad \text{or} \qquad C(n, r) = C(n, n - r)$$

Thus, for example,

$$\binom{15}{5} = \binom{15}{10}$$

2. $\binom{n}{n} = \binom{n}{0} = 1.$

**EXAMPLE 1** Compute $\binom{n+2}{n}$.

$$\binom{n+2}{n} = \frac{(n+2)!}{n!2!} = \frac{(n+2)(n+1)\not{n!}}{\not{n!} \cdot 2!} = \frac{(n+2)(n+1)}{2}$$

**EXAMPLE 2** How many ways can a committee of four people be chosen from a group of 10?

Here we are not concerned with the distinct arrangements of people, rather just the collections of four at a time. Hence we want the combinations of 10 taken four at a time. We want

$$C(10, 4) \quad \text{or} \quad \binom{10}{4} = \frac{10!}{6! \cdot 4!}$$

One need not carry out the entire multiplication intended by this result, rather simply "cancel" the factorials:

$$\frac{10!}{6! \cdot 4!} = \frac{10 \cdot 9 \cdot 8 \cdot 7 \cdot \not{6!}}{\not{6!} \cdot 4 \cdot 3 \cdot 2 \cdot 1} = \frac{10 \cdot \overset{3}{\not{9}} \cdot \not{8} \cdot 7}{\not{4} \cdot \not{3} \cdot \not{2}} = 210$$

**EXAMPLE 3** A student can take five courses next semester chosen from the computer science and mathematics curriculums. Two computer science courses can be chosen from the four that are offered and three mathematics courses can be chosen from the five offered courses. How many ways can a schedule be arranged, assuming no time conflicts?

The computer science choices are $\binom{4}{2}$ and the mathematics choices are $\binom{5}{3}$. By using the fundamental multiplication rule we see that the $\binom{4}{2}$ computer science courses can be "followed" by the $\binom{5}{3}$ mathematics courses. Thus we obtain $\binom{4}{2}\binom{5}{3}$ schedules.

$$\binom{4}{2} \cdot \binom{5}{3} = \frac{4!}{2! \cdot 2!} \frac{5!}{3! \cdot 2!} = \frac{\not{4} \cdot 3 \cdot 2 \cdot 1}{\not{2} \cdot 1 \cdot \not{2} \cdot 1} \cdot \frac{5 \cdot \overset{2}{\not{4}} \cdot \not{3!}}{\not{3!} \cdot \not{2} \cdot 1} = 60 \text{ schedules}$$

**Binomial Coefficients** The numbers $\binom{n}{r}$ are frequently referred to as *binomial coefficients* due to their role in the expansion of the binomial $(x + y)^n$. Specifically, one can prove that

$$(x+y)^n = \binom{n}{0} x^n y^0 + \binom{n}{1} x^{n-1} y^1 + \cdots + \binom{n}{k} x^{n-k} y^k + \cdots + \binom{n}{n} x^0 y^n$$

$$= \sum_{k=0}^{n} \binom{n}{k} x^{n-k} y^k$$

As special cases, we have

$$(x+y)^1 = \binom{1}{0} x^1 y^0 + \binom{1}{1} x^0 y^1 = x + y$$

$$(x+y)^2 = \binom{2}{0}x^2y^0 + \binom{2}{1}x^1y^1 + \binom{2}{2}x^0y^2 = x^2 + 2xy + y^2$$

$$(x+y)^3 = \binom{3}{0}x^3y^0 + \binom{3}{1}x^2y^1 + \binom{3}{2}x^1y^2 + \binom{3}{3}x^0y^3$$

$$= x^3 + 3x^2y + 3xy^2 + y^3$$

These binomial coefficients when laid out in tabular form give rise to the famous *Pascal triangle*, in which each number is obtained as the sum of the two neighboring numbers in the preceding row (Table 8.2). The way the entries are constructed in the table give rise to the formula

$$\binom{n+1}{k} = \binom{n}{k-1} + \binom{n}{k}$$

**TABLE 8.2** Table of Binomial Coefficients, or Pascal's Triangle [the $k$th number in row $n$ is $\binom{n}{k}$]

| | | | | | | | | | | | |
|---|---|---|---|---|---|---|---|---|---|---|---|
| $n = 0$ | | | | | | 1 | | | | | |
| $n = 1$ | | | | | 1 | | 1 | | | | |
| $n = 2$ | | | | 1 | | 2 | | 1 | | | |
| $n = 3$ | | | 1 | | 3 | | 3 | | 1 | | |
| $n = 4$ | | 1 | | 4 | | 6 | | 4 | | 1 | |
| $n = 5$ | 1 | | 5 | | 10 | | 10 | | 5 | | 1 |

## EXERCISES

1. Compute the following.

   (a) $\frac{12!}{4!}$ (b) $\frac{n!}{(n-2)!}$ (c) $\frac{7!}{11!}$

2. Compute the following.

   (a) $\binom{12}{6}$ (b) $\binom{7}{5}$ (c) $\binom{8}{8}$ (d) $\binom{10}{0}$ (e) $C(n+2, n)$

3. Compute the rows of Pascal's triangle for $n = 6$ and $n = 7$.

4. How many ways can a committee of three people be chosen from a group of five? From a group of 10?

5. An instructor decided to assign five problems from a set of 10. How many possible assignments can the instructor make?

6. A quality control inspector tests 10 microprocessor chips out of each batch of 100 produced. In how many ways can the inspector choose the sample to be inspected?

7. A university committee is to be composed of four faculty members and five students.

The faculty are to be chosen from 10 who are eligible for membership and the students will be selected from the 12 who volunteered. In how many ways can the committee be composed?

## Summary and Selected References

The subject of combinatorics is broad and expansive. As noted, the coverage in this chapter is meant only to give a flavor of some of the techniques used in discrete counting and is in no way complete. For a full theoretical treatment the reader may consult Ryser (1963), Hall (1967), Liu (1968), and Brualdi (1977).

For modern applied treatments, Tucker (1984), Bogart (1983), and Roberts (1984) are especially informative. The applicability of combinatorics to the design of computer algorithms is found in Hu (1982), Even (1973, 1979), Reingold, Nievergelt, Deo (1977), and Nijenhuis and Wilf (1978).

Several discrete mathematics textbooks include combinatorics as an integral part. Notable among these are Stanat and McAllister (1977), Preparata and Yeh (1973), and Mott, Kandel, and Baker (1983).

# Posets and Lattices

# 9

## 9.1 Partial Orders Revisited

In this chapter we extend the notion of a partial order on discrete units of a structure to examine some theoretical aspects of such a relation. Recall from Chapter 2 that we defined a relation as a partial order if it has the properties of reflexivity, antisymmetry, and transitivity. Any set that is so ordered under some relation we called a partially ordered set (poset).

The relation, usually designated by $\leq$, imposes an ordering on the elements of the set, which is considered partial in that there might not be any prescribed order between certain pairs of elements of the set. If two elements, $x$ and $y$, are ordered so that $x \leq y$, we state that $x$ *precedes* $y$ (and $y$ *succeeds* $x$) and $x$ and $y$ are said to be comparable. When two elements are not ordered, they are said to be *noncomparable.*

Our first example of a poset was in connection with Boolean algebra. There we saw that natural binary operations exist on elements of a poset. We defined the *meet* (denoted by $\wedge$ or $\cdot$) of two comparable elements by the following:

$$x \leq y \quad \text{iff} \quad x \wedge y = x \quad \text{or} \quad x \cdot y = x$$

Additionally, we defined the join ($\vee$ or $+$) of two comparable elements by

$$x \leq y \quad \text{iff} \quad x \vee y = y \quad \text{or} \quad x + y = y$$

It was shown that these definitions of meet and join do indeed satisfy the requirements of a partial order. (See Section 3.4 where we proved that the properties of reflexivity, antisymmetry, and transitivity resulted from these definitions.)

An important consequence is that if two elements are related by $\leq$ in that $x \leq y$ or $y \leq x$, then the meet and join, so defined, are *unique*. There is only one meet and only one join. Nothing was said about the meet and join of two noncomparable elements: whether they exist or not, and if so, whether they are unique. We will address this shortly.

In Chapter 7 we further encountered partial orders in connection with activity digraphs. Here the elements represented tasks and the relation was "is a prerequisite for." In this particular case the property of reflexivity did not hold and we stated that what we had was called an incompletely specified partial order. The ordering was still *partial* in that the prerequisite nature possibly did not exist between some two tasks.

In the case of both activity digraphs and diagrams of Boolean algebras we observed that there were *no cycles*. The figures were directed acyclic graphs. The property of antisymmetry is what guarantees that no directed cycles be present. If an event or node $A$ precedes node $B$ and if node $B$ precedes node $A$ as in Figure 9.1, we have a cycle. The property of antisymmetry states that this is clearly impossible: for if $A \leq B$ and at the same time $B \leq A$, then $A = B$ and we don't have two nodes at all!

Some of the diagrams of posets that we encountered earlier are reproduced as Figure 9.2 together with their previous figure numbers.

## 9.2 Upper and Lower Bounds

Consider a set $S$ partially ordered with respect to the relation $\leq$, and a subset $A \subseteq S$. An element $m \in S$ is said to be a *lower bound* of the subset $A$ if for every $a \in A$, we have $m \leq a$. That is, the element $m$ precedes every element of set $A$. We note that the only restriction on $m$ is that it be an element of set $S$. It may or may not be in the subset $A$.

**EXAMPLE 1** Let $S$ be the set of Figure 9.2(a) and let $A = \{4, 6, 12\}$. Then the element 2 constitutes the lower bound of $A$. Elements 2, 3, and 6 are lower bounds of the subset $\{6, 12\}$. In Figure 9.2(d), the elements $\varnothing$, $\{0\}$, and $\{1\}$ are all lower bounds for $\{0, 1\}$, whereas the elements $\varnothing$ and $\{1\}$ are lower bounds for the set $\{\{0, 1\}, \{1\}\}$.

*Upper bounds* are defined similarly. Let $B$ be a subset of a poset $S$. Then $n \in S$ is said to be an upper bound of $B$ if for every $b \in B$, we have $b \leq n$. That is, element $n$ *succeeds* every element of $B$. Here again, $n$ may or may not be in set $B$.

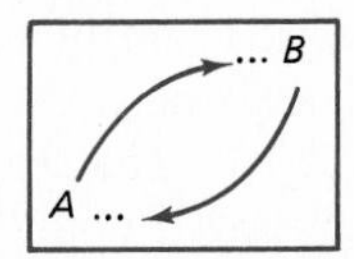


**FIGURE 9.1**

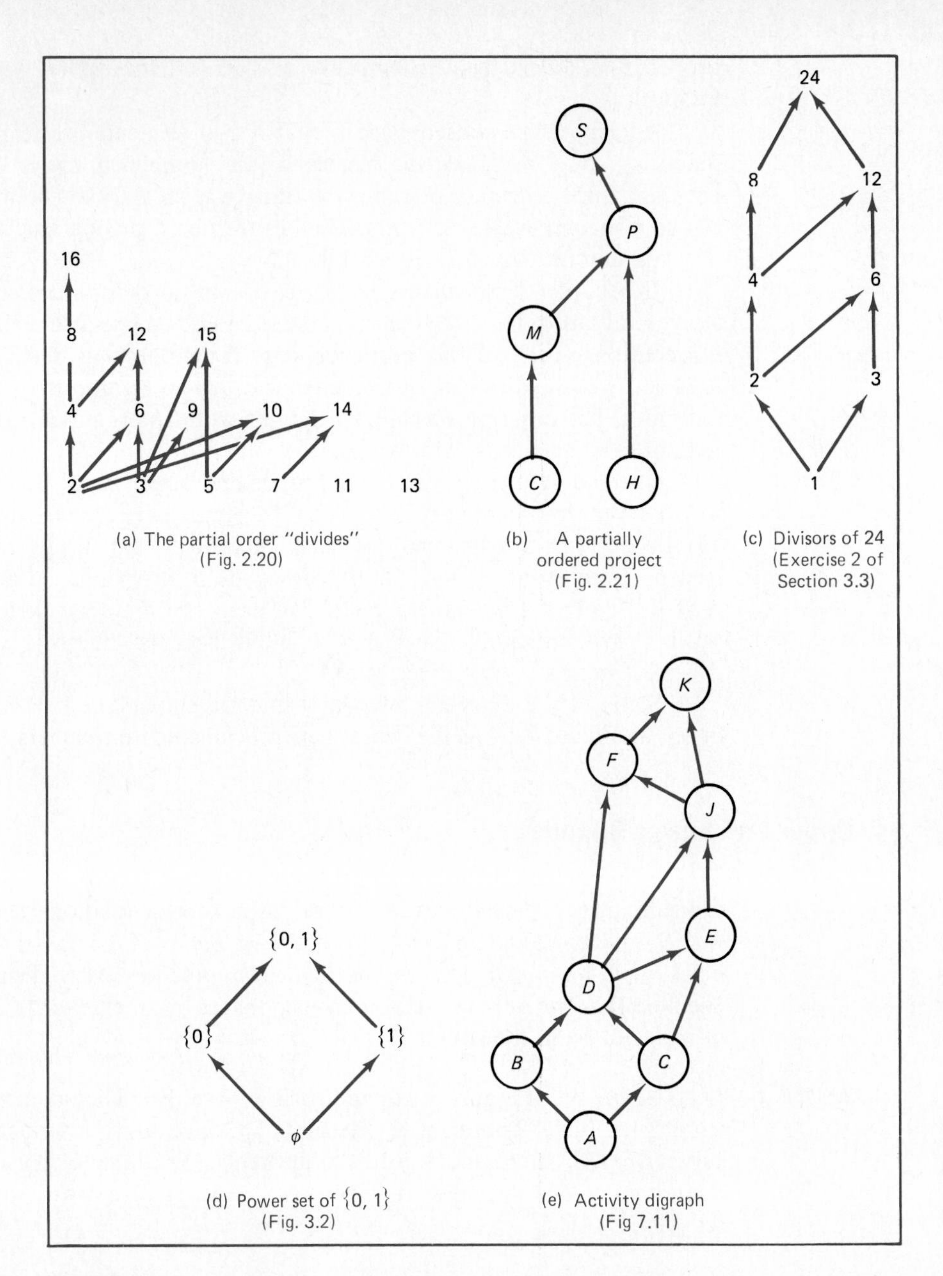


**FIGURE 9.2**

**EXAMPLE 2** To give an example of upper bounds, let $S$ be the poset of Figure 9.2(a), and consider the subset $B = \{4, 6\}$. It should be clear that the element $12 \in S$ is the only upper bound for $B$. Also, it should be observed that $\{4, 8, 12\}$ has *no* upper bounds. (There is no element in $S$ that succeeds all of $\{4, 8, 12\}$.) Neither is there an upper bound for $\{6, 10\}$ or for $\{5, 7, 10\}$. The set $\{M, H\}$ of Figure 9.2(b) has $\{P, S\}$ as upper bounds but has no lower bounds.

From the previous discussion it is seen that upper and lower bounds of sets need not exist.

## EXERCISES

1. Let $S = \{a, b, c, d, e, f\}$ be the poset ordered by the diagram

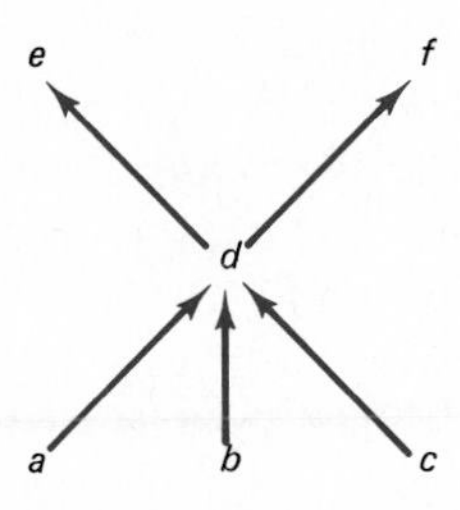


Find the upper bounds and lower bounds of the following subsets if they exist.

(a) $\{a, b, d\}$ (b) $\{a, f\}$ (c) $\{c, d\}$ (d) $\{d, e, f\}$

2. Let $S = \{a, b, c, d, e, f\}$ be the poset ordered by the diagram

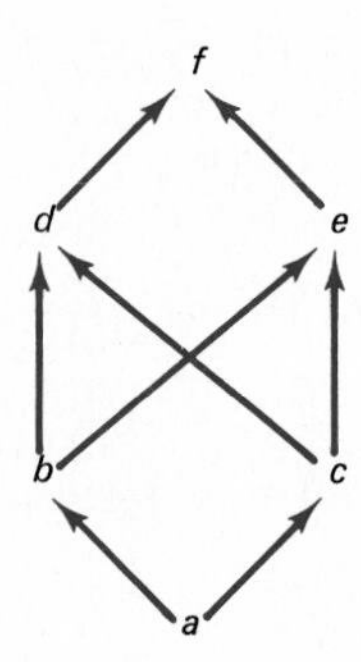


Find the upper and lower bounds of the following subsets if they exist.

(a) $\{a, b, c\}$ (b) $\{b, c, d\}$ (c) $\{b, e, f\}$ (d) $\{a, c, f\}$ (e) $\{c, d\}$

3. In Figure 9.2(c) find upper and lower bounds for the following subsets if they exist.

(a) $\{3, 4, 8\}$ (b) $\{4, 6\}$ (c) $\{6, 8\}$ (d) $\{1, 24\}$

## 9.3 Least Upper Bound and Greatest Lower Bound

In Figure 9.2(a) we saw that the set $\{6, 12\}$ has *three* lower bounds, namely the elements 2, 3, and 6. The element *6*, however, has the property that it succeeds all the lower bounds ($2 \leq 6$, $3 \leq 6$, and $6 \leq 6$). Consequently, we call it the *greatest lower bound* (*glb*). In general, an element $x$ of a poset $S$ is a *glb* of a subset $A \subseteq S$ iff $x$ is a lower bound of $A$ *and* for every lower bound $a$ of $A$, we have $a \leq x$. That is, the *glb* $x$ succeeds all lower bounds.

For Figure 9.2(a) the *glb* of $\{4, 8\}$ is 4. (The set of lower bounds is $\{2, 4\}$ and $2 \le 4$.) In Figure 9.2(d) the singleton set $\{\{0, 1\}\}$ has three lower bounds, namely $\{1\}$, $\{0\}$, and $\varnothing$. But since *no one* of them succeeds all others, there is *no glb*. In Figure 9.2(c), the subset $\{24, 12, 8\}$ has three lower bounds, namely the elements 4, 2, and 1; and element 4 is the *glb*.

An analogous situation exists for upper bounds. If there is one upper bound of a set that precedes all others, then it is called the *least upper bound*, or *lub*. For example, in Figure 9.2(b) the upper bounds of the subset $\{C, H, M\}$ are events $P$ and $S$. Since $P$ precedes $S$, then $P$ is the *lub*. In Figure 9.2(e), node $D$ is the *lub* for the subset $\{A, B, C, D\}$. In Figure 9.2(a), the set $\{2, 4\}$ has as upper bounds the set $\{4, 8, 12, 16\}$ and 4 is the *lub;* the set $\{8, 9, 10\}$ has no upper bounds at all.

We now prove a very important theorem concerning *glbs* and *lubs* of subsets of partially ordered sets.

□ Theorem 9.1 Given a partially ordered set $S$ and a subset $A \subseteq S$. If the *glb* (*lub*) of $A$ exists, then it is unique. That is, there is *only one*.

**Proof** We use the property of antisymmetry. Suppose that subset $A$ has two *glbs*, namely $x$ and $y$. These two are, of course, lower bounds. Since $x$ is a *glb* we know that $y \le x$, and since $y$ also is a *glb* we know that $x \le y$. Thus since $y \le x$ and $x \le y$ we know from antisymmetry that $x = y$, so there is only one *glb*. An analogous proof holds for *lubs*.

## EXERCISES

1. For the poset $S = \{a, b, c, d, e, f, g, h\}$ ordered by the diagram

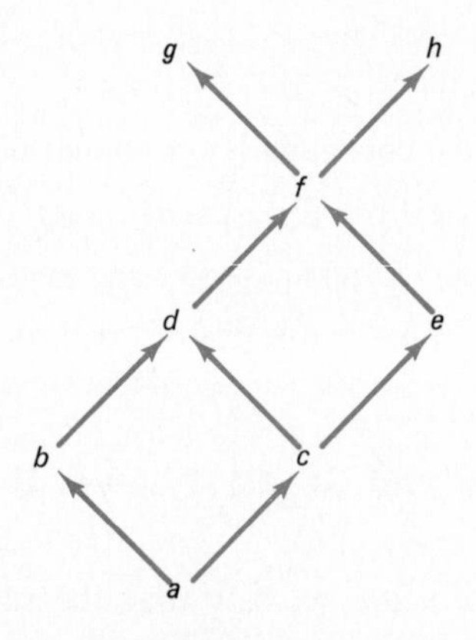


find the *glbs* and *lubs* for the following subsets.

(a) $\{b, c, d\}$ (b) $\{a, e, f\}$ (c) $\{d, h\}$ (d) $\{f, g, h\}$ (e) $\{g, h\}$

2. For the poset $S = \{a, b, c, d, e, f\}$ ordered by the diagram

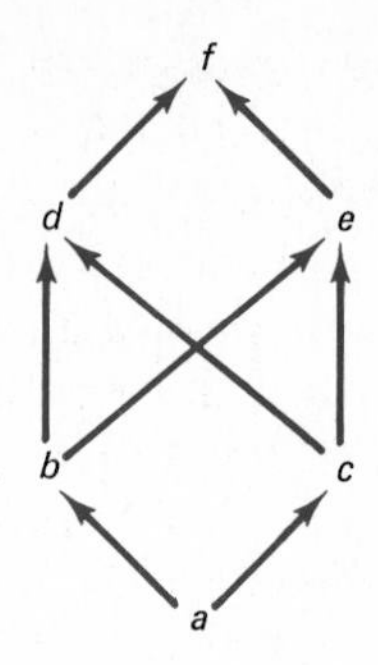


find the *glbs* and *lubs* for the following subsets.

(a) $\{b, c\}$ (b) $\{b, e, f\}$ (c) $\{d, e\}$ (d) $\{a, d, f\}$

3. Prove that the *lub* of a subset $A$ of a partially ordered set $S$ is unique.

## 9.4 Lattices

As pointed out earlier, not all subsets of posets have *lubs* and *glbs*. But posets that do have this characteristic have special meaning. To be more precise, for a given poset $S$, if *every* subset of $S$ has a *lub* and a *glb*, then $S$ is classified as a system known as a *lattice*. In particular, every *pair* of elements has a *glb* and a *lub*. We have shown that the *glb* and *lub* are unique, so we have again a case in which natural *binary* operations can be defined on elements of the poset.

The unique *glb* of two elements $x$ and $y$ of a lattice is known as the *meet* of the two and is usually designated by the familiar symbol, $x \wedge y$. The unique *lub* is known as the *join* and is symbolized by $x \vee y$.

From earlier discussions of partial orders we can stipulate the following about elements $x$, $y$, and $z$ of any lattice, $L$, without proof:

**L.1.** If $x$ and $y$ are ordered by $x \leq y$, then
(a) $x \wedge y = x$ (the meet is the predecessor)
(b) $x \vee y = y$ (the join is the successor)

**L.2.** Since $x \leq x$ (by reflexivity) we have
(a) $x \wedge x = x$ (idempotence of meet)
(b) $x \vee x = x$ (idempotence of join)

**L.3.** (a) $x \wedge y = y \wedge x$ (commutativity of meet)
(b) $x \vee y = y \vee x$ (commutativity of join)

**L.4.** (a) $x \leq x \vee y$ ($x$ precedes its join with any other element)
(b) $x \wedge y \leq x$ ($x$ succeeds its meet with any other element)

**L.5.** Absorption ($x = x \wedge (x \vee y)$) follows from L.4(a) and L.1(a). From L.4(a): $x \leq x \vee y$, so from L.1(a): $x = x \wedge (x \vee y)$. The dual absorption law can be established similarly.

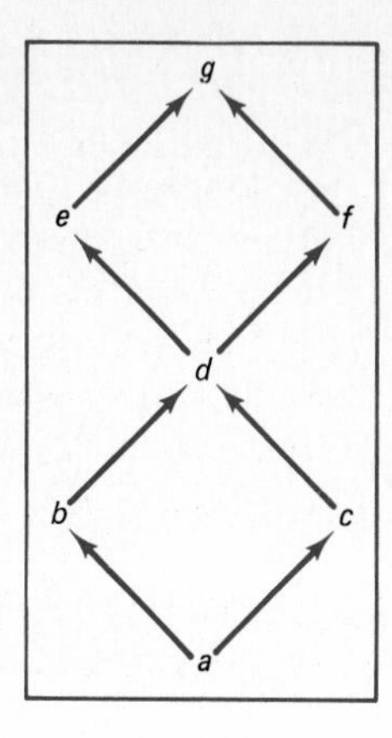


FIGURE 9.3

**EXAMPLE 1** For the lattice shown in Figure 9.3, *we have the meets*

$$b \wedge d = b \qquad f \wedge g = f$$

$$e \wedge f = d \qquad a \wedge g = a$$

and the *joins*

$$b \vee d = d \qquad f \vee g = g$$

$$e \vee f = g \qquad a \vee g = g$$

**EXAMPLE 2** The three diagrams in Figure 9.4 are *not* lattices.

FIGURE 9.4

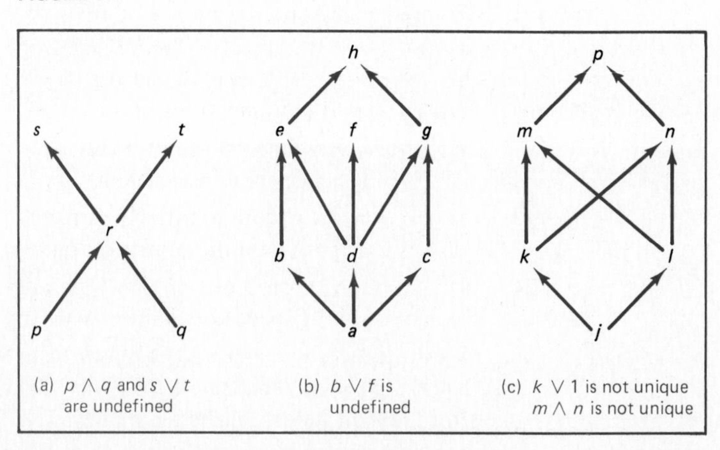


(a) $p \wedge q$ and $s \vee t$ are undefined

(b) $b \vee f$ is undefined

(c) $k \vee 1$ is not unique $m \wedge n$ is not unique

**Bounded Lattices** Every finite lattice $L$ has a unique upper bound for the entire set. It is usually designated as the unit element, written as 1. It has the property that for every element $a \in L$ we have $a \vee 1 = 1$. (This is the boundness axiom for Boolean algebras.) Similarly, a unique lower bound for the entire set exists. It is known as the zero element (0), and has the property that $a \wedge 0 = 0$ for all $a \in L$. We say that every finite lattice is bounded. This is obviously not the case for infinite lattices. Elements $m$ and $n$ are the lower and upper bounds, respectively, in Figure 9.5.

Since the unique unit element 1 of a bounded lattice has the property that $1 = a \vee 1$ for every $a \in L$, we can show that the *identity* law, $a \wedge 1 = a$ holds since

$$\begin{aligned} a \wedge 1 &= a \wedge (a \vee 1) && \text{by property of unit} \\ &= a && \text{by absorption} \end{aligned}$$

A dual argument can be shown for the other *identity* law, $a \vee 0 = a$, because of the boundness property for the zero element.

**Complemented Lattices** A bounded lattice, $L$ (with zero and unit elements), is said to be *complemented* if for *any* $x \in L$ there exists an element $y \in L$ such that $x \wedge y = 0$ and $x \vee y = 1$. Not all lattices are complemented and complements need not be unique.

The lattice of Figure 9.5 is not complemented, for elements $u$, $v$, $w$, $x$, and $y$ do not have complements. (If node $m$ were excluded from the figure, then node $u$ would be the zero element and $w$ and $x$ would be complements.)

The lattice of Figure 9.6 is complemented; however, the complements are not unique. Both elements $a$ and $b$ are complements of element $c$: $c \wedge a = c \wedge b = 0$; $c \vee a = c \vee b = 1$.

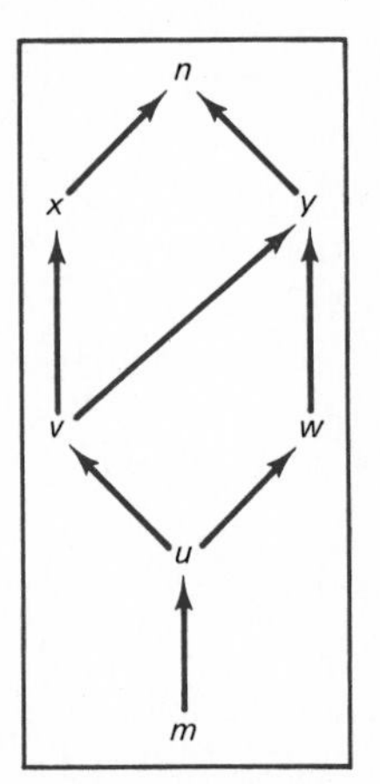


**FIGURE 9.5**

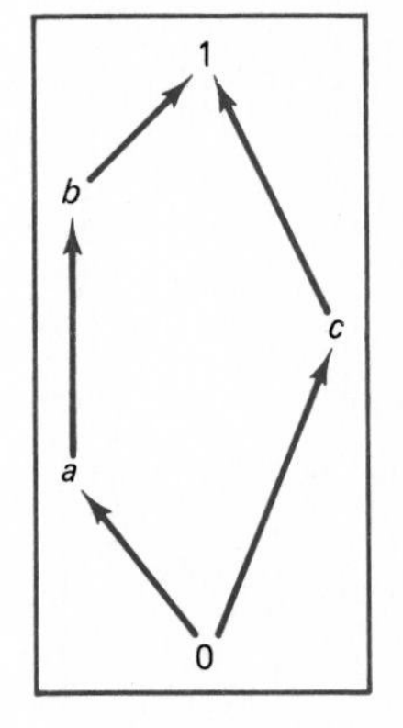


**FIGURE 9.6**

**Distributive Lattices** Lattices may or may not be distributive. When the distributive property holds, then for all $x$, $y$, and $z$ of a lattice, we have

$$x \wedge (y \vee z) = (x \wedge y) \vee (x \wedge z)$$

$$x \vee (y \wedge z) = (x \vee y) \wedge (x \vee z)$$

For example, in the lattice of Figure 9.6, we have

$$a \vee (b \wedge c) = a \vee 0 = a$$

but

$$(a \vee b) \wedge (a \vee c) = b \wedge 1 = b$$

Thus the distributive property does not hold.

Neither does it hold for the lattice of Figure 9.7, for

$$a \vee (b \wedge c) = a \vee 0 = a$$

$$(a \vee b) \wedge (a \vee c) = 1 \wedge 1 = 1$$

□ **Theorem 9.2** If lattice $L$ is bounded, distributive, and complemented, then complements are unique.

**Proof** Let 0 and 1 be the zero and unit elements of $L$. Since $L$ is complemented, then element $a \in L$ has at least one complement. Suppose that it has two, namely the elements $x$ and $y$. Then

$$a \wedge x = 0 \qquad a \vee x = 1$$

and

$$a \wedge y = 0 \qquad a \vee y = 1$$

Then we have

$$\begin{aligned} x &= x \vee 0 \\ &= x \vee (a \wedge y) \\ &= (x \vee a) \wedge (x \vee y) \\ &= 1 \wedge (x \vee y) \\ &= x \vee y \end{aligned} \qquad \text{and} \qquad \begin{aligned} y &= y \vee 0 \\ &= y \vee (a \wedge x) \\ &= (y \vee a) \wedge (y \vee x) \\ &= 1 \wedge (x \vee y) \\ &= x \vee y \end{aligned}$$

Thus $x = y$, and complements are unique.

□ **Theorem 9.3** A complemented (and bounded with $0 \neq 1$) distributive lattice determines a Boolean algebra.

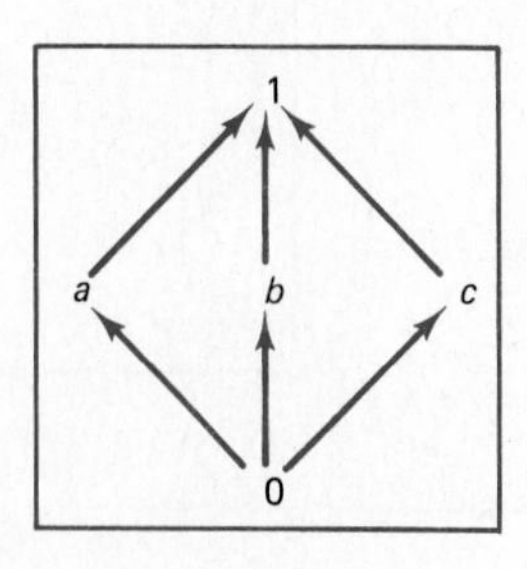


FIGURE 9.7

**Proof** We need to show that the eight axioms of a Boolean algebra stated in Section 3.2 hold. Commutativity, axioms la and 1b, has been shown above, as has distribution, axioms 2a and 2b. We have verified the complement laws (axioms 4a and 4b) by showing in Theorem 9.2 that an element $x$ has a unique complement, $x'$, such that $x \wedge x' = 0$ and $x \vee x' = 1$. The identity laws (axioms 3a and 3b) state that $x \wedge 1 = x$ and $x \vee 0 = x$. These were established earlier in this section.

## EXERCISES

1. Given the following diagrams of partial orders, which are lattices?

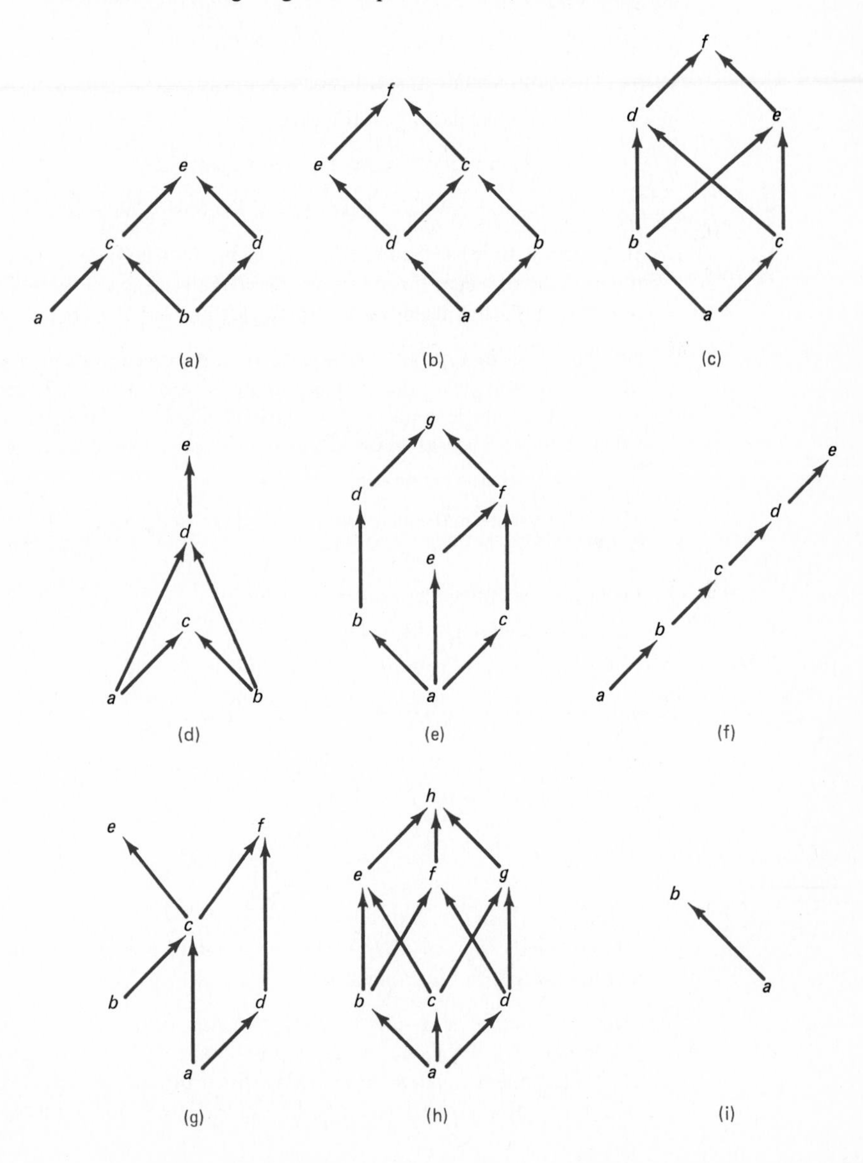

2. For figure (b) of Exercise 1, find

   (a) $c \wedge e$.

   (b) $a \vee e$.

   (c) $b \wedge f$.

   (d) $b \vee f$.

   (e) The complement of $b$, if it exists.

   (f) The complement of $f$, if it exists.

3. For figure (e) of Exercise 1, find

   (a) $c \wedge d$.

   (b) $c \vee d$.

   (c) $b \vee c$.

   (d) The complement of $c$, if it exists.

   (e) The complement of $f$, if it exists.

4. Show that figure (e) of Exercise 1 is not distributive.

5. Do elements $b$, $c$, and $d$ of figure ($f$) of Exercise 1 have complements?

6. In lattices with lower bound 0, the elements that immediately succeed 0 are called *atoms*. To be precise: If $a \in L$ is an atom, then if $b \neq a$ is such that $b \wedge a = b$, then $b = 0$. Identify the atoms of figures (b), (c), and (d) of Exercise 1.

7 An element $a$ in a poset $S$ is said to be a *maximal* element if no element (other than $a$) succeeds $a$. In other words, if there is an element $x \in S$ such that $a \leq x$, then $x = a$. Minimal elements are similarly defined. An element $b$ is *minimal* if no other element precedes $b$. Or, if $y \leq b$, then $y = b$. Find all maximal and minimal elements of figures (d) and (g) of Exercise 1.

8. How many maximal and minimal elements (see Exercise 7) do bounded lattices have?

9. Determine whether the lattice shown is

   (a) Complemented. (b) Distributive.

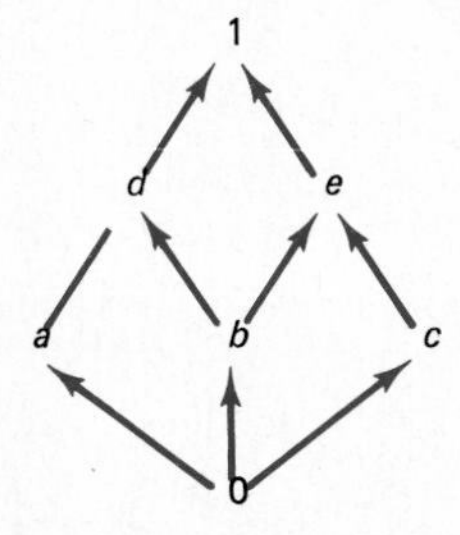


10. Which of the figures of Exercise 1 are complemented, distributive lattices, thus representing a Boolean algebra?

11. Suppose that $M$ is a subset of the elements of a lattice $L$. We say that $M$ is a *sublattice* of $L$ if the operations of meet and join as defined in $L$ are preserved in $M$. These operations are closed in $M$. For example, figure (a) below is a subset of the lattice of Exercise 9 and is a sublattice. However, the subset of figure (b) is not

a sublattice. ($b \vee c \neq e$, as it should.) Determine two other sublattices of the lattice of Exercise 9.

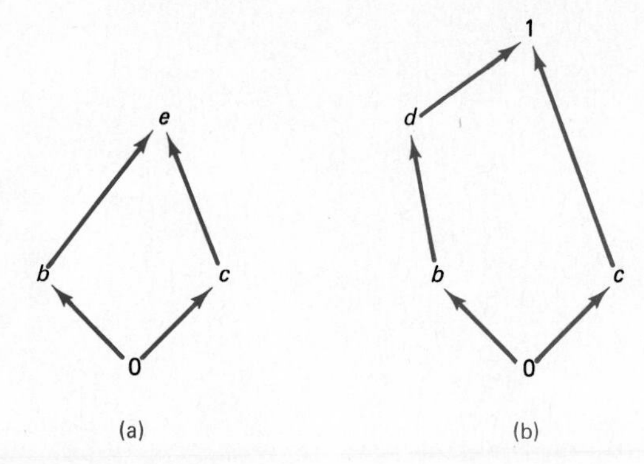

(a) (b)

**12.** Complete the proof of Theorem 9.2 by supplying reasons for the stated equalities.

## Summary and Selected References

The theory of lattices is an important theoretical mathematics topic. Its importance in computer science is notable as a foundation for posets and Boolean algebras. Theoretical aspects of lattice theory can be found in Birkhoff (1967) and Grätzer (1971). Applications to computer science can be found in Birkhoff and Bartee (1970), Gill (1976), Lipschutz (1976), Liu (1977), Stone (1973), and Preparata and Yeh (1973).

# The Binary Number System

# A

## A.1 Binary Numeration

We are accustomed to using the decimal number system. For a variety of reasons, but primarily because human beings have 10 fingers, our ancestors found it more convenient to perform groupings in terms of *tens* rather than in other numbers. The metric system of measurement is a prime example of this grouping, as is our monetary system: 10 cents in a dime, 10 dimes in a dollar, $10^2 =$ 100 cents in a dollar, and so on.

When applied to numbers, we use the familiar positional numeration system where the position of a digit in a numeral determines which power of 10 it represents. Any numeral in the decimal system is a string of decimal digits from 0 through 9. Each digit, by its position, represents a multiple of a power of 10.

Thus $N = 30{,}518$ can be written in its expanded notation:

$$\begin{aligned} N &= 3 \times 10^4 + 0 \times 10^3 + 5 \times 10^2 + 1 \times 10^1 + 8 \times 10^0 \\ &= 3 \times 10{,}000 + 0 \times 1000 + 5 \times 100 + 1 \times 10 + 8 \times 1 \\ &= 30{,}000 + 500 + 10 + 8 \\ &= 30{,}518 \end{aligned}$$

The digit "5" in $N$ does not represent the number "five" but, because of its position, represents 500. The locations of the digits in a numeral are referred to as their "place values." Notice that the highest power of the base used is one less than the number of digits in the number.

In general, any numeral can be written as the string $a_k a_{k-1} a_{k-2} \cdots a_2 a_1 a_0$, where, for the decimal representation, $0 \leq a_i \leq 9$ for $0 \leq i \leq k$, and $a_k \neq 0$. Any term, $a_i$, then, is a multiple of the chosen base, $b$. The string of digits then in expanded notation is $a_k b^k + a_{k-1} b^{k-1} + \cdots + a_2 b^2 + a_1 b^1 + a_0 b^0$.

Electronic components of digital computers are of a discrete nature. They can be in either of two states (on or off), usually represented by 1 and 0. Thus the arithmetic performed is done on these 1/0 values. These are the values of digits of the binary numeration system (base 2).

In the decimal system we can count from 0 through 9 and the next number is represented by 10 to mean that we have "1 of the base (10) and 0 other digits (ones)." In expanded notation, then, 10 is $1 \times 10^1 + 0$. The string 100 would mean $1 \times 10^2 + 0 \times 10^1 + 0$. 1000 is $1 \times 10^3 + 0 \times 10^2 + 0 \times 10 + 0$.

When we count in base 2 we can count from 0 to 1, and then we encounter the base, 2. This must be the next numeral, which will be represented by 10, "one of the base + zero." The next is one more, 11, the next, 100, and so on. Thus the first nine numerals in the binary system are

0   1   10   11   100   101   110   111   1000

A binary numeral is expressed as a string of binary digits (*bits* for short), each of which is either 0 or 1. Thus the binary number $a_k a_{k-1} a_{k-2} \cdots a_2 a_1 a_0$ in expanded notation is $a_k 2^k + a_{k-1} 2^{k-1} + \cdots + a_2 2^2 + a_1 2^1 + a_0$, with each $a_i$ being either a zero (0) or a one (1).

Consequently, the binary number $N = 10111$ in expanded notation is

$$\begin{aligned} N &= 1 \times 2^4 + 0 \times 2^3 + 1 \times 2^2 + 1 \times 2^1 + 1 \\ &= 1 \times 16 + 0 \times 8 + 1 \times 4 + 1 \times 2 + 1 \\ &= 16 + 4 + 2 + 1 \\ &= 23 \quad \text{(as the decimal equivalent)} \end{aligned}$$

## A.2 Conversions between Binary and Decimal

Conversion from Binary to Decimal   Any binary numeral, then, can be converted to its *decimal counterpart* by utilizing the expanded positional notation. The highest power of the base is one less than the number of digits since the rightmost digit is always a multiple of $2^0 = 1$.

To indicate that a number is in binary notation, usually a subscript 2 is appended to the string of digits. Thus $101_2$ indicates the number "five" rather than "one hundred one" in decimal.

**EXAMPLE 1**   To convert $N = 1011001_2$ to its decimal equivalent, we can construct a table similar to Table A.1. Thus we see that $1011001_2 = 64 + 16 + 8 + 1 = 89$ in decimal written as $89_{10}$.

**TABLE A.1**

| | $N$ | | | | | | |
|---|---|---|---|---|---|---|---|
| | 1 | 0 | 1 | 1 | 0 | 0 | 1 |
| Power of 2 | $2^6$ | $2^5$ | $2^4$ | $2^3$ | $2^2$ | $2^1$ | $2^0$ |
| Decimal Counterpart | $1 \times 64$ + | $0 \times 32$ + | $1 \times 16$ + | $1 \times 8$ + | $0 \times 4$ + | $0 \times 2$ + | $1 \times 1$ |

For purposes of reference we show in Table A.2 powers of 2 from 0 through 12.

Another procedure that can be used to convert a binary representation of a number to its decimal equivalent is known as *Horner's method*. This method uses what is known as a *nested polynomial* representation of a number.

To illustrate, consider the decimal $M = 4356$

$$M = 4 \times 10^3 + 3 \times 10^2 + 5 \times 10 + 6$$

As a nested polynomial, $M$ can be written as

$$\begin{aligned} M &= ((4 \times 10 + 3) \times 10 + 5) \times 10 + 6 \\ &= ((43) \times 10 + 5) \times 10 + 6 \\ &= (435) \times 10 + 6 \\ &= 4350 + 6 \\ &= 4356 \end{aligned}$$

Now consider the binary number $N = 10111_2$. From above,

$$N = 1 \times 2^4 + 0 \times 2^3 + 1 \times 2^2 + 1 \times 2 + 1$$

and written as a nested polynomial, we have

$$\begin{aligned} &\quad\ \ \underset{\downarrow}{\text{first digit}} \quad \underset{\downarrow}{\text{second digit}} \quad \underset{\downarrow}{\text{third digit}} \quad \underset{\downarrow}{\text{fourth digit}} \quad \underset{\downarrow}{\text{fifth digit}} \\ N &= (((1 \times 2 + 0) \times 2 + 1) \times 2 + 1) \times 2 + 1 \\ &= (((2) \times 2 + 1) \times 2 + 1) \times 2 + 1 \\ &= ((5) \times 2 + 1) \times 2 + 1 \\ &= (11) \times 2 + 1 \\ &= 23 \end{aligned}$$

**TABLE A.2** Powers of 2

| $N$ | 0 | 1 | 2 | 3 | 4 | 5 | 6 | 7 | 8 | 9 | 10 | 11 | 12 |
|---|---|---|---|---|---|---|---|---|---|---|---|---|---|
| $2^N$ | 1 | 2 | 4 | 8 | 16 | 32 | 64 | 128 | 256 | 512 | 1024 | 2048 | 4096 |

In general, the procedure is as follows: Let $N = a_k a_{k-1} \cdots a_1 a_0$ be a binary numeral.

*Step 1:* Multiply $a_k$ by 2.
*Step 2:* Add the next digit.
*Step 3:* If the last digit has been added, stop. We have the desired result.
*Step 4:* Otherwise, multiply by 2.
*Step 5:* Repeat steps 2, 3, and 4.

For $N = 10111_2$, we have the highest power of 2 being 4, so $k = 4$.

*Step 1:* $a_4 = 1$; $1 \times 2 = 2$.
*Step 2:* $a_3 = 0$; $2 + 0 = 2$
*Step 4:* $2 \times 2 = 4$.
*Step 2:* $a_2 = 1$; $4 + 1 = 5$.
*Step 4:* $5 \times 2 = 10$.
*Step 2:* $a_1 = 1$; $10 + 1 = 11$.
*Step 4:* $11 \times 2 = 22$.
*Step 2:* $a_0 = 1$; $22 + 1 = 23$.
*Step 3:* The last digit has been added; the result is 23.

An interesting way to visualize the evaluation of nested polynomials is that of synthetic division, in which the divisor is the base (2) with the dividend being the string of digits representing the binary numeral. The remainder obtained from the division will be the desired decimal equivalent.

$$2 \rfloor \; 1 \quad 0 \quad 1 \quad 1 \quad 1$$

We "bring down" the first 1 ($a_k$):

$$\begin{array}{r|ccccc} 2 & 1 & 0 & 1 & 1 & 1 \\ \hline & 1 & & & & \end{array}$$

Multiply 1 by 2, placing this product under the next digit (0), and add this next digit.

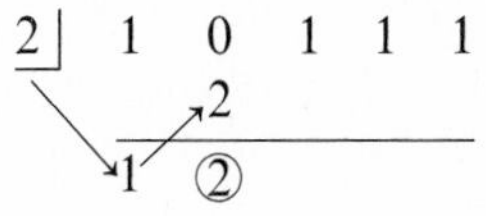

$$\begin{array}{r|ccccc} 2 & 1 & 0 & 1 & 1 & 1 \\ & & 2 & & & \\ \hline & 1 & \textcircled{2} & & & \end{array}$$

Repeat this process.

$$\begin{array}{r|ccccc} 2 & 1 & 0 & 1 & 1 & 1 \\ & & 2 & 4 & & \\ \hline & 1 & 2 & \textcircled{5} & & \end{array}$$

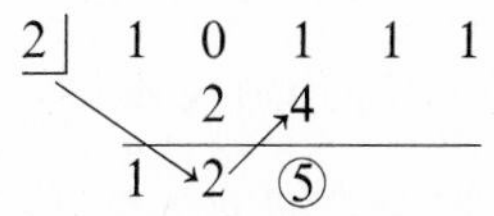

| 2 | 1 | 0 | 1 | 1 | 1 |
|---|---|---|---|---|---|
| | | 2 | 4 | 10 | |
| | 1 | 2 | 5 | (11) | |

| 2 | 1 | 0 | 1 | 1 | 1 | |
|---|---|---|---|---|---|---|
| | | 2 | 4 | 10 | 22 | |
| | 1 | 2 | 5 | 11 | (23) | ← the decimal equivalent |

**EXAMPLE 2** Convert $1011001_2$ to its decimal equivalent by Horner's method.

| 2 | 1 | 0 | 1 | 1 | 0 | 0 | 1 |
|---|---|---|---|---|---|---|---|
| | | 2 | 4 | 10 | 22 | 44 | 88 |
| | 1 | 2 | 5 | 11 | 22 | 44 | 89 |

So we have $1011001_2 = 89_{10}$.

### Conversion from Decimal to Binary

We show two techniques for converting a decimal numeral to its binary counterpart.

TECHNIQUE 1: To convert a decimal number $N$ to binary, we find the largest power of two contained in $N$ (use Table A.2). This largest power will be contained either *once* or not at all. We repeat the process on the remainder obtained by dividing it by this largest power. This process is repeated until we exhaust all powers of 2. We record the powers by which we divide with a 1, and those that are skipped by a 0. This string of 1's and 0's then will be the desired binary conversion.

**EXAMPLE 3** Convert $1900_{10}$ to binary.

From Table A.2 we find that 10 is the largest power of 2 ($2^{10} = 1024$) that divides 1900. We perform the division $1900 \div 1024$, getting a quotient of 1 and a remainder of 876. We then repeat the process on 876 and each successive remainder, recording the powers used with 1 and those skipped by 0. This process can be illustrated by Table A.3. The binary string obtained then from *top down* is 11101101100. We can check to see if this is correct:

| 1 | 1 | 1 | 0 | 1 | 1 | 0 | 1 | 1 | 0 | 0 |
|---|---|---|---|---|---|---|---|---|---|---|
| $2^{10}$ | $2^9$ | $2^8$ | $2^7$ | $2^6$ | $2^5$ | $2^4$ | $2^3$ | $2^2$ | $2^1$ | $2^0$ |

$$1024 + 512 + 256 + 0 + 64 + 32 + 0 + 8 + 4 + 0 + 0 = 1900$$

**EXAMPLE 4** Convert $2050_{10}$ to binary.

The highest power of 2 contained in 2050 is 11 ($2^{11} = 2048$).

$$2050 \div 2048 = 1 \qquad \text{with remainder of 2}$$

The next highest power of 2 contained in the remainder is 1 ($2^1 = 2$). The

**TABLE A.3**

| *Number (Original or Remainder)* | *Highest Power of 2* | *Division* | *Quotient* | *Remainder* |
|---|---|---|---|---|
| 1900 | $2^{10} = 1024$ | $1900 \div 1024$ | 1 | 876 |
| 876 | $2^9 = 512$ | $976 \div 512$ | 1 | 364 |
| 364 | $2^8 = 256$ | $364 \div 256$ | 1 | 108 |
| 108 | $2^7 = 128$ (not contained in 108) | | 0 | |
| | $2^6 = 64$ | $108 \div 64$ | 1 | 44 |
| 44 | $2^5 = 32$ | $44 \div 32$ | 1 | 12 |
| 12 | $2^4 = 16$ (not contained in 12) | | 0 | |
| | $2^3 = 8$ | $12 \div 8$ | 1 | 4 |
| 4 | $2^2 = 4$ | $4 \div 4$ | 1 | 0 |
| 0 | $2^1$ | | 0 | |
| 0 | $2^0$ | | 0 | |

other powers are each represented 0 times. Thus the binary equivalent is

$$\begin{array}{cccccccccccc} 1 & 0 & 0 & 0 & 0 & 0 & 0 & 0 & 0 & 0 & 1 & 0 \\ 2^{11} & 2^{10} & 2^9 & 2^8 & 2^7 & 2^6 & 2^5 & 2^4 & 2^3 & 2^2 & 2^1 & 2^0 \end{array}$$

TECHNIQUE 2: The second technique is to divide the decimal number $N$ and each successive *quotient* by 2 and record the remainders as either 1 or 0. We continue this until a 0 quotient is obtained. The string of 1/0 remainders, read in *reverse* order from which they were obtained, is the desired binary numeral. Table A.4 illustrates this procedure on the decimal under $N = 1900$. The *reverse* string of remainders is the binary representation, 11101101100.

**TABLE A.4** Decimal-to-Binary Conversion

| *Division* | *Quotient* | *Remainder* |
|---|---|---|
| $1900 \div 2$ | 950 | 0 ↑ |
| $950 \div 2$ | 475 | 0 |
| $475 \div 2$ | 237 | 1 |
| $237 \div 2$ | 118 | 1 |
| $118 \div 2$ | 59 | 0 |
| $59 \div 2$ | 29 | 1 |
| $29 \div 2$ | 14 | 1 |
| $14 \div 2$ | 7 | 0 |
| $7 \div 2$ | 3 | 1 |
| $3 \div 2$ | 1 | 1 |
| $1 \div 2$ | 0 | 1 |

**TABLE A.5**

| *Decimal* | *Binary* | | | | | | |
|---|---|---|---|---|---|---|---|
| | $2^6$ | $2^5$ | $2^4$ | $2^3$ | $2^2$ | $2^1$ | $2^0$ |
| 0 | | | | | | | 0 |
| 1 | | | | | | | 1 |
| 2 | | | | | | 1 | 0 |
| 3 | | | | | | 1 | 1 |
| 4 | | | | | 1 | 0 | 0 |
| 5 | | | | | 1 | 0 | 1 |
| 6 | | | | | 1 | 1 | 0 |
| 7 | | | | | 1 | 1 | 1 |
| 8 | | | | 1 | 0 | 0 | 0 |
| 9 | | | | 1 | 0 | 0 | 1 |
| 10 | | | | 1 | 0 | 1 | 0 |
| 11 | | | | 1 | 0 | 1 | 1 |
| 12 | | | | 1 | 1 | 0 | 0 |
| 13 | | | | 1 | 1 | 0 | 1 |
| 14 | | | | 1 | 1 | 1 | 0 |
| 15 | | | | 1 | 1 | 1 | 1 |
| 16 | | | 1 | 0 | 0 | 0 | 0 |
| 17 | | | 1 | 0 | 0 | 0 | 1 |
| 20 | | | 1 | 0 | 1 | 0 | 0 |
| 23 | | | 1 | 0 | 1 | 1 | 1 |
| 25 | | | 1 | 1 | 0 | 0 | 1 |
| 30 | | | 1 | 1 | 1 | 1 | 0 |
| 50 | | 1 | 1 | 0 | 0 | 1 | 0 |
| 75 | 1 | 0 | 0 | 1 | 0 | 1 | 1 |
| 100 | 1 | 1 | 0 | 0 | 1 | 0 | 0 |

For later reference Table A.5 gives the binary equivalent of some decimal numerals.

## A.3 Binary Arithmetic

In this section we will show how to add and multiply in binary. The algorithms used are the same as those used for numbers expressed in any base, and in particular, those learned in elementary school for decimal numbers. Additionally, a technique for binary subtraction will be shown.

Addition In decimal addition of two numbers, we proceed as follows: (1) add the rightmost digits; (2) record the units digit and if the sum exceeds 9, we regroup in terms of tens and "carry" the tens digit to be added to the next column of digits; and (3) repeat this process for all the columns of digits.

We can write the carry down in the appropriate column or note it mentally. The procedure is as follows:

Add $576 + 718$ in base 10.

$$\begin{array}{rcccl} & 1 & 0 & 1 & \leftarrow \text{ carry from previous addition} \\ & & 5 & 7 & 6 \\ + & & 7 & 1 & 8 \\ \hline & 1 & 2 & 9 & 4 \end{array}$$

The procedure is the same for any base. We regroup our sums in terms of the base and "carry" our regrouping to the next column.

In order to add in binary we need a binary addition table (Table A.6). Notice that three of the sums in the table involve no carry digit. However, in the sum $1 + 1 = 10$, we have a *sum of 0* and a *carry of 1*.

In adding two binary numbers, all we need to know how to add are

$$\begin{array}{rl} 0 + 0 = 0 & \text{with a carry of } 0 \\ 0 + 1 = 1 + 0 = 1 & \text{with a carry of } 0 \\ 1 + 1 = 0 & \text{with a carry of } 1 \\ 1 + 1 + 1 = 1 & \text{with a carry of } 1 \end{array}$$

To illustrate, find the sum of $110_2 + 11_2$.

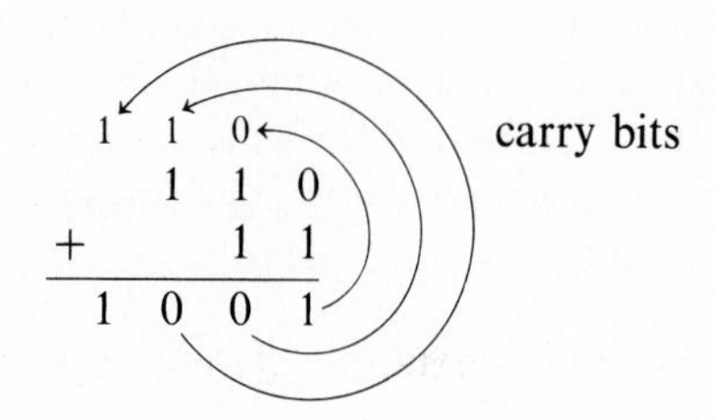


As another illustration, add $1110010_2 + 1100111_2$.

$$\begin{array}{cccccccccl} & 1 & 1 & 0 & 0 & 1 & 1 & 0 & \leftarrow & \text{carry bits} \\ & & 1 & 1 & 1 & 0 & 0 & 1 & 0 & \\ + & & 1 & 1 & 0 & 0 & 1 & 1 & 1 & \\ \hline & 1 & 1 & 0 & 1 & 1 & 0 & 0 & 1 & \end{array}$$

**TABLE A.6** Addition Table in Binary

| + | 0 | 1 |
|---|---|---|
| 0 | 0 | 1 |
| 1 | 1 | 10 |

To add more than two numbers, some of the carrys may extend over more than one column, as illustrated in performing the addition of

$$
\begin{array}{r}
1\ 1\ 1\ 1 \\
1\ 0\ 0\ 1 \\
1\ 1\ 1 \\
+1\ 0\ 1\ 0 \\
\hline
\end{array}
$$

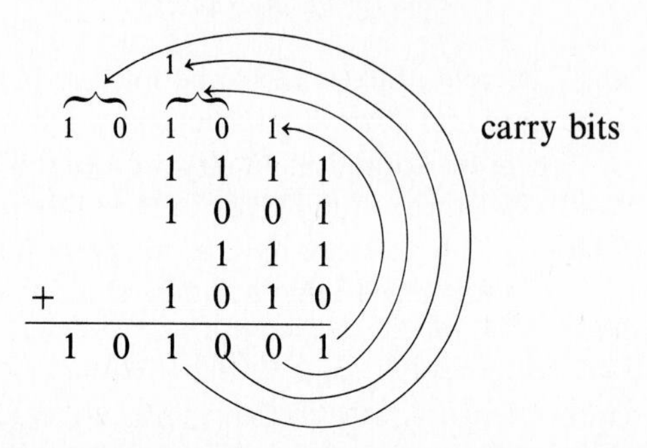


The procedures are identical to decimal addition. With a little practice you will get used to carrying "twos" and powers of 2 instead of "tens" and powers of 10. The regrouping is in terms of *two* rather than *ten.*

Multiplication Multiplication in binary is based on Table A.7. It is seen that the only time a nonzero product is obtained is when the multiplication is by 1, and then all that is obtained is a recopy of the factor being multiplied.

In carrying out the multiplication of two numbers, one lines up the partial products as done in ordinary decimal multiplication and then binary addition is performed. To illustrate, let us multiply 1101 by 101.

$$
\begin{array}{rl}
1\ 1\ 0\ 1 & \\
+\quad 1\ 0\ 1 & \\
\hline
1\ 1\ 0\ 1 & (1101 \times 1 = 1101) \\
0\ 0\ 0\ 0\ \phantom{0} & (1101 \times 0 = 0000) \\
1\ 1\ 0\ 1\ \phantom{0\ 0} & (1101 \times 1 = 1101) \\
\hline
1\ 0\ 0\ 0\ 0\ 0\ 1 & \leftarrow \text{binary addition of partial products}
\end{array}
$$

It can be seen that multiplication in binary is nothing more than a series of "shifts" and "adds," and this is how most computers actually perform multiplication.

**TABLE A.7** Multiplication Table in Binary

| × | 0 | 1 |
|---|---|---|
| 0 | 0 | 0 |
| 1 | 0 | 1 |

**Subtraction** Subtraction in binary can be carried out as is done in ordinary decimal subtraction, the borrowing process being fundamentally the same. The process at times is somewhat awkward since the borrowing can spill over several places. This is illustrated in subtracting $y = 1011$ from $x = 11010$.

$$\begin{array}{rccccc} & & 0 & 1 & 0 & \\ x = & 1 & \not{1} & \not{0} & \not{1} & 0 \\ -y = & & 1 & 0 & 1 & 1 \\ \cline{2-6} & & 1 & 1 & 1 & 1 \end{array}$$

The borrowing process can be avoided completely and subtraction can be performed in terms of addition only. Most computers utilize this procedure in carrying out subtraction. The procedure is known as *two's-complement* subtraction.

The first requirement is that both numbers have the same number of digits, and this is perfectly reasonable for fixed-length computer memory locations. For our algorithm one simply introduces 0's at the beginning of a number which is shorter in length than the other one. Thus our subtraction above becomes

$$\begin{array}{rccccc} x = & 1 & 1 & 0 & 1 & 0 \\ -y = & 0 & 1 & 0 & 1 & 1 \\ \cline{2-6} \end{array}$$

The two's complement of a binary number is the following: All 1's are changed to 0's and all 0's are changed to 1's and then we add 1. Thus for

$$y = 0 \quad 1 \quad 0 \quad 1 \quad 1$$

the two's complement is

$$1 \quad 0 \quad 1 \quad 0 \quad 0 + 1 = 1 \quad 0 \quad 1 \quad 0 \quad 1$$

This value is then *added* to the $x$ and the leading 1 in the sum is deleted. The result is the required subtraction $x - y$.

The two's-complement subtraction for the values of $x$ and $y$ above becomes

$$-\begin{array}{rccccc} x = & 1 & 1 & 0 & 1 & 0 \\ y = & 0 & 1 & 0 & 1 & 1 \end{array} \xrightarrow{} \text{two's complement} \quad \begin{array}{rcccccc} x = & & 1 & 1 & 0 & 1 & 0 \\ = & + & 1 & 0 & 1 & 0 & 1 \\ \cline{3-7} & \textcircled{1} & 0 & 1 & 1 & 1 & 1 \\ & \underbrace{}_{\text{delete}} & & & & & \end{array}$$

The result is 1111, and we see that this is the same difference as we obtained above.

---

**EXAMPLE 1** Subtract by two's complement:

$$\begin{array}{cccccccccc} & 1 & 1 & 0 & 0 & 1 & 0 & 1 & 1 & x \\ - & & & & 1 & 0 & 1 & 1 & 1 & y \\ \cline{1-9} \end{array}$$

*Step 1:* Left-pad $y$ with 0's.

$$\begin{array}{rl} 1\ 1\ 0\ 0\ 1\ 0\ 1\ 1 & x \\ -0\ 0\ 0\ 1\ 0\ 1\ 1\ 1 & y \\ \hline \end{array}$$

*Step 2:* Compute the two's complement of $y$, obtaining

$$1\ 1\ 1\ 0\ 1\ 0\ 0\ 0 + 1 = 1\ 1\ 1\ 0\ 1\ 0\ 0\ 1$$

*Step 3:* Add $x$ + two's complement of $y$.

$$\begin{array}{rl} 1\ 1\ 0\ 0\ 1\ 0\ 1\ 1 & x \\ +1\ 1\ 1\ 0\ 1\ 0\ 0\ 1 & \text{two's complement of } y \\ \hline \textcircled{1}\ 1\ 0\ 1\ 1\ 0\ 1\ 0\ 0 & \end{array}$$

*Step 4:* Delete the leading 1, obtaining 10110100. Thus

$$\begin{array}{r} 1\ 1\ 0\ 0\ 1\ 0\ 1\ 1 \\ -\quad 1\ 0\ 1\ 1\ 1 \\ \hline 1\ 0\ 1\ 1\ 0\ 1\ 0\ 0 \end{array}$$

How does this procedure work for $x - y$? Let the number of digits in $x$ be $n$. Then consider $2^n$, namely a 1 followed by $n$ 0's. Now $2^n - 1$ is a string of $n$ 1's. Subtract $y$ from $2^n - 1$. This results in changing 0's to 1's and 1's to 0's in $y$. We now have

$$2^n - 1 - y = \underbrace{111 \cdots 1}_{n \text{ digits}} - y$$

Adding this to $x$, we get

$$x + 2^n - 1 - y = x - y + 2^n - 1$$

The addition of the 1 yields

$$x - y + 2^n - 1 + 1 = x - y + 2^n$$

The deletion of the leading 1 (which is the $2^n$) then yields $x - y$. We have

$$x - y = x + \underbrace{(2^n - 1 - y) + 1}_{\substack{\uparrow \\ \text{two's complement} \\ \text{of } y}} - \underset{\substack{\uparrow \\ \text{leading 1}}}{2^n}$$

What happens if $y > x$ in the subtraction of $x - y$? This can be performed as $y - x$ and then the result is the negative of the desired result. [We see this from: $x - y = -(y - x)$.]

To illustrate, subtract $y = 11010$ from $x = 100$. Since $y > x$, perform $y - x$.

$$
\begin{array}{rcccccl}
y = & 1 & 1 & 0 & 1 & 0 & \\
-x = & 0 & 0 & 1 & 0 & 0 & \longrightarrow \text{two's complement} =
\end{array}
\quad
\begin{array}{rcccccc}
x = & & 1 & 1 & 0 & 1 & 0 \\
+ & & 1 & 1 & 1 & 0 & 0 \\ \hline
& \underset{\underset{\text{delete}}{\nwarrow}}{\textcircled{1}} & 1 & 0 & 1 & 1 & 0
\end{array}
$$

The result is 10110, which is the *negative* of $x - y$.

*Note:* The subtraction in decimal above was $x - y = 4 - 26 = -22$, which is $-10110_2$.

## A.4 Further Computer Arithmetic

Octal For purposes of storage in a computer's memory, the digits of a binary number (the bits) are usually grouped by either threes or fours. If the grouping is in terms of threes, then each memory location can contain 3 bits and the numbers that can be stored range in value as shown in Table A.8. In the figure we have indicated the decimal counterpart for each of the 3-bit groupings.

The fact that there are only eight possible combinations of 3-bit groups gives rise to the octal (base 8) system of numeration. Each 3-bit group of digits represents one digit of octal (in decimal: 0 through 7). Thus to convert a binary string of digits to its octal representation, all we have to do is to group the bits by 3's from the *right* and add 0's to the left if necessary (*zero-pad* to the left). The following two examples show how this is done.

**EXAMPLE 1** Convert $101011100111_2$ to octal.

If we group the digits by threes from the right, we obtain

$$101\quad 011\quad 100\quad 111$$

Now each of the 3-bit groups can be converted to octal by the correspondence in Table A.8.

$$\underbrace{101}_{5}\quad \underbrace{011}_{3}\quad \underbrace{100}_{4}\quad \underbrace{111}_{7}$$

Thus the result is $5347_8$.

**TABLE A.8**

| 3-bit groups | 000 | 001 | 010 | 011 | 100 | 101 | 110 | 111 |
|---|---|---|---|---|---|---|---|---|
| Decimal Counterpart | 0 | 1 | 2 | 3 | 4 | 5 | 6 | 7 |

**EXAMPLE 2** Convert $1010110000011_2$ to octal.

In grouping by threes we notice that two zeros have to be added to the left, giving

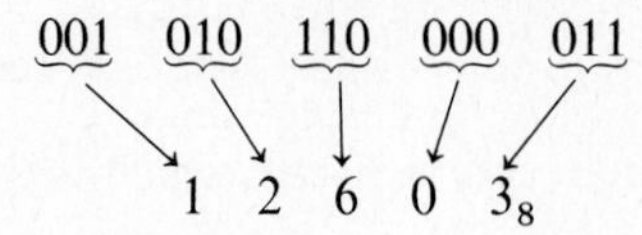


To perform conversion from octal to binary the process discussed above is just reversed, as shown in Example 3.

**EXAMPLE 3** Convert $37105_8$ to binary.

By using Table A.8 we obtain

$$\begin{array}{ccccc} 3 & 7 & 1 & 0 & 5 \\ \downarrow & \downarrow & \downarrow & \downarrow & \downarrow \\ 011 & 111 & 001 & 000 & 101_2 \end{array}$$

To convert from an octal representation to decimal, merely express the octal number in its place-value representation, as was done earlier in this appendix for conversion from binary to decimal.

**EXAMPLE 4** Convert $5342_8$ to decimal.

$$\begin{aligned} 5342_8 &= 5 \times 8^3 + 3 \times 8^2 + 4 \times 8 + 2 \\ &= 5(512) + 3(64) + 4(8) + 2 \\ &= 2560 + 192 + 32 + 2 \\ &= 2786_{10} \end{aligned}$$

To convert a decimal representation to its equivalent in octal, one divides the number and resulting quotients by 8 until the final quotient is zero. The octal representation is the *reverse* string of remainders. This procedure is exactly the same as was done earlier for conversion from decimal to binary.

**EXAMPLE 5** To convert $7693_{10}$ to octal, we construct Table A.9. The reverse string of remainders is the required octal representation: $17015_8$.

**TABLE A.9**

| *Division* | *Quotient* | *Remainder* |
|---|---|---|
| 7693 ÷ 8 | 961 | 5 ↑ |
| 961 ÷ 8 | 120 | 1 |
| 120 ÷ 8 | 15 | 0 |
| 15 ÷ 8 | 1 | 7 |
| 1 ÷ 8 | 0 | 1 |

**Hexadecimal** As mentioned above, the grouping of bits in a computer's memory is sometimes done by fours. Here we have a total of 16 possibilities for the arrangement of bits, giving rise to the hexadecimal (base 16) system of numeration. In order to work strictly within base 16, we need to have access to 16 different digits. There is no problem with the first 10 of these. We simply use the digits 0 to 9 of the decimal system. The first six letters of the alphabet, A to F, are then used to represent the next six digits needed.

In Table A.10 we give this hexadecimal numeration system together with the decimal and binary counterparts.

The next four examples will illustrate conversions between hexadecimal, binary, and decimal. These conversions parallel those shown above relative to the octal system. The only changes we will see will be in regard to the extra six symbols used (A to F).

**EXAMPLE 6** Convert $10010111011010_2$ to hexadecimal.

Group the bits by *four* from the right and zero-pad the leftmost to obtain a 4-bit grouping by using Table A.10.

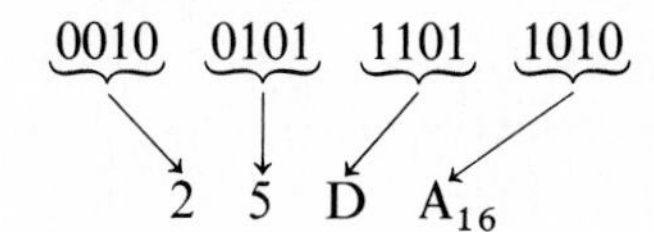


**EXAMPLE 7** Convert $7C09_{16}$ to binary.

$$\begin{array}{cccc} 7 & C & 0 & 9 \\ \downarrow & \downarrow & \downarrow & \downarrow \\ 0111 & 1100 & 0000 & 1001_2 \end{array}$$

**TABLE A.10**

| *Hexadecimal Digit* | *4-Bit Binary Equivalent* | *Decimal Value* |
|---|---|---|
| 0 | 0000 | 0 |
| 1 | 0001 | 1 |
| 2 | 0010 | 2 |
| 3 | 0011 | 3 |
| 4 | 0100 | 4 |
| 5 | 0101 | 5 |
| 6 | 0110 | 6 |
| 7 | 0111 | 7 |
| 8 | 1000 | 8 |
| 9 | 1001 | 9 |
| A | 1010 | 10 |
| B | 1011 | 11 |
| C | 1100 | 12 |
| D | 1101 | 13 |
| E | 1110 | 14 |
| F | 1111 | 15 |

**EXAMPLE 8** Convert $23DA_{16}$ to decimal.

$$\begin{aligned} 23DA &= 2 \times 16^3 + 3 \times 16^2 + D \times 16 + A \\ &= 2(4096) + 3(256) + 13(16) + 10 \\ &= 8192 + 768 + 208 + 10 \\ &= 9178_{10} \end{aligned}$$

**EXAMPLE 9** To convert $2793_{10}$ to hexadecimal, we construct Table A.11.

**TABLE A.11**

| *Division* | *Quotient* | *Remainder* |
|---|---|---|
| 2793 ÷ 16 | 174 | 9 ↑ |
| 174 ÷ 16 | 10 | 14 = E |
| 10 ÷ 16 | 0 | 10 = A |

The result is $AE9_{16}$

## EXERCISES

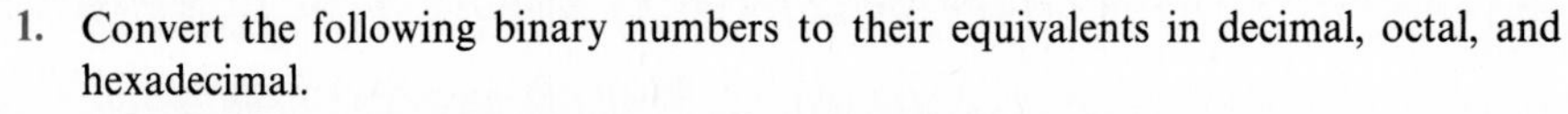

1. Convert the following binary numbers to their equivalents in decimal, octal, and hexadecimal.
   (a) 1010101 (b) 1111001011 (c) 11001001000
2. Convert the following octal numbers to their equivalents in decimal and binary.
   (a) 40725 (b) 12345 (c) 7654
3. Convert the following hexadecimal numbers to their equivalents in decimal and binary.
   (a) 50AB (b) A7C (c) 592D
4. Convert the following decimal numbers to their equivalents in binary, octal, and hexadecimal.
   (a) 5723 (b) 896 (c) 1023
5. Add in binary.
   (a) 10110 + 101001 (b) 11111 + 100010 (c) 1101 + 10010 + 1111
6. Multiply in binary.
   (a) 10110 × 10101 (b) 1110 × 1110 (c) 11011 × 1001
7. Use two's complement to subtract in binary.
   (a) 1110101 – 10110 (b) 1010110 – 1000 (c) 10101 – 111011
8. Devise schemes for adding and multiplying in octal and hexadecimal.

# Matrices: An Introduction

# B

This appendix will serve to outline some of the elementary results from matrix theory. It is not intended to be a definitive study of matrices, which is extensive, but rather to give some important definitions and manipulations which prove useful to our coverage of discrete mathematics.

## B.1 What Is a Matrix?

A matrix (plural: matrices) is, very simply, a rectangular array of elements, usually numbers, which stand in a special relationship to one another. Individually, these elements have significance, but their totality constitutes the object—the matrix—that will be interpreted and manipulated according to certain rules.

Suppose that you as a student have obtained the following five scores on tests in a given course:

$$93 \quad 89 \quad 98 \quad 90 \quad 97$$

Individually, each score is important in its own right, but the entire set has special significance in that collectively these scores represent your grade in the course.

If we let these scores be represented by the letter $s$, then the first score, 93, can be designated, by using a subscript, as $s_1$. The second score, $s_2$, is 89, and so on. The subscripts refer to the position of the individual entries in the list.

Those of you with programming experience are accustomed to working with such lists and should have facility with subscripts. As stated above, the subscript signifies the *position* in the list. Such a list as the one above is referred to as a *one-dimensional matrix* since it consists of only one row of numbers. The term *vector* is also used for such a list, which has only one row or one column.

More commonly, matrices refer to two-dimensional arrays which are characterized by having both rows and columns. To extend our example above, suppose that there were four students in the class. Then the matrix of scores could be written as follows:

$$
\begin{array}{l} \\ \text{Student 1} \\ \text{Student 2} \\ \text{Student 3} \\ \text{Student 4} \end{array}
\begin{array}{c}
\begin{array}{ccccc} \text{Test 1} & \text{Test 2} & \text{Test 3} & \text{Test 4} & \text{Test 5} \end{array} \\
\begin{bmatrix} 93 & 89 & 98 & 90 & 97 \\ 82 & 96 & 78 & 83 & 87 \\ 70 & 100 & 92 & 75 & 78 \\ 65 & 76 & 82 & 90 & 95 \end{bmatrix}
\end{array}
$$

This matrix has *four rows* and *five columns* and we say that its *order* is $4 \times 5$. Notice that the number of rows is listed first and columns second. The score of the third student on the fourth test is 75. This entry is said to be in the (3, 4) position of the matrix. If the entire matrix is designated by the letter $S$, then the 75 can be represented by the double subscript $S_{3,4}$.

In general, a $4 \times 5$ matrix $A$ can be written as follows:

$$
A = \begin{array}{l} \\ \text{Row 1} \\ \text{Row 2} \\ \text{Row 3} \\ \text{Row 4} \end{array}
\begin{array}{c}
\begin{array}{ccccc} \text{Col. 1} & \text{Col. 2} & \text{Col. 3} & \text{Col. 4} & \text{Col. 5} \end{array} \\
\begin{bmatrix} a_{11} & a_{12} & a_{13} & a_{14} & a_{15} \\ a_{21} & a_{22} & a_{23} & a_{24} & a_{25} \\ a_{31} & a_{32} & a_{33} & a_{34} & a_{35} \\ a_{41} & a_{42} & a_{43} & a_{44} & a_{45} \end{bmatrix}
\end{array}
$$

In this matrix, the doubly subscripted entries are indicative of their positions, with the row being given first and the column second. Thus the element $a_{23}$ refers to the element in the second row and the third column. In general, the term $a_{ij}$ will refer to the element in the $i$th row and the $j$th column (see Figure B.1). This $a_{ij}$ notation has many advantages in that the elements of a

$$
\begin{array}{l} \\ \text{Row 1} \\ \text{Row 2} \\ \vdots \\ \text{Row } i \\ \vdots \\ \text{Row } m \end{array}
\begin{array}{c}
\begin{array}{ccccccc} \text{Col. 1} & \text{Col. 2} & & \text{Col. } j & & \text{Col. } n \end{array} \\
\begin{bmatrix} a_{11} & a_{12} & \cdots & a_{1j} & \cdots & a_{1n} \\ a_{21} & a_{22} & \cdots & a_{2j} & \cdots & a_{2n} \\ & & \cdots & & & \\ a_{i1} & a_{i2} & \cdots & a_{ij} & \cdots & a_{in} \\ & & \cdots & & & \\ a_{m1} & a_{m2} & \cdots & a_{mj} & \cdots & a_{mn} \end{bmatrix}
\end{array}
$$

**FIGURE B.1** General matrix of order $m \times n$.

matrix can be determined by giving a "formula" for a typical element. This is explained in the following example.

**EXAMPLE 1** Show the $3 \times 3$ matrix whose typical element $a_{ij}$ is given by $a_{ij} = i + 1$.

The matrix will have the general form as

$$\begin{bmatrix} a_{11} & a_{12} & a_{13} \\ a_{21} & a_{22} & a_{23} \\ a_{31} & a_{32} & a_{33} \end{bmatrix}$$

and specifically will be

$$\begin{bmatrix} 2 & 2 & 2 \\ 3 & 3 & 3 \\ 4 & 4 & 4 \end{bmatrix}$$

**EXAMPLE 2** Show the $3 \times 4$ matrix whose typical element is given by

$$a_{ij} = 1 \quad \text{if} \quad i = j$$
$$a_{ij} = 0 \quad \text{if} \quad i \neq j$$

*Note:* This can be restated as

$$a_{ij} = \begin{cases} 1 & i = j \\ 0 & i \neq j \end{cases}$$

Here the row subscript, $i$, will take on values $i = 1, 2, 3$; and the column values are $j = 1, 2, 3, 4$. The places where $i = j$ are the (1, 1), (2, 2), and (3, 3) positions. For all the others we have $i \neq j$. Hence the desired matrix is

$$\begin{bmatrix} 1 & 0 & 0 & 0 \\ 0 & 1 & 0 & 0 \\ 0 & 0 & 1 & 0 \end{bmatrix}$$

## B.2 Special Matrices

Square Matrices A matrix is said to be *square* if it has the same number of rows as columns. Some square matrices are shown in Figure B.2. Matrices that are not square are usually referred to as *rectangular*.

The elements $a_{11}, a_{22}, \ldots, a_{nn}$ of a square matrix are said to constitute the *main diagonal*. These elements run from "northwest to southeast" in the

$$[6] \qquad \begin{bmatrix} 1 & 2 & 3 \\ 9 & 5 & 4 \\ 8 & 6 & 7 \end{bmatrix} \qquad \begin{bmatrix} a_{11} & a_{12} & \cdots & a_{1n} \\ a_{21} & a_{22} & \cdots & a_{2n} \\ \cdots & \cdots & \cdots & \cdots \\ a_{n1} & a_{n2} & \cdots & a_{nn} \end{bmatrix}$$

(a) $1 \times 1$ matrix (b) $3 \times 3$ matrix (c) $n \times n$ matrix

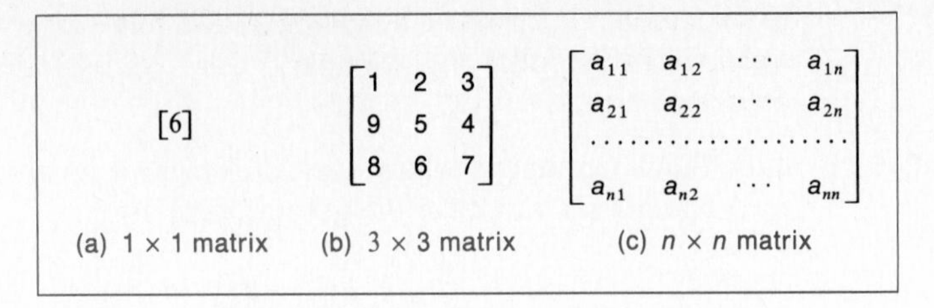

**FIGURE B.2** Some square matrices.

matrix. Thus the elements 3, −1, 7 are the main diagonal elements of the $3 \times 3$ matrix:

$$\begin{bmatrix} 3 & 0 & 9 \\ 5 & -1 & 8 \\ 6 & 2 & 7 \end{bmatrix}$$

Zero Matrices The *zero matrix* is one whose elements all are zero (see Figure B.3). These can be of any order.

Identity Matrices The *identity* (or unit) matrix is square and has 1's on the main diagonal and 0's everywhere else. There is an identity matrix of every order. The identity matrix of order $m$ is referred to as $I_m$. The typical element of the identity matrix is given as

$$a_{ij} = \begin{cases} 1 & \text{if } i = j \\ 0 & \text{if } i \neq j \end{cases}$$

Some typical identity matrices are given in Figure B.4.

**FIGURE B.3** Some zero matrices.

$$[0 \;\; 0 \;\; 0] \qquad \begin{bmatrix} 0 \\ 0 \\ 0 \\ 0 \end{bmatrix} \qquad \begin{bmatrix} 0 & 0 \\ 0 & 0 \end{bmatrix}$$

(a) $1 \times 3$ zero matrix (b) $4 \times 1$ zero matrix (c) $2 \times 2$ zero matrix

$$[1] \qquad \begin{bmatrix} 1 & 0 \\ 0 & 1 \end{bmatrix} \qquad \begin{bmatrix} 1 & 0 & 0 & 0 \\ 0 & 1 & 0 & 0 \\ 0 & 0 & 1 & 0 \\ 0 & 0 & 0 & 1 \end{bmatrix}$$

(a) $I_1$ (b) $I_2$ (c) $I_4$

**FIGURE B.4** Identity matrices.

**Transpose of a Matrix** The *transpose* of a matrix $A$ is a matrix whose rows are the columns of $A$, in order, and whose columns are the rows of $A$, in order. It is denoted by $A^T$. For example, if

$$A = \begin{bmatrix} 2 & 3 & 4 \\ -1 & 6 & 0 \end{bmatrix}$$

then

$$A^T = \begin{bmatrix} 2 & -1 \\ 3 & 6 \\ 4 & 0 \end{bmatrix}$$

If matrix $A$ is of order $n \times m$, then $A^T$ will be of order $m \times n$.

**Symmetric Matrices** A *symmetric* matrix is a *square* matrix in which the element $a_{ij}$ equals element $a_{ji}$. These matrices remain unchanged when all the columns are substituted for rows, and vice versa. A condition for symmetry is that the transpose of a matrix is the same as the matrix, or $A = A^T$. The following are symmetric matrices:

$$\begin{bmatrix} 1 & 2 & 3 \\ 2 & 0 & 6 \\ 3 & 6 & 5 \end{bmatrix} \quad \begin{bmatrix} 1 & 0 & 1 & 1 \\ 0 & 1 & 0 & 1 \\ 1 & 0 & 0 & 1 \\ 1 & 1 & 1 & 0 \end{bmatrix}$$

**Matrix Equality** Two matrices $A$ and $B$ with typical elements $a_{ij}$ and $b_{ij}$ are said to be *equal*, written as $A = B$, if they are of the same order and if corresponding elements are equal throughout. Thus $a_{ij} = b_{ij}$ for all $i$ and $j$.

For example, we have

$$\begin{bmatrix} 2 & 1 \\ 8 & 4 \end{bmatrix} = \begin{bmatrix} 1+1 & 6-5 \\ 2^3 & 2 \times 2 \end{bmatrix}$$

**EXAMPLE 1** The matrix equality:

$$\begin{bmatrix} x+w & 2y-w \\ x-z & y-3z \end{bmatrix} = \begin{bmatrix} 0 & 10 \\ 5 & 13 \end{bmatrix}$$

is equivalent to the system of equations

$$\begin{aligned} x + \quad w &= 0 \\ 2y - \quad w &= 10 \\ x - \quad z &= 5 \\ y - 3z &= 13 \end{aligned}$$

The solution to this system is $x = 2$, $y = 4$, $z = -3$, $w = -2$.

## EXERCISES

1. How many elements does an $m \times n$ matrix have?

2. What are the possible orders of a matrix with
   (a) 100 elements? (b) 7 elements?

3. Given an $n \times n$ matrix, how many elements are there that are *not* on the main diagonal?

4. Use the double-subscript notation to write the elements of the next-to-last row of an $m \times n$ matrix.

5. Construct the $4 \times 4$ matrix, $A$, whose $a_{ij}$ elements are given by
$$a_{ij} = i + j$$

6. Construct the $3 \times 3$ matrix, $A$, whose $a_{ij}$ elements are given by
$$a_{ij} = (i - j)^2$$

7. Construct the $4 \times 4$ matrix, $A$, whose $a_{ij}$ elements are given by
$$\text{(a) } a_{ij} = \begin{cases} 0 & \text{if } i \neq j \\ i & \text{if } i = j \end{cases} \qquad \text{(b) } a_{ij} = \begin{cases} i^2 j & \text{if } i > j \\ j - i & \text{if } i \leq j \end{cases}$$

8. Use inspection to derive a general formula for the elements of the following matrices.
$$\text{(a)} \begin{bmatrix} 1 & 1 & 1 & 1 \\ 2 & 4 & 8 & 16 \\ 3 & 9 & 27 & 81 \\ 4 & 16 & 64 & 256 \end{bmatrix} \qquad \text{(b)} \begin{bmatrix} 1 & 1 & 1 & 1 & 1 & \cdots \\ 1 & 2 & 3 & 4 & 5 & \cdots \\ 1 & 4 & 9 & 16 & 25 & \cdots \\ 1 & 8 & 27 & 64 & 125 & \cdots \end{bmatrix}$$

9. Construct $A^T$ for the following matrices.
$$\text{(a) } A = \begin{bmatrix} 1 & 2 \\ 3 & 4 \\ 6 & 5 \end{bmatrix} \qquad \text{(b) } A = [1 \quad 4 \quad 9 \quad 16 \quad 25]$$

10. Show that $A$ is symmetric if $a_{ij} = i^2 + j^2$.

11. Find the values of $x$, $y$, $z$, and $w$ if
$$\begin{bmatrix} x + z & w - 2y \\ w - z & y - 3x \end{bmatrix} = \begin{bmatrix} 4 & 0 \\ 1 & -1 \end{bmatrix}$$

## B.3 Matrix Arithmetic

Addition For two matrices, $A$ and $B$, the sum $A + B$ is possible only if $A$ and $B$ are of the *same order*. The sum is a new matrix of the same order as $A$ and $B$ and is obtained by adding corresponding elements from $A$ and $B$. Addition of two matrices of different order is not defined and they are said to be non-conformable for addition.

In general,

$$\begin{bmatrix} a_{11} & a_{12} & \cdots & a_{1n} \\ a_{21} & a_{22} & \cdots & a_{2n} \\ \cdots & \cdots & \cdots & \cdots \\ \cdots & \cdots & \cdots & \cdots \\ a_{m1} & a_{m2} & \cdots & a_{mn} \end{bmatrix} + \begin{bmatrix} b_{11} & b_{12} & \cdots & b_{1n} \\ b_{21} & b_{22} & \cdots & b_{2n} \\ \cdots & \cdots & \cdots & \cdots \\ \cdots & \cdots & \cdots & \cdots \\ b_{m1} & b_{m2} & \cdots & b_{mn} \end{bmatrix}$$

$$= \begin{bmatrix} a_{11} + b_{11} & \cdots & a_{1n} + b_{1n} \\ \cdots & \cdots & \cdots \\ \cdots & \cdots & \cdots \\ a_{m1} + b_{m1} & \cdots & a_{mn} + b_{mn} \end{bmatrix}$$

**EXAMPLE 1**

(a) $\begin{bmatrix} 3 & 2 & -1 \\ 4 & 0 & 7 \end{bmatrix} + \begin{bmatrix} -2 & 7 & 2 \\ -4 & 6 & -5 \end{bmatrix} = \begin{bmatrix} 1 & 9 & 1 \\ 0 & 6 & 2 \end{bmatrix}$

(b) $\begin{bmatrix} 1 \\ 2 \\ 4 \end{bmatrix} + \begin{bmatrix} 6 \\ -5 \\ 0 \end{bmatrix} = \begin{bmatrix} 7 \\ -3 \\ 4 \end{bmatrix}$

(c) $\begin{bmatrix} 1 & 2 \\ 3 & 4 \end{bmatrix} + \begin{bmatrix} 2 \\ -2 \\ 6 \end{bmatrix}$ is not defined

We can observe that matrix addition is commutative and associative. That is, $A + B = B + A$ and $A + (B + C) = (A + B) + C$.

Subtraction Subtraction of two matrices is defined similarly to addition. The difference, $A - B$, can be carried out only if $A$ and $B$ are of the same order and the result is obtained by subtracting corresponding elements. For example,

$$\begin{bmatrix} 2 & -3 \\ 4 & 0 \end{bmatrix} - \begin{bmatrix} 3 & -7 \\ 4 & 2 \end{bmatrix} = \begin{bmatrix} -1 & 4 \\ 0 & -2 \end{bmatrix}$$

If $A$ is an $m \times n$ matrix whose general element is $a_{ij}$, then its negative, $-A$, is the matrix whose general element is given by $-a_{ij}$. Thus subtraction is usually defined in terms of addition: $A - B = A + (-B)$.

Several observations can be made about matrix subtraction:

1. $A - A = A + (-A) =$ the zero matrix of the appropriate order
2. $-(-A) = A$
3. $-(A + B) = -A + (-B) = -A - B$

Scalar Multiplication The product of a scalar (an ordinary real number), $k$, and a matrix $A$ is written as $kA$, and the product is a new matrix obtained by multiplying each $a_{ij}$ element of $A$ by $k$. Thus

$$k \cdot \begin{bmatrix} a_{11} & a_{12} & \cdots & a_{1n} \\ a_{21} & a_{22} & \cdots & a_{2n} \\ \cdots & \cdots & \cdots & \cdots \\ a_{m1} & a_{m2} & \cdots & a_{mn} \end{bmatrix} = \begin{bmatrix} ka_{11} & ka_{12} & \cdots & ka_{1n} \\ ka_{21} & ka_{22} & \cdots & ka_{2n} \\ \cdots & \cdots & \cdots & \cdots \\ ka_{m1} & ka_{m2} & \cdots & ka_{mn} \end{bmatrix}$$

We have, for example,

$$3 \cdot \begin{bmatrix} 1 & 2 \\ 4 & 3 \end{bmatrix} = \begin{bmatrix} 3 & 6 \\ 12 & 9 \end{bmatrix}$$

$$(-1) \cdot \begin{bmatrix} 6 \\ 8 \\ -5 \end{bmatrix} = \begin{bmatrix} -6 \\ -8 \\ 5 \end{bmatrix}$$

From the latter example we see that $(-1)A = -A$.

Several observations can be made concerning scalar multiplication. (Assume conformability throughout for addition.)

1. $k(A + B) = kA + kB$
2. $(k_1 + k_2)A = k_1A + k_2A$
3. $k_1(k_2A) = (k_1k_2)A$

Matrix Multiplication By analogy with addition of matrices, one might think that the multiplication of two matrices can be carried out in the same way, that is, by forming the product of corresponding entries. As it happens, a more useful form of multiplication is used for matrices. Without going into the practicalities here, we present the procedure.

We say that matrices $A$ and $B$ are conformable for multiplication, in the order $AB$, only if the number of *columns* of $A$ is the same as the number of *rows* of $B$. Thus it is possible to form the product $AB$ if $A$ is of order $3 \times 4$ and if $B$ is of order $4 \times 5$. Notice that the product $BA$ is not possible.

If $A$ is an $m \times p$ matrix and $B$ is a $p \times n$ matrix, then the product, $AB = C$, is a new matrix of order $m \times n$. The general element $c_{ij}$ of the new matrix $C$ is obtained by forming the following sum of products:

$$\begin{aligned} c_{ij} &= \sum_{k=1}^{p} a_{ik}b_{kj} \\ &= a_{i1}b_{1j} + a_{i2}b_{2j} + a_{i3}b_{3j} + \cdots + a_{ip}b_{pj} \end{aligned}$$

To illustrate, let us form the product of

$$\overset{A}{\begin{bmatrix} 1 & 2 & 4 \\ 3 & 4 & -1 \end{bmatrix}_{2 \times 3}} \times \overset{B}{\begin{bmatrix} 2 & 1 \\ 4 & 0 \\ 6 & 7 \end{bmatrix}_{3 \times 2}} = C$$

Form $C_{11}$

$$\begin{array}{c} \qquad\qquad j=1 \qquad\qquad\qquad\qquad C_{11} \\ i=1\rightarrow \begin{bmatrix} 1 & 2 & 4 \\ 3 & 4 & -1 \end{bmatrix} \begin{bmatrix} 2 & 1 \\ 4 & 0 \\ 6 & 7 \end{bmatrix} = \begin{bmatrix} 1\cdot 2 + 2\cdot 4 + 4\cdot 6 & \end{bmatrix} = \begin{bmatrix} 34 & \end{bmatrix} \end{array}$$

Form $C_{12}$

$$\begin{array}{c} \qquad\qquad j=2 \qquad\qquad\qquad\qquad C_{12} \\ i=1\rightarrow \begin{bmatrix} 1 & 2 & 4 \\ 3 & 4 & -1 \end{bmatrix} \begin{bmatrix} 2 & 1 \\ 4 & 0 \\ 6 & 7 \end{bmatrix} = \begin{bmatrix} 34 & 1\cdot 1 + 2\cdot 0 + 4\cdot 7 \end{bmatrix} = \begin{bmatrix} 34 & 29 \end{bmatrix} \end{array}$$

Form $C_{21}$

$$\begin{array}{c} \qquad\qquad j=1 \qquad\qquad\qquad\qquad C_{21} \\ i=2\rightarrow \begin{bmatrix} 1 & 2 & 4 \\ 3 & 4 & -1 \end{bmatrix} \begin{bmatrix} 2 & 1 \\ 4 & 0 \\ 6 & 7 \end{bmatrix} = \begin{bmatrix} 34 & 29 \\ 3\cdot 2 + 4\cdot 4 + (-1)\cdot 6 & \end{bmatrix} = \begin{bmatrix} 34 & 29 \\ 16 & \end{bmatrix} \end{array}$$

Form $C_{22}$

$$\begin{array}{c} \qquad\qquad j=2 \qquad\qquad\qquad\qquad C_{22} \\ i=2\rightarrow \begin{bmatrix} 1 & 2 & 4 \\ 3 & 4 & -1 \end{bmatrix} \begin{bmatrix} 2 & 1 \\ 4 & 0 \\ 6 & 7 \end{bmatrix} = \begin{bmatrix} 34 & 29 \\ 16 & 3\cdot 1 + 4\cdot 0 + (-1)\cdot 7 \end{bmatrix} = \begin{bmatrix} 34 & 29 \\ 16 & -4 \end{bmatrix} \end{array}$$

so that

$$\begin{bmatrix} 1 & 2 & 4 \\ 3 & 4 & -1 \end{bmatrix}_{2\times 3} \times \begin{bmatrix} 2 & 1 \\ 4 & 0 \\ 6 & 7 \end{bmatrix}_{3\times 2} = \begin{bmatrix} 34 & 29 \\ 16 & -4 \end{bmatrix}_{2\times 2}$$

It can be seen that in general the product of two matrices is not commutative. For example, even though matrices $A$ and $B$ may be conformable in one direction, they may not be in the other.

$$\begin{bmatrix} 2 & 1 \\ 3 & 2 \end{bmatrix} \begin{bmatrix} 2 \\ 1 \end{bmatrix} = \begin{bmatrix} 2\cdot 2 + 1\cdot 1 \\ 3\cdot 2 + 2\cdot 1 \end{bmatrix} = \begin{bmatrix} 5 \\ 8 \end{bmatrix}$$

but

$$\begin{bmatrix} 2 \\ 1 \end{bmatrix} \begin{bmatrix} 2 & 1 \\ 3 & 2 \end{bmatrix}$$

is not possible.

Several observations can be made relating multiplication and addition of matrices.

1. $A \cdot B \neq B \cdot A$, in general
2. $A(BC) = (AB)C$
3. $A(B + C) = AB + AC$
4. $(A + B)C = AC + BC$

When a matrix $A$ is multiplied by itself, $AA$, we designate the product as $A^2$, $AAA = A^3$, and so on.

*Note:* Because of associativity, $A^3 = AAA = A^2A = AA^2$.

We also observe that the identity matrix, $I_n$, has the property that for all matrices $A$ that are conformable for multiplication, we have

$$A \times I = I \times A = A$$

Thus $I_n$ serves as the identity element for multiplication just as 1 serves as the identity for ordinary scalar multiplication.

**EXAMPLE 2**

(a) $$\begin{bmatrix} 3 & 2 & 1 \\ 1 & 0 & 7 \\ -1 & 2 & 4 \end{bmatrix}\begin{bmatrix} 1 & 0 & 0 \\ 0 & 1 & 0 \\ 0 & 0 & 1 \end{bmatrix} = \begin{bmatrix} 3 & 2 & 1 \\ 1 & 0 & 7 \\ -1 & 2 & 4 \end{bmatrix}$$

(b) $$\begin{bmatrix} 3 & 2 & -6 \end{bmatrix}\begin{bmatrix} 1 & 0 & 0 \\ 0 & 1 & 0 \\ 0 & 0 & 1 \end{bmatrix} = \begin{bmatrix} 3 & 2 & -6 \end{bmatrix}$$

*but* $\begin{bmatrix} 1 & 0 & 0 \\ 0 & 1 & 0 \\ 0 & 0 & 1 \end{bmatrix}\begin{bmatrix} 3 & 2 & -6 \end{bmatrix}$ is not defined.

## EXERCISES

1. In the following, let

$$A = \begin{bmatrix} 2 & 7 \\ 4 & 2 \\ 8 & 6 \end{bmatrix} \quad B = \begin{bmatrix} 3 \\ 2 \\ 4 \end{bmatrix} \quad C = \begin{bmatrix} 5 & 2 & -1 \end{bmatrix} \quad D = \begin{bmatrix} 1 & 1 & 0 \\ 1 & 0 & 0 \\ 1 & 1 & 1 \end{bmatrix}$$

$$E = \begin{bmatrix} 1 & -1 \\ -2 & 3 \end{bmatrix}$$

Perform the following operations, if possible.

(a) $A + B$ (b) $E + I_2$ (c) $BI_2$ (d) $AB$
(e) $BA$ (f) $BC$ (g) $CB$ (h) $I_3D$
(i) $DI_3$ (j) $D - I_3$ (k) $E^2$ (l) Show that $E^2E = EE^2$.
(m) $EA^{\mathrm{T}}$ (n) $E^{\mathrm{T}}E$ (o) $7A$ (p) $(-3)B$

2. For an arbitrary matrix, $A$, under what conditions can $AA^{\mathrm{T}}$ be performed? $A + A^{\mathrm{T}}$?

# Selected Bibliography


AHO, A., J. E. HOPCROFT, and J. D. ULLMAN [1975]. *The Design and Analysis of Computer Algorithms*, Addison-Wesley, Reading, Mass.

AHO, A., J. E. HOPCROFT, and J. D. ULLMAN [1983]. *Data Structures and Algorithms*, Addison-Wesley, Reading, Mass.

ARNOLD, B. H. [1962]. *Logic and Boolean Algebra*, Prentice-Hall, Englewood Cliffs, N.J.

AUGENSTEIN, M., and A. TENENBAUM [1979]. *Data Structures and PL/I Programming*, Prentice-Hall, Englewood Cliffs, N.J.

BAASE, S. [1978]. *Computer Algorithms: Introduction to Design and Analysis*, Addison-Wesley, Reading, Mass.

BARON, R. J., and L. G. SHAPIRO [1980]. *Data Structures and Their Implementation*, Van Nostrand Reinhold, New York.

BECKMAN, F. S. [1980]. *Mathematical Foundations of Programming*, Addison-Wesley, Reading, Mass.

BEHZAD, M., and G. CHARTRAND [1971]. *Introduction to the Theory of Graphs*, Allyn and Bacon, Boston.

BEIDLER, J. [1982]. *An Introduction to Data Structures*, Allyn and Bacon, Boston.

BERZTISS, A. T. [1975]. *Data Structures: Theory and Practice*, 2nd ed., Academic Press, New York.

BIRKHOFF, G. [1967]. *Lattice Theory*, 3rd ed., American Mathematical Society, Providence, R.I.

BIRKHOFF, G., and T. C. BARTEE [1970]. *Modern Applied Algebra*, McGraw-Hill, New York.

BOGART, K. P. [1983]. *Introductory Combinatorics*, Pitman, Marshfield, Mass.

BONDY, J. A., and U. S. R. MURTY [1976]. *Graph Theory with Applications*, North-Holland, New York.

BOOTH, T. L. [1978]. *Digital Networks and Computer Systems*, Wiley, New York.

BRUALDI, R. A. [1977]. *Introductory Combinatorics*, Elsevier North-Holland, New York.

BUSACKER, R. G., and T. L. SAATY [1965]. *Finite Graphs and Networks: An Introduction with Applications*, McGraw-Hill, New York.

CHARTRAND, G. [1977]. *Graphs as Mathematical Models*, Prindle, Weber & Schmidt, Boston.

CHRISTOFIDES, N. [1975]. *Graph Theory: An Algorithmic Approach*, Academic Press, New York.

DEO, N. [1974]. *Graph Theory with Applications to Engineering and Computer Science*, Prentice-Hall, Englewood Cliffs, N.J.

DIETMEYER, D. L. [1978]. *Logic Design of Digital Systems*, 2nd ed., Allyn and Bacon, Boston.

EVEN, S. [1973]. *Algorithmic Combinatorics*, Macmillan, New York.

EVEN, S. [1979]. *Graph Algorithms*, Computer Science Press, Potomac, Md.

FISHER, J. L. [1977]. *Application-Oriented Algebra*, Thomas Y. Crowell, New York.

GERSTING, J. L. [1982]. *Mathematical Structures for Computer Science*, W. H. Freeman, San Francisco.

GILL, A. [1976]. *Applied Algebra for the Computer Sciences*, Prentice-Hall, Englewood Cliffs, N.J.

GOODMAN, S. E., and S. T. HEDETNIEMI [1977]. *Introduction to the Design and Analysis of Algorithms*, McGraw-Hill, New York.

GRÄTZER, G. [1971]. *Lattice Theory*, W. H. Freeman, San Francisco.

HALL, M., JR. [1967]. *Combinatorial Theory*, Blaisdell, Waltham, Mass.

HALMOS, P. R. [1960]. *Naive Set Theory*, D. Van Nostrand, New York.

HALMOS, P. R. [1963]. *Lectures on Boolean Algebras*, D. Van Nostrand, Princeton, N.J.

HARARY, F. [1969]. *Graph Theory*, Addison-Wesley, Reading, Mass.

HARARY, F., R. Z. NORMAN, and D. CARTWRIGHT [1965]. *Structural Models: An Introduction to the Theory of Directed Graphs*, Wiley, New York.

HOFSTADTER, D. R. [1979]. *Gödel, Escher, Bach: An Eternal Golden Braid*, Basic Books, New York.

HOHN, F. E. [1966]. *Applied Boolean Algebra*, 2nd ed., Macmillan, New York.

HOROWITZ, E., and S. SAHNI [1978]. *Fundamentals of Computer Algorithms*, Computer Science Press, Potomac, Md.

HOROWITZ, E., and S. SAHNI [1982]. *Fundamentals of Data Structures*, Computer Science Press, Potomac, Md.

HOROWITZ, E., and S. SAHNI [1984]. *Fundamentals of Data Structures in Pascal*, Computer Science Press, Potomac, Md.

HU, T. C. [1982]. *Combinatorial Algorithms*, Addison-Wesley, Reading, Mass.

KAPPS, C. A., and S. BERGMAN [1975]. *Introduction to the Theory of Computing*, Charles E. Merrill, Columbus, Ohio.

KOHAVI, Z. [1978]. *Switching and Finite Automata Theory*, 2nd ed., McGraw-Hill, New York.

KORFHAGE, R. R. [1966]. *Logic and Algorithms*, Wiley, New York.

KORFHAGE, R. R. [1974]. *Discrete Computational Structures*, Academic Press, New York.

LEVY, L. S. [1980]. *Discrete Structures of Computer Science*, Wiley, New York.

LEWIN, M. H. [1983]. *Logic Design and Computer Organization*, Addison-Wesley, Reading, Mass.

LEWIS, T. G., and M. Z. SMITH [1982]. *Applying Data Structures*, 2nd ed., Houghton Mifflin, Boston.

LIN, Y.-F., and S. LIN [1981]. *Set Theory with Applications*, 2nd ed., Mariner, Tampa, Fla.

LIPSCHUTZ, S. [1976]. *Discrete Mathematics*, McGraw-Hill (Schaum), New York.

LIU, C. L. [1968]. *Introduction to Combinatorial Mathematics*, McGraw-Hill, New York.

LIU, C. L. [1977]. *Elements of Discrete Mathematics*, McGraw-Hill, New York.

MAKI, D. P., and M. THOMPSON [1973]. *Mathematical Models and Applications*, Prentice-Hall, Englewood Cliffs, N.J.

MALKEVITCH, J., and W. MEYER [1974]. *Graphs, Models, and Finite Mathematics*, Prentice-Hall, Englewood Cliffs, N.J.

MANO, M. M. [1979]. *Digital Logic and Computer Design*, Prentice-Hall, Englewood Cliffs, N.J.

MENDELSON, E. [1964]. *Introduction to Mathematical Logic*, Van Nostrand Reinhold, New York.

MENDELSON, E. [1970]. *Boolean Algebra and Switching Circuits*, McGraw-Hill (Schaum), New York.

MONK, J. D. [1969]. *Introduction to Set Theory*, McGraw-Hill, New York.

MOTT, J. L., A. KANDEL, and T. P. BAKER [1983]. *Discrete Mathematics for Computer Scientists*, Reston, Reston, Va.

NIJENHUIS, A., and H. S. WILF [1978]. *Combinatorial Algorithms for Computers and Calculators*, 2nd ed., Academic Press, New York.

ORE, O. [1963]. *Graphs and Their Uses*, Random House, New York.

PRATHER, R. E. [1976]. *Discrete Mathematical Structures for Computer Science*, Houghton Mifflin, Boston.

PREPARATA, F. P., and R. T. YEH [1973]. *Introduction to Discrete Structures*, Addison-Wesley, Reading, Mass.

REINGOLD, E. M., and W. J. HANSEN [1983]. *Data Structures*, Little, Brown, Boston.

REINGOLD, E. M., J. NIEVERGELT, and N. DEO [1977]. *Combinatorial Algorithms: Theory and Practice*, Prentice-Hall, Englewood Cliffs, N.J.

ROBERTS, F. S. [1976]. *Discrete Mathematical Models*, Prentice-Hall, Englewood Cliffs, N.J.

ROBERTS, F. S. [1984]. *Applied Combinatorics*, Prentice-Hall, Englewood Cliffs, N.J.

RYSER, H. J. [1963]. *Combinatorial Mathematics*, The Mathematical Association of America (A Carus Monograph), distributed by Wiley, New York.

SAHNI, S. [1981]. *Concepts in Discrete Mathematics*, Camelot, Fridley, Minn.

SHOENFIELD, J. R. [1967]. *Mathematical Logic*, Addison-Wesley, Reading, Mass.

STANAT, D. F., and D. F. MCALLISTER [1977]. *Discrete Mathematics in Computer Science*, Prentice-Hall, Englewood Cliffs, N.J.

STANDISH, T. A. [1980]. *Data Structure Techniques*, Addison-Wesley, Reading, Mass.

STOLL, R. R. [1963]. *Set Theory and Logic*, W. H. Freeman, San Francisco.

STOLL, R. R. [1974]. *Sets, Logic and Axiomatic Theories*, 2nd ed., W. H. Freeman, San Francisco.

STONE, H. S. [1973]. *Discrete Mathematical Structures and Their Applications*, Science Research Associates, Chicago.

TENENBAUM, A., and M. AUGENSTEIN [1981]. *Data Structures Using Pascal*, Prentice-Hall, Englewood Cliffs, N.J.

TOCCI, R. J. [1980]. *Digital Systems, Principles and Applications*, rev. ed., Prentice-Hall, Englewood Cliffs, N.J.

TREMBLAY, J-P., and R. MONAHAR [1975]. *Discrete Mathematical Structures with Applications to Computer Science*, McGraw-Hill, New York.

TREMBLAY, J-P., and P. G. SORENSON [1984]. *An Introduction to Data Structures with Applications*, 2nd ed., McGraw-Hill, New York.

TUCKER, A. [1984]. *Applied Combinatorics*, 2nd ed. Wiley, New York.

WAND, M. [1980]. *Induction, Recursion and Programming*, Elsevier North-Holland, New York.

WILDER, R. L. [1965]. *Introduction to the Foundations of Mathematics*, 2nd ed., Wiley, New York.

WIRTH, N. [1976]. *Algorithms + Data Structures = Programs*, Prentice-Hall, Englewood Cliffs, N.J.

# Solutions to Selected Exercises

## CHAPTER 1

### SECTION 1.2 Pages 3–4

Propositions are at numbers 6, 9, 10, 12, 13, 17.

### SECTION 1.3 Page 7

1. (b)

| $p$ | $q$ | $p \wedge \neg(p \wedge q)$ |
|---|---|---|
| T | T | T F F T |
| T | F | T T T F |
| F | T | F F T F |
| F | F | F F T F |

(c)

| $p$ | $q$ | $(\neg p \wedge q) \wedge \neg p$ |
|---|---|---|
| T | T | F F T F F |
| T | F | F F F F F |
| F | T | T T T T T |
| F | F | T F F F T |

(f)

| $p$ | $q$ | $\neg(\neg p \wedge \neg(\neg q))$ |
|---|---|---|
| T | T | T F F T F |
| T | F | T F F F T |
| F | T | F T T T F |
| F | F | T T F F T |

### Page 9

2. (b)

| $p$ | $q$ | $\neg p \vee (q \wedge p)$ |
|---|---|---|
| T | T | F T T |
| T | F | F F F |
| F | T | T T F |
| F | F | T T F |

(c)

| $p$ | $q$ | $p \wedge (p \vee q)$ |
|---|---|---|
| T | T | T T T |
| T | F | T T T |
| F | T | F F T |
| F | F | F F F |

(g)

| $p$ | $q$ | $r$ | $p \wedge (q \wedge r)$ |
|---|---|---|---|
| T | T | T | T T T |
| T | T | F | T F F |
| T | F | T | T F F |
| T | F | F | T F F |
| F | T | T | F F T |
| F | T | F | F F F |
| F | F | T | F F F |
| F | F | F | F F F |

(i)

| $p$ | $q$ | $p \vee (p \oplus q)$ |
|---|---|---|
| T | T | T T F |
| T | F | T T T |
| F | T | F T T |
| F | F | F F F |

3. (a) $2^4 = 16$ (c) $2^n$ 4. (a) (i) $r \vee \neg q$ (iii) $\neg(p \vee q) \vee r$

(b) (ii) It is not the case that the Braves won the pennant and Abraham Lincoln is alive, or that snow fell in Anchorage and the Braves won the pennant.

**Page 16**

5. (a)

| $p$ | $q$ | $\neg p \to (p \vee q)$ |
|---|---|---|
| T | T | F T T |
| T | F | F T T |
| F | T | T T T |
| F | F | T F F |

(b)

| $p$ | $q$ | $q \to \neg(p \wedge q)$ |
|---|---|---|
| T | T | T F F T |
| T | F | F T T F |
| F | T | T T T F |
| F | F | F T T F |

(d)

| $p$ | $q$ | $r$ | $p \to (q \oplus \neg r)$ |
|---|---|---|---|
| T | T | T | T T T T F |
| T | T | F | T F T F T |
| T | F | T | T F F F F |
| T | F | F | T T F T T |
| F | T | T | F T T T F |
| F | T | F | F T T F T |
| F | F | T | F T F F F |
| F | F | F | F T F T T |

| $p$ | $q$ | $r$ | (h) $p \to (q \to r)$ | (i) $(p \to q) \to r$ |
|---|---|---|---|---|
| T | T | T | T T T | T T T |
| T | T | F | T F F | T F F |
| T | F | T | T T T | F T T |
| T | F | F | T T T | F T F |
| F | T | T | F T T | T T T |
| F | T | F | F T F | T F F |
| F | F | T | F T T | T T T |
| F | F | F | F T T | T F F |

Note the difference between these two results. We state that the conditional operator ($\to$) is not associative. See Section 1.5 following.

6. Letting the propositions be represented by key words in the sentences, we have:
(a) Tuesday $\to$ Belgium (b) John $\vee$ Fred $\vee$ Mary (c) Grass $\to$ Sunshine (d) $\neg$Aid $\to$ Starve
(f) $(2 + 3 = 5) \to \neg$Sunshine (g) Reelect $\to$ District 11

7. (a) (i) This sentence is ambiguously stated. It can be interpreted as:

"Either doctors are rich, or lawyers are smart when teachers are not foolish", in which case we have: $p \vee (\neg q \to r)$.

"Either doctors are rich or lawyers are smart, when teachers are not foolish" has the symbolic form: $\neg q \to (p \vee r)$.

(b) (i) If doctors are rich or lawyers are smart then teachers are foolish.

8. (a) $\neg p \wedge \neg q$ (b) $(p \wedge \neg q \wedge r) \vee (p \wedge \neg q \wedge \neg r) \vee (\neg p \wedge q \wedge \neg r)$

**Page 18**

9. (a)

| $p$ | $q$ | $(p \to q) \leftrightarrow (q \to \neg p)$ |
|---|---|---|
| T | T | T F T F F |
| T | F | F F F T F |
| F | T | T T T T T |
| F | F | T T F T T |

(c)

| $p$ | $q$ | $r$ | $(p \vee r) \wedge (\neg q \leftrightarrow p)$ |
|---|---|---|---|
| T | T | T | T F F F T |
| T | T | F | T F F F T |
| T | F | T | T T T T T |
| T | F | F | T T T T T |
| F | T | T | T T F T F |
| F | T | F | F F F T F |
| F | F | T | T F T F F |
| F | F | F | F F T F F |

## SECTION 1.4 Page 19

2.

| $p$ | $q$ | $((p \to q) \to p) \to p$ |
|---|---|---|
| T | T | T T T T T |
| T | F | F T T T T |
| F | T | T F F T F |
| F | F | T F F T F |

a tautology

4.

| $p$ | $q$ | $(p \wedge (p \to q)) \wedge (p \to \neg q)$ |
|---|---|---|
| T | T | TT T F T F F |
| T | F | TF F F T T T |
| F | T | FF T F F T F |
| F | F | FF T F F T T |

a contradiction

6. (b) and (e) are tautologies; (a) is a contradiction; (c) and (d) are neither.

## SECTION 1.5 Page 24

1. (c)

| $p$ | $q$ | $p \leftrightarrow (p \vee (p \wedge q))$ |
|---|---|---|
| T | T | T T TT T |
| T | F | T T TT F |
| F | T | F T FF F |
| F | F | F T FF F |

(d) (rule 9)

| $p$ | $q$ | $r$ | $[(p \wedge (q \vee r)] \leftrightarrow [(p \wedge q) \vee (p \wedge r)]$ |
|---|---|---|---|
| T | T | T | TT T T T T T |
| T | T | F | TT T T T T F |
| T | F | T | TT T T F T T |
| T | F | F | TF F T F F F |
| F | T | T | FF T T F F F |
| F | T | F | FF T T F F F |
| F | F | T | FF T T F F F |
| F | F | F | FF F T F F F |

(e) (rule 11)

| $p$ | $q$ | $\neg(p \wedge q) \leftrightarrow (\neg p \vee \neg q)$ |
|---|---|---|
| T | T | F T T F FF |
| T | F | T F T F TT |
| F | T | T F T T TF |
| F | F | T F T T TT |

(g)

| $p$ | $q$ | $(p \wedge q) \leftrightarrow \neg(\neg p \vee \neg q)$ |
|---|---|---|
| T | T | T T T F FF |
| T | F | F T F F TT |
| F | T | F T F T TF |
| F | F | F T F T TT |

(i)

| $p$ | $q$ | $(\neg p \wedge (p \vee q)) \leftrightarrow (\neg p \wedge q)$ |
|---|---|---|
| T | T | F F T T F |
| T | F | F F T T F |
| F | T | T T T T T |
| F | F | T F F T F |

**Page 26**

2. Start with more complex side.

(a) $(p \wedge \neg q) \vee q \Leftrightarrow q \vee (p \wedge \neg q)$ commutativity, rule (2)

$\Leftrightarrow (q \vee p) \wedge (q \vee \neg q)$ distribution, rule (10)

$\Leftrightarrow (q \vee p) \wedge \mathrm{T}$ complement, rule (14f)

$\Leftrightarrow (q \vee p)$ identity, rule (14a)

$\Leftrightarrow p \vee q$ commutativity, rule (2)

(c) $p \vee (p \wedge q) \Leftrightarrow (p \wedge \mathrm{T}) \vee (p \wedge q)$ identity, rule (14a)

$\Leftrightarrow p \wedge (\mathrm{T} \vee q)$ distribution, rule (9)

$\Leftrightarrow p \wedge (q \vee \mathrm{T})$ commutativity, rule (2)

$\Leftrightarrow p \wedge \mathrm{T}$ boundness, rule (14d)

$\Leftrightarrow p$ identity, rule (14a)

(e) $\neg p \vee (q \rightarrow \neg r) \Leftrightarrow \neg p \vee (\neg q \vee \neg r)$ Figure 1.11

$\Leftrightarrow \neg p \vee \neg q \vee \neg r$ associativity, rule (4)

3. (b) and (d) are equivalent; (a) and (c) are not. 6. $p \rightarrow q \Leftrightarrow \neg p \vee q \Leftrightarrow \neg(\neg\neg p \wedge \neg q) \Leftrightarrow \neg(p \wedge \neg q)$

7. $p \leftrightarrow q \Leftrightarrow (p \wedge q) \vee (\neg p \wedge \neg q)$

$\Leftrightarrow \neg(\neg p \vee \neg q) \vee \neg(\neg\neg p \vee \neg\neg q)$

$\Leftrightarrow \neg(\neg p \vee \neg q) \vee \neg(p \vee q)$

9. $p \oplus q \Leftrightarrow \neg(p \leftrightarrow q)$

(a) $\neg(\neg(p \wedge \neg q) \wedge \neg(\neg p \wedge q))$

(b) $\neg(\neg p \vee q) \vee \neg(p \vee \neg q)$

**SECTION 1.6 Pages 27–28**

1. (a) $(p \wedge q) \wedge \neg(\mathrm{F} \vee q)$ (c) $(p \vee q) \wedge (\neg p \vee q) \wedge (p \vee \neg q) \wedge (\neg p \vee \neg q)$ 3. Duals are (c) only.

**SECTION 1.7 Pages 30–31**

| | conditional | converse | inverse | contrapositive |
|---|---|---|---|---|
| 1. (a) | $H \rightarrow M$ | $M \rightarrow H$ | $\neg H \rightarrow \neg M$ | $\neg M \rightarrow \neg H$ |
| (b) | $T \rightarrow G$ | $G \rightarrow T$ | $\neg T \rightarrow \neg G$ | $\neg G \rightarrow \neg T$ |
| (d) | $(W \vee D) \rightarrow P$ | $P \rightarrow (W \vee D)$ | $\neg(W \vee D) \rightarrow \neg P$ | $\neg P \rightarrow \neg(W \vee D)$ |
| 2. (b) | $(p \vee \neg q) \rightarrow \neg r$ | $\neg r \rightarrow (p \vee \neg q)$ | $\neg(p \vee \neg q) \rightarrow r$ | $r \rightarrow \neg(p \vee \neg q)$ |

3. Assume both factors are even, and thus can be written as $2k$ and $2m$. Now form the product.

**SECTION 1.8 Pages 33–34**

3. Premises are: $H \wedge W \rightarrow S$

$H \wedge \neg W$

Conclusion: $\neg S$.

| $H$ | $W$ | $S$ | $\{[(H \wedge W)$ | $\rightarrow S]$ | $\wedge$ | $[H \wedge \neg W]\}$ | $\rightarrow$ | $\neg S$ |
|---|---|---|---|---|---|---|---|---|
| T | T | T | T | T | F | F | T | F |
| T | T | F | T | F | F | F | T | T |
| T | F | T | F | T | T | T | (F) | F |
| T | F | F | F | T | T | T | T | T |
| F | T | T | F | T | F | F | T | F |
| F | T | F | F | T | F | F | T | T |
| F | F | T | F | T | F | F | T | F |
| F | F | F | F | T | F | F | T | T |

not valid

5. Premises are: $W \to M$

$\neg W \to T$

Conclusion: $M \vee T$

| $W$ | $M$ | $T$ | $[(W \to M)$ | $\wedge$ | $(\neg W \to T)]$ | $\to$ | $(M \vee T)$ |
|---|---|---|---|---|---|---|---|
| T | T | T | T | T | T | T | T |
| T | T | F | T | T | T | T | T |
| T | F | T | F | F | T | T | T |
| T | F | F | F | F | T | T | T |
| F | T | T | T | T | T | T | T |
| F | T | F | T | F | F | T | F |
| F | F | T | T | F | T | T | T |
| F | F | F | T | F | F | T | F |

valid

**SECTION 1.9 Page 35**

1. $\neg p \to q \Leftrightarrow \neg\neg p \vee q \Leftrightarrow p \vee q$; column 2

2.

| $p$ | $q$ | $\neg p$ | $\leftrightarrow$ | $\neg q$ |
|---|---|---|---|---|
| T | T | F | T | F |
| T | F | F | F | T |
| F | T | T | F | F |
| F | F | T | T | T |

is column 7. Equivalent to $p \leftrightarrow q$

4. (a) $2^8$ (b) $2^{16}$ (c) $2^{2^n}$

**Page 36**

5. $\neg p \Leftrightarrow \neg(p \wedge p)$ by idempotence

7. (a) $p \wedge \neg p \Leftrightarrow p \wedge \neg(p \wedge p)$

$\Leftrightarrow \neg(\neg(p \wedge \neg(p \wedge p)))$

$\Leftrightarrow \neg(\neg(p \wedge \neg(p \wedge p)) \wedge \neg(p \wedge \neg(p \wedge p)))$

which is $[p|(p|p)]|[p|(p|p)]$

(b) $\neg(\neg(p \wedge p) \wedge p)$ or $(p|p)|p$

**Page 37**

9. $p \to q \Leftrightarrow \neg p \vee q$

$\Leftrightarrow \neg(p \vee p) \vee q$

$\Leftrightarrow \neg(\neg(\neg(p \vee p) \vee q))$

$\Leftrightarrow \neg(\neg(\neg(p \vee p) \vee q) \vee \neg(\neg(p \vee p) \vee q))$

$\Leftrightarrow [(p \downarrow p) \downarrow q] \downarrow [(p \downarrow p) \downarrow q]$

10. (a) $\neg(\neg(p \vee p) \vee p)$ or $(p \downarrow p) \downarrow p$

## CHAPTER 2

**SECTION 2.2 Pages 44–45**

1. Those that can be classified as sets are (b), (e), (f), and (g).
2. $U = \{$Washington, Adams, . . . , Carter, Reagan$\}$

Post-WW II Presidents = {Truman, Eisenhower, Kennedy, Nixon, Ford, Carter, Reagan}.
Post-WW II Presidents whose name begins with N, R, T = {Nixon, Reagan, Truman}
(a) {Washington, Adams, . . . , F. D. Roosevelt, Eisenhower, Kennedy, Ford, Carter} or $U$ without {Nixon, Reagan, Truman}
(b) $U$ (c) $\varnothing$ (d) not defined **3.** (a) {10, 11, 12, 13, 14, 15, 16, 17, 18, 19, 20}
(b) {1, 3, 5, 7, 9, 11} (c) $\varnothing$ (d) $U$ (e) {3, 6, 9, 12, . . .} (f) {1, 4, 9, 16, 25, 36, 49, 64, 81, 100, 121, 144}
(g) {100, 400, 900, 1600, . . .} (h) {2, 4, 6, 8, 10, 12, 13, 14, 15, 16, 17, . . .}
**4.** (b) $\{x \mid x \text{ is even } \vee x > 12\} = \{2, 4, 6, 8, 10, 12, 13, 14, 15, \ldots\}$

## SECTION 2.3 Pages 50–51

**1.** If $B \subseteq A$ and $C \subseteq B$ then $C \subseteq A$ can be rewritten as $[(B \to A) \wedge (C \to B)] \to (C \to A)$

| $A$ | $B$ | $C$ | $[(B \to A)$ | $\wedge$ | $(C \to B)]$ | $\to$ | $(C \to A)$ |
|---|---|---|---|---|---|---|---|
| T | T | T | T | T | T | T | T |
| T | T | F | T | T | T | T | T |
| T | F | T | T | F | F | T | T |
| T | F | F | T | T | T | T | T |
| F | T | T | F | F | T | T | F |
| F | T | F | F | F | T | T | T |
| F | F | T | T | F | F | T | T |
| F | F | F | T | T | T | T | T |

a tautology

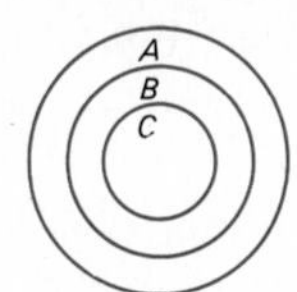


**2.** is not a tautology **3.** (a) $A = \varnothing$ (b) $A = \{\varnothing, a, b\}$ (e) $A = \{3\}$ $B = \{2, 3, \{3\}, 4\}$
**5.** (a) False (c) True (e) True **6.** True ones are (b), (e), (f), (g), (h), (j), (k), and (p).

## SECTION 2.4 Page 55

**1.** (a) $\{a, d\}$ (c) $\{a, e, f\}$ (f) $\{a, f\}$ (h) $\{a, b, c, g, h, j\}$
**2.** Truth tables should be applied to: $[(A \wedge B) \leftrightarrow (C \wedge B)] \to (A \leftrightarrow C)$ and $[(A \vee B) \leftrightarrow (C \vee B) \to (A \leftrightarrow C)$.
Neither are tautologies.
**4.** (a) If $A \subseteq B$ then $A \cup B = B$ and $A \cap B = A$. (c) (i) 4 and 6 (iv) 4 and 6 (v) $U$

## SECTION 2.5 Page 58

**1.** (a)

| $A$ | $B$ | $(A \wedge B)$ | $\to$ | $B$ |
|---|---|---|---|---|
| T | T | T | T | T |
| T | F | F | T | F |
| F | T | F | T | T |
| F | F | F | T | F |

(d)

| $A$ | $B$ | $C$ | $[(A \to B)$ | $\wedge$ | $(A \to C)]$ | $\to$ | $[A$ | $\to$ | $(B \vee C)]$ |
|---|---|---|---|---|---|---|---|---|---|
| T | T | T | T | T | T | T | T | T | T |
| T | T | F | T | F | F | T | T | T | T |
| T | F | T | F | F | T | T | T | T | T |
| T | F | F | F | F | F | T | T | F | F |
| F | T | T | T | T | T | T | F | T | T |
| F | T | F | T | T | T | T | F | T | T |
| F | F | T | T | T | T | T | F | T | T |
| F | F | F | T | T | T | T | F | T | F |

(g)

| $A$ | $B$ | $[(A \vee B) \leftrightarrow F] \leftrightarrow [(A \leftrightarrow F) \wedge (B \leftrightarrow F)]$ |
|---|---|---|
| T | T | T F F T F F F |
| T | F | T F F T F F T |
| F | T | T F F T T F F |
| F | F | F T F T T T T |

2. False statements are (a), (c), and (g).

## SECTION 2.6 Page 65

1. (a) $A \cap (\bar{A} \cup B) = (A \cap \bar{A}) \cup (A \cap B)$ distribution
$= \varnothing \cup (A \cap B)$ complement
$= A \cap B$ identity

(c) $\overline{\bar{A} \cap \bar{B}} = \bar{\bar{A}} \cup \bar{\bar{B}}$ De Morgan's law
$= A \cup B$ involution

(e) $A \cup (\overline{(\bar{B} \cup A) \cap B}) = A \cup (\overline{\bar{B} \cup A} \cup \bar{B})$ De Morgan's law
$= A \cup ((\bar{\bar{B}} \cap \bar{A}) \cup \bar{B})$ De Morgan's law
$= A \cup (\bar{B} \cup (B \cap \bar{A}))$ commutative, involution
$= A \cup ((\bar{B} \cup B) \cap (\bar{B} \cup \bar{A}))$ distribution
$= A \cup (U \cap (\bar{B} \cup \bar{A})$ complement
$= A \cup (\bar{B} \cup \bar{A})$ identity
$= (A \cup \bar{A}) \cup \bar{B}$ commutative, associative
$= U \cup \bar{B}$ complement
$= U$ boundness

(g) $A \oplus A = (A \cap \bar{A}) \cup (\bar{A} \cap A)$ definition of $X$-OR
$= (A \cap \bar{A}) \cup (A \cap \bar{A})$ commutative
$= A \cap \bar{A}$ idempotence
$= \varnothing$ complement

2. (b) (iii) $A - (B \cup C) = A \cap (\overline{B \cup C})$ relative complement def.
$= A \cap (\bar{B} \cap \bar{C})$ De Morgan's law
$= (A \cap \bar{B}) \cap \bar{C}$ associative
$= (A - B) \cap \bar{C}$ relative complement def.
$= (A - B) - C$ relative complement def.

## SECTION 2.7 Page 69

2. (a) 20 (b) 15 (c) 14 (d) 12 3. 6
6. Brands A, B, and C alone, respectively: 18, 0, and 3. Brands A or C but not brand B: 28. None: 42

## SECTION 2.8 Page 71

1. (d) $\{a\}, \{b\}, \{\{a\}\}, \{a, b\}, \{a, \{a\}\}, \{b, \{\{a\}\}, \{a, b, \{a\}\}, \varnothing$. 2. yes 3. no. 4. yes
5. (a) is false, others are true. 6. If $A \cap B = \varnothing$ then $P(A) \cap P(B)$ has one element: $\varnothing$.

## SECTION 2.9 Pages 78–79

1. (b) antisymmetric only 2. reflexive, antisymmetric, transitive
3. (b) transitive only (d) symmetric only
4. (b) see 3(d) (e)

| $R$ | $a$ | $b$ | $c$ |
|---|---|---|---|
| $a$ | | × | × |
| $b$ | | | |
| $c$ | | | |

is antisymmetric and irreflexive

5. irreflexive, symmetric, antisymmetric, transitive, and asymmetric

### SECTION 2.10 Pages 81–82

**1.** (a) not antisymmetric (d) is a partial order

**2.** The set containment relation, $\subseteq$, is reflexive, antisymmetric, and transitive.

**6.** Given any two distinct integers $a$ and $b$, it is true that either $a < b$ or $b < a$. There will always be a relation between the two. In a partial order it is possible for two elements not to be related by the relation.

**9.** (e) Perpendicularity is not transitive. (f) $\geq$ is not symmetric.
(g) is neither reflexive nor symmetric. (h) Friendship is not necessarily transitive.

## CHAPTER 3

### SECTION 3.1 Page 86

**1.** (a) $(x\bar{y}) + (\overline{y\bar{z}\bar{x}})$ (c) $xyz + \bar{x}\bar{y}\bar{z} + x\bar{y}\bar{z}$

**2.** (b) $(y + 1)(x + 0)(\overline{xy1})$ (d) $(x \cdot 1) + (\bar{x} \cdot 0) + (y \cdot \bar{1})$

### SECTION 3.2 Pages 92–93

**1.** (b)

$$\begin{aligned} x(\bar{x} + y) &= x\bar{x} + xy && \text{distribution} \\ &= 0 + xy && \text{complement} \\ &= xy + 0 && \text{commutative} \\ &= xy && \text{identity} \end{aligned}$$

(f)

$$\begin{aligned} (x + y)(\overline{xy}) &= \overline{xy}(x + y) && \text{commutative} \\ &= \overline{xy}x + \overline{xy}y && \text{distribution} \\ &= x\overline{xy} + y\overline{xy} && \text{commutative} \\ &= x(\bar{x} + \bar{y}) + y(\bar{x} + \bar{y}) && \text{De Morgan's law} \\ &= x\bar{x} + x\bar{y} + y\bar{x} + y\bar{y} && \text{distribution} \\ &= 0 + x\bar{y} + y\bar{x} + 0 && \text{complement} \\ &= x\bar{y} + y\bar{x} && \text{identity} \\ &= x\bar{y} + \bar{x}y && \text{commutative} \end{aligned}$$

(f) (alternative solution)

$$\begin{aligned} x\bar{y} + \bar{x}y &= (x\bar{y} + \bar{x})(x\bar{y} + y) && \text{distribution} \\ &= (\bar{x} + x\bar{y})(y + \bar{y}x) && \text{commutative} \\ &= (\bar{x} + \bar{y})(y + x) && \text{from 1(c)} \\ &= \overline{xy}(y + x) && \text{De Morgan's law} \\ &= (x + y)(\overline{xy}) && \text{commutative} \end{aligned}$$

(g)

$$\begin{aligned} (\overline{\bar{x}y}) + x\bar{y} &= (\bar{\bar{x}} + \bar{y}) + x\bar{y} && \text{De Morgan's law} \\ &= (x + \bar{y}) + x\bar{y} && \text{involution} \\ &= x + (\bar{y} + x\bar{y}) && \text{associativity} \\ &= x + (\bar{y} + \bar{y}x) && \text{commutative} \\ &= x + \bar{y} && \text{absorption} \end{aligned}$$

**2.** $x \cdot 0 = x \cdot 0 + 0 = x \cdot 0 + x\bar{x} = x(0 + \bar{x}) = x\bar{x} = 0$

**3.** A counterexample is: Let set $A = \{2, 3\}$, set $B = \{3, 4\}$ and set $C = \{3, 5\}$. Then $A \cap B = A \cap C = \{3\}$ but $B \neq C$. Also, the truth table for $[(x \wedge y) \leftrightarrow (x \wedge z)] \rightarrow (y \leftrightarrow z)$ is not a tautology.

**7.** Prove $\overline{xy} = \bar{x} + \bar{y}$. *Proof:* Show that the complement of $xy$ is $\bar{x} + \bar{y}$.
(a) The meet should be 0: $(xy)(\bar{x} + \bar{y}) = 0$. (b) The join should be 1: $(xy) + (\bar{x} + \bar{y}) = 1$.

(a)

$$\begin{aligned} xy(\bar{x} + \bar{y}) &= xy\bar{x} + xy\bar{y} \\ &= x\bar{x}y + xy\bar{y} \\ &= 0 \cdot y + x \cdot 0 \\ &= 0 && \text{Thus the meet is 0.} \end{aligned}$$

(b)

$$\begin{aligned} xy + (\bar{x} + \bar{y}) &= (\bar{x} + \bar{y}) + xy \\ &= (\bar{x} + \bar{y} + x) \cdot (\bar{x} + \bar{y} + y) && \text{distribution} \\ &= (x + \bar{x} + \bar{y}) \cdot (\bar{x} + y + \bar{y}) \\ &= (1 + \bar{y}) \cdot (\bar{x} + 1) \\ &= 1 \cdot 1 \\ &= 1 && \text{Thus the join is 1.} \end{aligned}$$

**8.** (c) $x \oplus \bar{x} = x\bar{\bar{x}} + \bar{x}\bar{x} = xx + \bar{x}\bar{x} = x + \bar{x} = 1$

(f)

$$\begin{aligned} xy \oplus xz &= xy(\overline{xz}) + xz(\overline{xy}) = xy(\bar{x} + \bar{z}) + xz(\bar{x} + \bar{y}) \\ &= xy\bar{x} + xy\bar{z} + xz\bar{x} + xz\bar{y} = 0 + xy\bar{z} + 0 + xz\bar{y} \\ &= x(y\bar{z} + \bar{y}z) = x(y \oplus z). \end{aligned}$$

(g)

$$\begin{aligned} x + (x \oplus y) &= x + (x\bar{y} + \bar{x}y) = (x + x\bar{y}) + \bar{x}y = x + \bar{x}y \\ &= (x + \bar{x})(x + y) = 1 \cdot (x + y) = x + y \end{aligned}$$

## SECTION 3.3 Page 99

1. $S = \{1, 2, 5, 7, 10, 14, 35, 70\}$

Complements: $\dfrac{70}{x}$

meet: gcd of two elements
join: lcm of two elements

$$\bar{1} = \frac{70}{1} = 70 \qquad \overline{10} = \frac{70}{10} = 7$$
$$\bar{2} = \frac{70}{2} = 35 \qquad \overline{14} = \frac{70}{14} = 5$$
$$\bar{5} = \frac{70}{5} = 14 \qquad \overline{35} = \frac{70}{35} = 2$$
$$\bar{7} = \frac{70}{7} = 10 \qquad \overline{70} = \frac{70}{70} = 1$$

*Axiom 1* (commutativity) is shown by observing that gcd and lcm are commutative.

For example,
$$\gcd(2, 35) = \gcd(35, 2) = 1$$
$$\text{lcm}(5, 10) = \text{lcm}(10, 5) = 10$$

*Axiom 2* (distribution)

For example:
$$2 \text{ meet } (5 \text{ join } 7) = \gcd(2, \text{lcm}(5, 7)) = \gcd(2, 35) = 1$$
and
$$(2 \text{ meet } 5) \text{ join } (2 \text{ meet } 7) = \text{lcm}(\gcd(2, 5), \gcd(2, 7)) = \text{lcm}(1, 1) = 1$$

*Axiom 3* (identity)

There exist unit and zero elements, namely 70 and 1, respectively such that: $\gcd(x, 70) = x$ and $\text{lcm}(x, 1) = x$ for all $x \in S$.

*Axiom 4* (complement laws)

Show that $\gcd(x, \bar{x}) = 0$ and $\text{lcm}(x, \bar{x}) = 1$ for all $x \in S$.

$$\gcd(1, 70) = 1 \qquad \text{lcm}(1, 70) = 70$$
$$\gcd(2, 35) = 1 \qquad \text{lcm}(2, 35) = 70$$
$$\gcd(5, 14) = 1 \qquad \text{lcm}(5, 14) = 70$$
$$\gcd(7, 10) = 1 \qquad \text{lcm}(7, 10) = 70$$

All the axioms of a Boolean algebra hold. Thus the integral divisors of 70 form a Boolean algebra.

2.

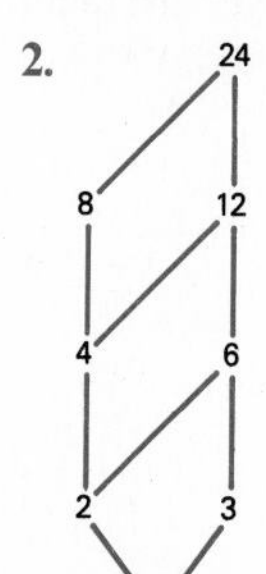


$$\bar{1} = 24 \qquad \bar{6} = 4$$
$$\bar{2} = 12 \qquad \bar{8} = 3$$
$$\bar{3} = 8 \qquad \overline{12} = 2$$
$$\bar{4} = 6 \qquad \overline{24} = 1$$

Axiom 4 does not hold:

$$\gcd(6, \bar{6}) = \gcd(6, 4) = 2 \neq 1$$
and
$$\text{lcm}(6, \bar{6}) = \text{lcm}(6, 4) = 12 \neq 24.$$

4.

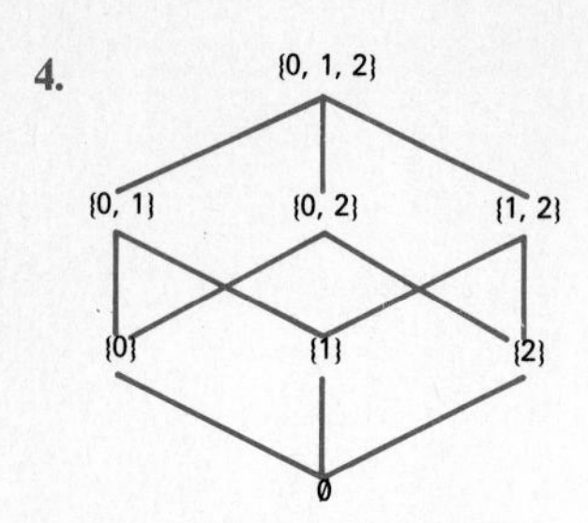


## SECTION 3.4 Page 102

1. Assume $x + y = y$, Show $x = xy$.

$$\begin{aligned} xy &= x(x + y) && \text{by substitution} \\ &= x && \text{by absorption} \\ \text{so} \quad x &= xy. \end{aligned}$$

## SECTION 3.5 Page 103

1. $x \le y$ means $x = xy$.

$$\begin{aligned} \text{so} \quad xz &= (xy)z && \text{by substitution} \\ &= x(yz) && \text{by associativity} \\ &= x(zy) && \text{by commutativity} \\ &= (xz)y && \text{by associativity} \\ \text{and} \quad xz &= (xz)y && \text{means that} \quad xz \le y. \end{aligned}$$

2. $x \le y$ means $x = xy$, so $\bar{x} = \bar{x} + \bar{y}$ by De Morgan's law, written as $\bar{y} + \bar{x} = \bar{x}$ means that $\bar{y} \le \bar{x}$.
5. $x \le y$ means $x = xy$, so $\bar{x} + y = \overline{xy} + y = (\bar{x} + \bar{y}) + y = \bar{x} + (\bar{y} + y) = \bar{x} + 1 = 1$.
6. $x = x \cdot 1 = x(y + \bar{y}) = xy + x\bar{y} = xy + 0 = xy$. Thus $x \le y$.
9. Note that $xy + x\bar{y} + xy + \bar{x}y = xy + (x \oplus y)$.

## SECTION 3.6 Page 104

1. Let $B = \{0, 1, a, b, c\}$. We know that $\bar{0} = 1$ and $\bar{1} = 0$ and that they are unique. $\bar{a}$ cannot equal $a$, for then we should have $a \cdot \bar{a} = a \cdot a = 0$, but $a \cdot a = a$ by idempotence and $a \ne 0$. Suppose $\bar{a} = b$, then $ab = 0$ and $a + b = 1$. This means additionally that $\bar{b} = a$ and that these complements are unique. What about $\bar{c}$? Because of uniqueness of complements, $\bar{c} \ne 0, \bar{c} \ne 1, \bar{c} \ne a$, and $\bar{c} \ne b$. Thus we must have $\bar{c} = c$. But then $c \cdot \bar{c} = c \cdot c = 0$, which it can't because $c \cdot c = c$. So a five-element Boolean algebra cannot exist. This argument can be extended to show that a Boolean algebra cannot have an odd number of elements.

# CHAPTER 4

## SECTION 4.2 Pages 111–112

1. (b) $(\bar{x} + z)y + x\bar{y}$

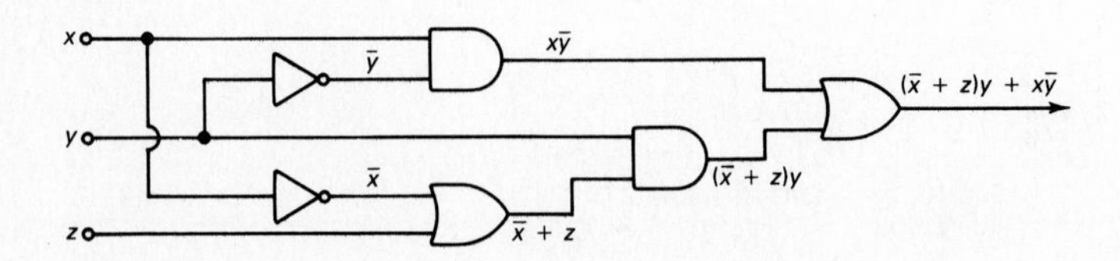

3. $x \leftrightarrow y = \overline{x\bar{y} + \bar{x}y} = (\overline{x\bar{y}})(\overline{\bar{x}y}) = (\bar{x} + y)(x + \bar{y})$

4. You are to determine whether:

$$\overline{x\overline{yz}} = \overline{\overline{xy}z} \quad \text{and} \quad \overline{x + \overline{y + z}} = \overline{\overline{x + y} + z}$$

By using De Morgan's laws and simplification you will find that neither NAND gates nor NOR gates are associative.

5. (b) $\bar{x}z(y + z) + (\bar{x} + xy)$

## SECTION 4.3 Page 114

1.

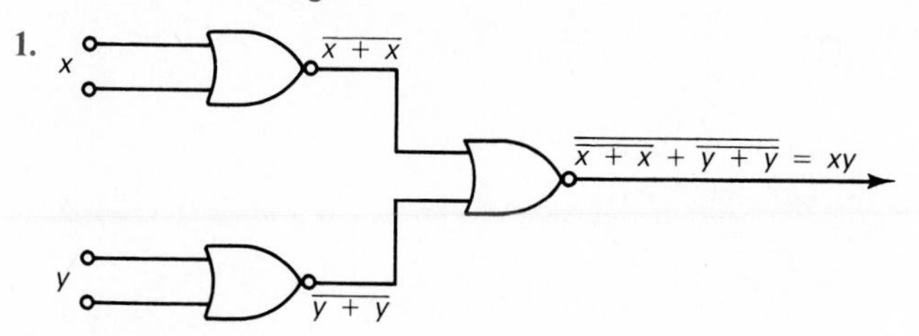


2. $xy = (xy) + (xy) = \overline{\overline{xy} \cdot \overline{xy}}$

4. $x \rightarrow y = \bar{x} + y = \overline{x\bar{y}} = \overline{(\overline{\overline{xx}\,\overline{xx}})\overline{yy}}$

$x \rightarrow y = \bar{x} + y = \overline{\overline{\bar{x} + y}} = \overline{(\overline{\bar{x} + y}) + (\overline{\bar{x} + y})} = \overline{(\overline{\overline{x + x} + y}) + (\overline{\overline{x + x} + y})}$

### Page 117

8. (a) $xyz + \bar{x}yz + x\bar{y}z + xy\bar{z}$ can be written by idempotence and commutativity as $(xyz + \bar{x}yz) + (xyz + x\bar{y}z) + (xyz + xy\bar{z})$. Distribution gives $yz(x + \bar{x}) + xz(y + \bar{y}) + xy(z + \bar{z})$. By the complement rule we obtain $xy + xz + yz$.

(b) $xyz + \bar{x}yz + x\bar{y}z + xy\bar{z} = xyz + xy\bar{z} + \bar{x}yz + x\bar{y}z$

$= xy(z + \bar{z}) + (\bar{x}y + x\bar{y})z$

$= xy + (x \oplus y)z$

## SECTION 4.4 Page 120

1. All are Boolean expressions. (a), (b), (e), (g), and (i) are in S.O.P. form.

### Page 121

2. (b) $(\overline{\bar{x} + z})(y + x) + x\bar{y} = (x\bar{z})(y + x) + x\bar{y}$

$= x\bar{z}y + x\bar{z}x + x\bar{y}$

$= x\bar{z}y + x\bar{z} + x\bar{y}$

$= x\bar{z} + x\bar{y}$ by absorption

(e) $wxyz + \bar{w}\bar{x}\bar{y}\bar{z} + x + \bar{w} = (x + xwyz) + (\bar{w} + \bar{w}\bar{x}\bar{y}\bar{z})$

$= x + \bar{w}$ by absorption twice

### Page 129

3. (b) $\bar{x}\bar{y} + xy$

| | $y$ | $\bar{y}$ |
|---|---|---|
| $x$ | ① | |
| $\bar{x}$ | | ① |

cannot be reduced.

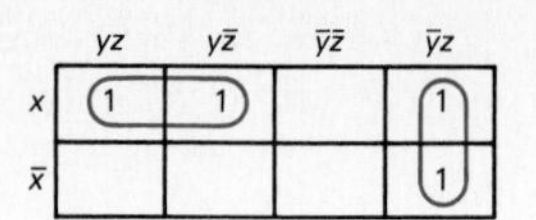


(c) $xy + xz + \bar{y}z = xy(z + \bar{z}) + xz(y + \bar{y}) + \bar{y}z(x + \bar{x})$
$= xyz + xy\bar{z} + xyz + x\bar{y}z + x\bar{y}z + \bar{x}\bar{y}z$
$= xyz + xy\bar{z} + x\bar{y}z + \bar{x}\bar{y}z$
which reduces to $xy + \bar{y}z$ (see accompanying map)

(f) $xyz + (\overline{\bar{x} + \bar{y}})x = xyz + (xy)x$
$= xyz + xy(z + \bar{z})$
$= xyz + xyz + xy\bar{z}$
$= xyz + xy\bar{z}$
$= xy$

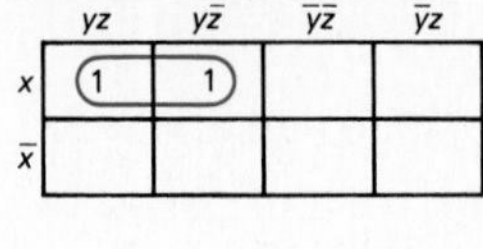


(g) $x + yz$ (h) $\bar{x} + \bar{y} + \bar{z}$ (i) One minimal sum is $\bar{w}y + xy + w\bar{y}\bar{z} + \bar{x}\bar{y}\bar{z}$. **4.** $\bar{x} + z$

**SECTION 4.5 Pages 134–135**

**1.** $xz + \bar{x}\bar{z} + xy$ or $xz + \bar{x}\bar{z} + y\bar{z}$ **3.** Is already in minimal form **5.** $wx + w\bar{z} + \bar{w}\bar{x}z$ **7.** $xz + \bar{x}\bar{z}$

# CHAPTER 5

**SECTION 5.2 Pages 143–145**

**1.** (a), (b) and (f) are functions. **3.** (d) is impossible.
**6.** (b) This function is an injection. It cannot be a surjection because all the languages cannot be used.
(c) $4 \cdot 3 \cdot 2 = 24$ **8.** (b) $5 \cdot 4 \cdot 3 \cdot 2 = 120$

**SECTION 5.3 Page 147**

**1.** (a) $f \circ g(x) = f(g(x)) = f(x^2 + 3x + 4) = \dfrac{1}{\sqrt{(x^2 + 3x + 4) + 3}} = \dfrac{1}{\sqrt{x^2 + 3x + 7}}$

(b) $g \circ f(x) = \dfrac{1}{x + 3} + \dfrac{3}{\sqrt{x + 3}} + 4$

**3.** $g^{-1}$ does not exist since the element $10 \in C$ is unused.

**SECTION 5.4 Page 151**

| | infix | prefix | postfix |
|---|---|---|---|
| **1.** (a) | $p \to (q \wedge (r \vee p))$ | $\to p \wedge q \vee rp$ | $pqrp \vee \wedge \to$ |
| (b) | $((p \leftrightarrow q) \to (r \wedge (p \vee q))$ | $\to\leftrightarrow pq \wedge r \vee pq$ | $pq \leftrightarrow rpq \vee \wedge \to$ |
| (d) | $(A + (B/C)) - (D \uparrow A)$ | $- + A/BC \uparrow DA$ | $ABC/ + DA\uparrow -$ |

**2.** (a) $(p \wedge (q \vee r)) \to p$ **3.** (a) $(A + B) \uparrow (C * D)$

**SECTION 5.5 Pages 156–157**

**1.** $F(7) = 7! = 5040$ **4.** $h(7) = 17$ **6.** (a) $FV(5) = \$1610.51$ (b) $FV(4) = \$3147.04$
**7.** $P(4) = 12$, $Q(4) = 12$

**SECTION 5.7 Pages 165–166**

**1.** (b) (i) *basis:* $\overline{x_1 x_2} = \overline{x_1} + \overline{x_2}$ is true by De Morgan's law.
*IH:* $\overline{x_1 x_2 \cdots x_{k-1}} = \overline{x_1} + \overline{x_2} + \cdots + \overline{x_{k-1}}$
*Show:* $\overline{x_1 x_2 \cdots x_{k-1} x_k} = \overline{x_1} + \overline{x_2} + \cdots + \overline{x_{k-1}} + \overline{x_k}$

Start with $\overline{x_1} + \overline{x_2} + \cdots + \overline{x_{k-1}} + \overline{x_k}$.

$$\begin{aligned}\overline{x_1} + \overline{x_2} + \cdots + \overline{x_{k-1}} + \overline{x_k} &= (\overline{x_1} + \overline{x_2} + \cdots + \overline{x_{k-1}}) + \overline{x_k} && \text{by associativity}\\ &= \overline{x_1 x_2 \cdots x_{k-1}} + \overline{x_k} && \text{by IH}\\ &= \overline{(x_1 x_2 \cdots x_{k-1}) x_k} && \text{by De Morgan's law and involution}\\ &= \overline{x_1 x_2 \cdots x_{k-1} x_k} && \text{by associativity}\end{aligned}$$

**3.** (b) $100 \cdot 3^n$

**4.** (b) *basis:* $k = 1: 1^3 + 5 \cdot 1 = 6$ which is divisible by 6.
*IH:* $k^3 + 5k$ is divisible by 6.
*Show:* $(k+1)^3 + 5(k+1)$ is divisible by 6.

$$\begin{aligned}(k+1)^3 + 5(k+1) &= k^3 + 3k^2 + 3k + 1 + 5k + 5\\ &= k^3 + 5k + 3k^2 + 3k + 6\\ &= \underbrace{(k^3 + 5k)}_{\substack{\text{is divisible}\\ \text{by 6 from IH}}} + \underbrace{3(k^2 + k)}_{\substack{\text{is divisible}\\ \text{by 2 from 4(a)}}} + \underbrace{6}_{\substack{\text{is divisible}\\ \text{by 6}}}\end{aligned}$$

Since each of the above terms is divisible by 6, their sum will also be divisible by 6, thus proving the induction step.

(c) *basis:* $k = 1: 2^1 > 1$ is true.
*IH:* $2^k > k$
*Show:* $2^{k+1} > k + 1$

$$2^{k+1} = 2 \cdot 2^k > 2 \cdot k \quad \text{by the IH} \qquad \text{and} \qquad 2k \geq k + 1 \quad \text{for all} \quad k \geq 1$$

Thus $2^{k+1} > k + 1$, which is what we were to show.

**5.** (b) *basis:* $k = 1: 2 \cdot 1 = 1(1+1)$
$2 = 2$ is true.
*IH:* $2 + 4 + \cdots + 2k = k(k+1)$
*Show:* $2 + 4 + \cdots + 2k + 2(k+1) = (k+1)(k+2)$

$$\begin{aligned}2 + 4 + \cdots + 2k + 2(k+1) &= (2 + 4 + \cdots + 2k) + 2(k+1)\\ &= k(k+1) + 2(k+1) \qquad \text{by the IH}\\ &= k^2 + k + 2k + 2\\ &= k^2 + 3k + 2\\ &= (k+1)(k+2), \text{ thus proving the inductive step.}\end{aligned}$$

Alternative solution: Multiply both sides of 5(a) by 2.

$$2(1 + 2 + \cdots + n) = 2 \cdot \frac{1}{2} n(n+1)$$

$$2 + 4 + \cdots + 2n = n(n+1)$$

## CHAPTER 6

### SECTION 6.2 Pages 173–175

**1.** (b)

| *for graph H* | *for graph J* |
|---|---|
| $\deg(v_1) = \deg(v_6) = 2$ | $\deg(A) = \deg(D) = \deg(F) = 2$ |
| $\deg(v_2) = \deg(v_3) = 3$ | $\deg(C) = 1$ |
| $\deg(v_4) = \deg(v_5) = 1$ | $\deg(E) = 5$ |
| total degree = 12 | $\deg(G) = 4$ |
| | total degree = 18 |

(c) $\text{order}(H) = 6$ $\quad \text{order}(J) = 7$
$\text{size}(H) = 6$ $\quad \text{size}(J) = 9$

2. (b) $\deg(v_1) = 0$
$\deg(v_2) = \deg(v_6) = \deg(v_7) = 1$
$\deg(v_3) = \deg(v_9) = \deg(v_{11}) = 3$
$\deg(v_4) = \deg(v_8) = \deg(v_{10}) = \deg(v_{12}) = 2$
$\deg(v_5) = 4$
total degree $= 24$
order $=$ size $= 12$
There are six odd vertices.

4. no, no

6. (b)

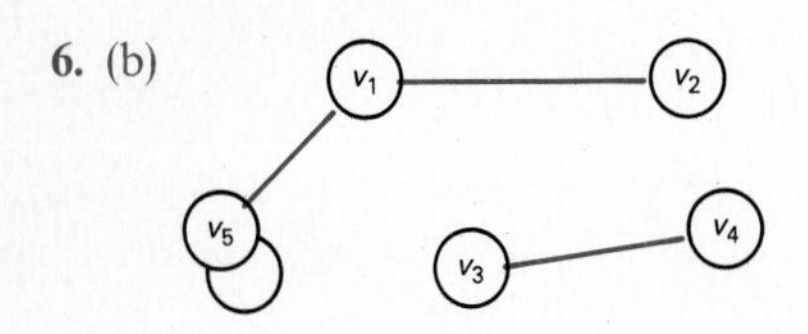


8. (a)

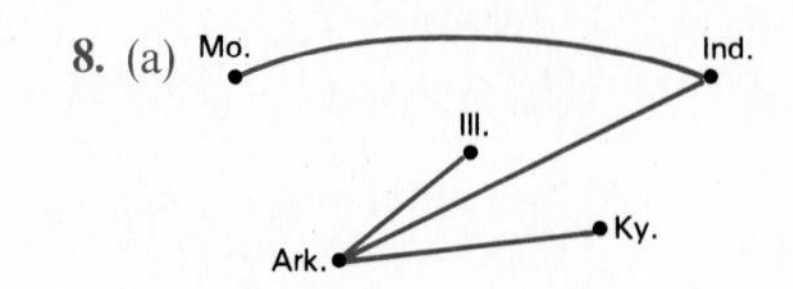


(c) Show the number of edges in $E \cup \bar{E} = n(n-1)/2$. When the two graphs are joined (the *union* as explained in Exercise 9), every vertex is adjacent to every other vertex. Thus the degree of each of the $n$ vertices is $n - 1$, giving a total degree of $n(n-1)$. From Theorem 6.1, we know that this is twice the number of edges. Therefore, the number of edges in $E \cup \bar{E}$ is $\frac{1}{2}n(n-1)$.

11. The complement of 10(a) is:

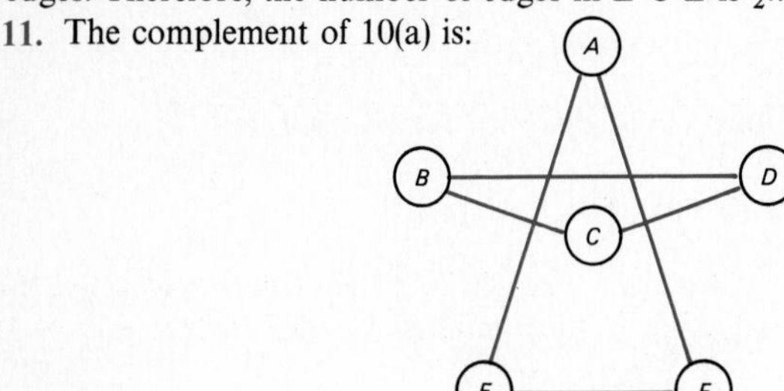


**SECTION 6.3 Page 180**

2. Paths are trails, but trails are not necessarily paths.

4. cut-points: $v_4$, $v_5$, $v_7$

5. size = order − 1 (see Section 6.5)

9. 3 bridges, 1 cut-point

10. no, no

11. $$M^5 = \begin{bmatrix} 0 & 9 & 0 & 0 & 0 & 0 \\ 9 & 0 & 9 & 9 & 0 & 0 \\ 0 & 9 & 0 & 0 & 0 & 0 \\ 0 & 9 & 0 & 0 & 0 & 0 \\ 0 & 0 & 0 & 0 & 0 & 1 \\ 0 & 0 & 0 & 0 & 1 & 0 \end{bmatrix}$$

**SECTION 6.4 Pages 184–185**

1. Show that $K_p$, the complete graph of order $p$, is $(p-1)$ regular, that is, that every vertex has degree $p - 1$. Proof by induction:
*basis:* Let $p = 1$. $K_1$ is a single point with $1 - 1 = 0$ edges.
*IH:* $K_n$ is $(n-1)$ regular. There are $n$ vertices each with degree $n - 1$.
*Prove for* $p = n + 1$. Show $K_{n+1}$ is $n$ regular. From the IH, every vertex of a $K_n$ graph has $n - 1$ edges. Add one more vertex to $K_n$. For this to be a complete graph, this new vertex must be adjacent to each of the $n$ vertices of $K_n$. This gives a new edge to each of them, giving each vertex a degree of $n - 1 + 1 = n$. The new vertex will be of degree $n$, also. Thus $K_{n+1}$ is $n$-regular.

**5.** Each of the $n$ vertices has degree $n - 1$, giving a total degree of $n(n - 1)$. So the number of edges is $\frac{1}{2}n(n - 1)$ from Theorem 6.1.

**6.** Assume $n - k + r = 2$ for $k$ edges. Prove true for $k + 1$ edges.

*Proof:* If one more edge is added, then we have two cases. *Case 1:* One more vertex is added, giving no new regions. Thus we have

$$\begin{aligned} (n + 1) - (k + 1) + r &= \underbrace{n - k + r} + 1 - 1 \\ &= 2 + 1 - 1 \qquad \text{from IH} \\ &= 2 \end{aligned}$$

*Case 2:* No new vertices are added, yielding an additional region. Thus we have

$$\begin{aligned} n - (k + 1) + (r + 1) &= \underbrace{n - k + r} - 1 + 1 \\ &= 2 - 1 + 1 \qquad \text{from IH} \\ &= 2 \end{aligned}$$

## SECTION 6.5 Pages 191–195

**2.** (c) 4 (f) 7 **3.** Size $= n - 1$. Total degree $= 2(n - 1) = 2n - 2$ from Theorem 6.1.

**4.** No. Consider the nodes of degree 1. Removal of these edges does not disconnect the graph.

**12.** *Basis:* $n = 2$. Each of the *two* nodes will be of degree 1.

*IH:* A tree of order $k > 2$ has at least two nodes of degree 1.

*Induction step:* (Prove for a $(k + 1)$ order tree).

Attach a new node to either (a) a node of degree 1, in which case the old node will be of degree 2, but the new node will be of degree 1, thus the $(k + 1)$ tree will still have at least two nodes of degree 1; or (b) a node of degree $> 1$, in which case we merely add another node of degree 1.

**13.** (b) Given a graph of order $n$, size $= n - 1$ and connected. Prove it is a tree. *Proof:* Assume it is not a tree. Then there will be one or more cycles. Consider a cycle of length $p$. $p$ nodes and $p$ edges of the graph will be accounted for by the cycle. Now consider the other $n - p$ nodes of the graph. There must be a path from each of these nodes to one or more nodes of the cycle. This means at least $n - p$ additional edges in the graph. Adding these edges to the $p$ edges of the cycle gives a total of at least $n - p + p = n$, which contradicts the assumption of $n - 1$ edges. Thus we conclude that there are no cycles. A connected acyclic graph is a tree.

**16.** **(a)** (i) 5 (ii) 12 (iii) 3 (iv) 7 (v) 8

## SECTION 6.6 Pages 198–200

**2.** Node $v_4$ is a source. There are no sinks.

**6.** $$M^4 = \begin{array}{c} \\ a \\ b \\ c \\ d \end{array} \begin{array}{c} \begin{matrix} a & b & c & d \end{matrix} \\ \begin{bmatrix} 1 & 0 & 1 & 0 \\ 1 & 1 & 1 & 0 \\ 0 & 0 & 1 & 1 \\ 0 & 1 & 0 & 1 \end{bmatrix} \end{array}$$ There are 9 walks of length 4.

**7.** Only (c) and (f) are strongly connected. Figures (a) through (f) are weakly connected. Figures (b) through (f) are unilaterally connected.

## CHAPTER 7

### SECTION 7.2 Pages 216–219

**2.** (a) (i), (iii), (iv) (b) (v), (vi) (c) (i), (iii), (iv), (vi) (d) (i), (ii), (iii), (iv), (vi)
**4.** 100 **8.** Both graphs can be drawn with no crossovers.
**9.** Orient the tree as rooted. Since all nodes at a level are not adjacent to one another, they can be assigned the same color. Then alternate the two colors from level to level.
**11.** chromatic number = 4

### SECTION 7.3 Pages 233–235

| | *Preorder* | *Inorder* | *Postorder* |
|---|---|---|---|
| **2.** (a) | *ABDEFCGHIJ* | *EDFBACGIHJ* | *EFDBIJHGCA* |
| (b) | *ABCDE* | *EDCBA* | *EDCBA* |

**4.** (a) preorder = $+-*+abcd\uparrow e-f*gh$
postorder = $ab+c*d-efgh*-\uparrow+$ **5.** Evaluation = 76
**10.** (a) 15 (b) 7 (c) 12 variables, 11 operators

### SECTION 7.4 Pages 240–241

**4.** Let the ordered pairs be (number of liters in jug 1, number of liters in jug 2). The following sequence is a minimal: (0, 0) → (5, 0) → (2, 3) → (2, 0) → (0, 2) → (5, 2) → (4, 3) → (4, 0)
**5.** Let the coins be: *a*, *b*, *c*, *d*, *e*, *f*, *g*, *h*; with one *different*, either heavier or lighter. Start by weighing three coins against three others.

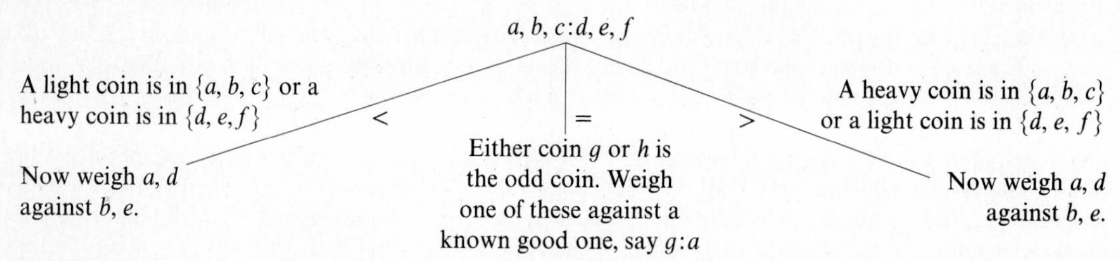


**8.** Weigh coins *a*, *b*, *c* against *d*, *e*, *f*.

## CHAPTER 8

### SECTION 8.2 Pages 244–245

**1.** The license plates take the form ddlll. There are 9 choices for the first digit, 10 for the second digit, and 26 choices for each of the three letters. This gives $9 \cdot 10 \cdot 26 \cdot 26 \cdot 26 = 1{,}581{,}840$.
**2.** (c) The eight digits have a 1 in the first three places. There will be $2^5 = 32$ choices for the other five.
**4.** (a) 6 (b) 12 (c) 8 (d) 10 **5.** (b) $9 \cdot 10 \cdot 5 = 450$ **6.** $3^5 = 243$

### SECTION 8.3 Page 247

**2.** If all programming courses are taken, there will be $1 \cdot 3 \cdot 2 \cdot 1 = 6$ orders.

**3.** (b) $P(5, 1) = \frac{5!}{4!} = 5$ (d) $P(n, 2) = \frac{n!}{(n-2)!} = \frac{n(n-1)(n-2)!}{(n-2)!} = n(n-1)$ **4.** 7

### SECTION 8.4 Pages 250–251

1. (c) $\frac{7!}{11!} = \frac{7!}{11 \cdot 10 \cdot 9 \cdot 8 \cdot 7!} = \frac{1}{7920}$ 2. (b) $\binom{7}{5} = \frac{7!}{5! \cdot 2!} = 21$ 5. $\binom{10}{5} = 252$ 6. $\binom{100}{10}$

## CHAPTER 9

### SECTION 9.2 Page 255

1. (a) upper bounds = $\{d, e, f\}$; lower bounds = $\varnothing$ (c) upper bounds = $\{d, e, f\}$; lower bounds = $\{c\}$
2. (c) upper bounds = $\{f\}$; lower bounds = $\{a, b\}$ 3. (d) upper bound = $\{24\}$; lower bound = $\{1\}$

### SECTION 9.3 Pages 256–257

1. (b) glb = $\{a\}$; lub = $\{f\}$ (c) glb = $\{d\}$; lub = $\{h\}$ 2. (c) glb = $\varnothing$; lub = $\{f\}$

### SECTION 9.5 Pages 261–263

1. Figures (b), (e), (f), (h), and (i) are lattices. 2. (a) $d$ (d) $f$ (e) $e$ 3. (b) $g$ (e) $b, d$ 6. atoms of figure (b): $\{b, d\}$; figure (d) has no atoms.
9. Element $b$ has no complement. The lattice is not distributive. 10. Figures (h) and (i)

## APPENDIX A

### SECTION A.4 Page 278

| | decimal | octal | hexadecimal |
|---|---|---|---|
| 1. (a) | 85 | 125 | 55 |
| (b) | 971 | 1713 | 3CB |
| (c) | 1608 | 3110 | 648 |

| | decimal | binary |
|---|---|---|
| 2. (a) | 16853 | 100, 000, 111, 010, 101 |
| (b) | 5349 | 1, 010, 011, 100, 101 |
| (c) | 4012 | 111, 110, 101, 100 |

| | decimal | binary |
|---|---|---|
| 3. (a) | 20651 | 101, 0000, 1010, 1011 |
| (b) | 2684 | 1010, 0111, 1100 |
| (c) | 22829 | 101, 1001, 0010, 1101 |

| | binary | octal | hexadecimal |
|---|---|---|---|
| 4. (a) | 1011001011011 | 13133 | 165B |
| (b) | 1110000000 | 1600 | 380 |
| (c) | 1111111111 | 1777 | 3FF |

5. (a) 111111 (b) 1000001 (c) 101110 6. (a) 111001110 (b) 11000100 (c) 11110011
7. (a) 1011111 (b) 1001110 (c) 100110 (negative)

## APPENDIX B

### SECTION B.2 Page 284

1. $mn$ elements
2. (a) $1 \times 100, 100 \times 1, 2 \times 50, 50 \times 2, 4 \times 25, 25 \times 4, 5 \times 20, 20 \times 5, 10 \times 10.$ (b) $1 \times 7, 7 \times 1$
3. The $n \times n$ matrix has a total of $n^2$ elements. There are $n$ elements on the main diagonal. Thus there are $n^2 - n$ elements off the main diagonal.
4. $a_{m-1,1}; a_{m-1,2}; a_{m-1,3}; \ldots; a_{m-1,n}$.

5. $\begin{bmatrix} 2 & 3 & 4 & 5 \\ 3 & 4 & 5 & 6 \\ 4 & 5 & 6 & 7 \\ 5 & 6 & 7 & 8 \end{bmatrix}$ 6. $\begin{bmatrix} 0 & 1 & 4 \\ 1 & 0 & 1 \\ 4 & 1 & 0 \end{bmatrix}$ 7. (a) $\begin{bmatrix} 1 & 0 & 0 & 0 \\ 0 & 1 & 0 & 0 \\ 0 & 0 & 1 & 0 \\ 0 & 0 & 0 & 1 \end{bmatrix}$ (b) $\begin{bmatrix} 0 & 1 & 2 & 3 \\ 4 & 0 & 1 & 2 \\ 9 & 18 & 0 & 1 \\ 16 & 32 & 48 & 0 \end{bmatrix}$

8. (a) $a_{ij} = i^j$ (b) $a_{ij} = j^{i-1}$ 10. $a_{ji} = j^2 + i^2 = i^2 + j^2 = a_{ij}$ 11. $x = 1, y = 2, z = 3, w = 4$

**SECTION B.3 Page 288**

1. (a) impossible (b) $\begin{bmatrix} 2 & -1 \\ -2 & 4 \end{bmatrix}$ (c) impossible (d) impossible (e) impossible (f) $\begin{bmatrix} 15 & 6 & -3 \\ 10 & 4 & -2 \\ 20 & 8 & -4 \end{bmatrix}$

(g) $[15]$ (h) $\begin{bmatrix} 1 & 1 & 0 \\ 1 & 0 & 0 \\ 1 & 1 & 1 \end{bmatrix}$ (i) $\begin{bmatrix} 1 & 1 & 0 \\ 1 & 0 & 0 \\ 1 & 1 & 1 \end{bmatrix}$ (j) $\begin{bmatrix} 0 & 1 & 0 \\ 1 & -1 & 0 \\ 1 & 1 & 0 \end{bmatrix}$ (k) $\begin{bmatrix} 3 & -4 \\ -8 & 11 \end{bmatrix}$ (l) $\begin{bmatrix} 11 & -15 \\ -30 & 41 \end{bmatrix}$

(m) $\begin{bmatrix} -5 & 2 & 2 \\ 17 & -2 & 2 \end{bmatrix}$ (n) $\begin{bmatrix} 5 & -7 \\ -7 & 10 \end{bmatrix}$ (o) $\begin{bmatrix} 14 & 49 \\ 28 & 14 \\ 56 & 42 \end{bmatrix}$ (p) $\begin{bmatrix} -9 \\ -6 \\ -12 \end{bmatrix}$

2. $A$ has to be a square matrix.

# Index of Symbols

| Symbol | Explanation |
|---|---|

## BOOLEAN ALGEBRA

| Symbol | Explanation |
|---|---|
| $x + y$ | Join operator of elements $x$ and $y$, 85 |
| $x \cdot y$ or $xy$ | Meet operator of elements $x$ and $y$, 85 |
| $x \leq y$ | "precedes" relation: $x$ precedes $y$, 100 |
| $x \oplus y$ | X-OR operator of element $x$ and $y$, 92 |

## COUNTING AND COMBINATORICS

| Symbol | Explanation |
|---|---|
| $C(n, r)$ | Combinations, $r$ from $n$ objects, 247 |
| $\binom{n}{r}$ | Combinations, $r$ from $n$ objects, 247 |
| $P(n, r)$ | Permutations, $r$ from $n$ objects, 246 |

## FUNCTIONS

| Symbol | Explanation |
|---|---|
| $f \circ g$ | Composition of functions $f$ and $g$, 145 |
| $f: A \rightarrow B$ | $f$ is a function from domain $A$ to codomain $B$, 137 |
| $f^{-1}$ | Inverse of a function $f$, 143 |
| $\sum$ | Summation notation, 164 |

## GRAPH THEORY AND TREES

| Symbol | Explanation |
|---|---|
| Lg | Base 2 logarithm, 194 |
| $\chi(M)$ | Chromatic number of a map $M$, 213 |
| $\bar{G}$ | Complement of a graph $G$, 174 |
| $K_{m,n}$ | Complete bipartite graph, 183 |
| $\deg(v)$ | Degree of a vertex, $v$, of a graph, 169 |
| $K_p$ | Regular graph of order $p$, 181 |
| $G(V, E)$ | Representation of a graph $G$ with vertex set $V$ and edge set $E$, 169 |

## LATTICES AND POSETS

| Symbol | Explanation |
|---|---|
| glb | Greatest lower bound of elements of posets, 255 |
| $a \vee b$ | Join operator of lattice elements $a$ and $b$, 257 |

## LOGIC AND PROPOSITIONS



## MATRICES



## SETS



## SWITCHING CIRCUITS AND LOGIC GATES

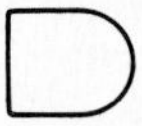

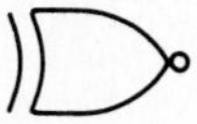

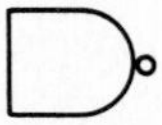

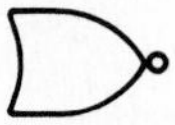

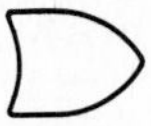

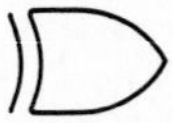

## A



## B

## C

**D**



**E**

F



G

## H



## I



## J, K



## L

## M



## N



## O

## P



## Q, R



## S

**T**

## U, V



## W, X, Z